IN THE SCOPE OF LOGIC, METHODOLOGY AND PHILOSOPHY OF SCIENCE

SYNTHESE LIBRARY

STUDIES IN EPISTEMOLOGY,

LOGIC, METHODOLOGY, AND PHILOSOPHY OF SCIENCE

VOLUME 315

IN THE SCOPE OF LOGIC, METHODOLOGY AND PHILOSOPHY OF SCIENCE

Volume One of the 11th International Congress of Logic, Methodology and Philosophy of Science, Cracow, August 1999

edited by

PETER GÄRDENFORS
Lund University,
Lund, Sweden

JAN WOLEŃSKI
Jagiellonian University,
Cracow, Poland

and

KATARZYNA KIJANIA-PLACEK
Jagiellonian University,
Cracow, Poland

KLUWER ACADEMIC PUBLISHERS
DORDRECHT / BOSTON / LONDON

A C.I.P. Catalogue record for this book is available from the Library of Congress.

ISBN 1-4020-0929-1 (HB)
ISBN 1-4020-0931-3 (set)

Published by Kluwer Academic Publishers,
P.O. Box 17, 3300 AA Dordrecht, The Netherlands.

Sold and distributed in North, Central and South America
by Kluwer Academic Publishers,
101 Philip Drive, Norwell, MA 02061, U.S.A.

In all other countries, sold and distributed
by Kluwer Academic Publishers,
P.O. Box 322, 3300 AH Dordrecht, The Netherlands.

Printed on acid-free paper

Printed in the Netherlands.

Contents

VI LOGIC, LANGUAGE AND COGNITION

VII METHODOLOGY

VIII PROBABILITY, INDUCTION, AND DECISION THEORY

IX PHILOSOPHY OF LOGIC, MATHEMATICS, AND COMPUTER SCIENCE

X PHILOSOPHY OF THE PHYSICAL SCIENCES

EDITORIAL

This volume contains invited lectures delivered during the 11th International Congress of Logic, Methodology and Philosophy of Science, held in Cracow, August 1999. The title of the volume alludes to the general idea behind the series of all Congresses of Logic, Methodology and Philosophy of Science (starting in Stanford 1960) which has been to review the current state of logic in the widest sense. Another aim of the Congress was to display the mutual interrelations between the variety of fields explored by logical research. Although the general structure of the program was similar to most of the previous congresses, some alterations were introduced in Cracow. The number of sections was increased by two as compared to the Congress in Florence 1995. In particular, the circle of problems defined by the heading "Ethical, Historical and Social Perspectives on the Philosophy of Science" formerly constituting a single section, was transformed into a larger unit as the fourth main area (beside Logic, General Philosophy of Science, and Philosophical Problems Concerning the Sciences). Moreover, the section "Philosophical Logic" was replaced with several more specific ones. Three plenary lectures were given by Professor Andrzej K. Wróblewski (opening lecture), Professor John Maynard Smith (closing lecture) and Stanisław Lem, a famous Polish science-fiction novelist.

Apart from the regular sections the program included three special symposia "A Hundred Years of the Philosophy of Science", "Cognitive Science Meets Philosophy of Science" and "Special Round-Table Symposium on Ethics and Responsibility in Science". There were also two affiliated symposia "Many Valued Logics for Computer Science Applications" and "The Scientific Meeting of Polish Association of Logic and Philosophy of Science". The papers delivered at two first special symposia are published in this volume. Many other contributions submitted to the Congress will be included in a separate volume. As usual the Congress was an occasion to meet people of various interests in logic and related disciplines for a successful exchange of ideas, views and proposals.

Kluwer Academic Publishers offered the opportunity to produce the Proceedings of the Congress. It was initiated by Rudolf Rijgersberg and continued by Floor Oosting. The editors are pleased to express their gratitude to these persons. The editorial work was done by Katarzyna Kijania-Placek.

Peter Gärdenfors, Chair of the General Program Committee (Lund)
Jan Woleński, Chair of the Local Organizing Committee (Kraków)
Katarzyna Kijania-Placek (Kraków)

SECTIONS, PLENARY LECTURES, AND SPECIAL SYMPOSIA

LOGIC

Section 1: Proof Theory

Section Program Committee: S. Buss (Chair), J.Y. Girard, J. Krajicek.

Invited Speakers:

V. Danos, *Sequent calculus and continuation passing style compilation*
P. Pudlák, *On the computational content of propositional intuitionistic calculus*
V. Orevkov, *Lower bounds on proof length in axiomatic theories*

Section 2: Model Theory

Section Program Committee: W. Hodges (Chair), J. Väänänen, B. Zilber.

Invited Speakers:

L. Newelski, *Constructions of countable models*,
S. Starchenko, *'Complex Analysis' over O-minimal fields*,
S. Shelah, *Categoricity of abstract elementary classes*

Section 3: Recursion Theory

Section Program Committee: Y. Moschovakis (Chair), R. Soare, H. Schwichtenberg.

Invited Speakers:

A. Nies, *Model theoretic properties of degree structures*
R.A. Shore, *Computable structures: presentations matter*

Section 4: Axiomatic Set Theory

Section Program Committee: A. Louveau (Chair), H. Woodin, S. Todorcevic.

Invited Speakers:

J. Steel, *Why not Supercompacts?*
G. Hjorth, *Efective cardinality*
R. Jensen, About Gödel's views on the foundations of set theory

Section 5: Logic and Computation

Section Program Committee: J. Tiuryn (Chair), D. Harel, M. Okada.

Invited Speakers:

D. Kozen, *On Hoare logic, Kleen algebra, and types*
Y. Gurevich, *Formalization of the notion of algorithm*
T.A. Henzinger, *Logics for reasoning about real-time computation*
J.P. Jouannaud, *Towards engineering proofs*
W. Thomas, *Determinacy results in automata theory*

Section 6: Logic, Language and Cognition

Section Program Committee: M. Davies (Chair), T. Iida, F. Veltman.

Invited Speakers:

F.J. Pelletier, *Is irrationality impossible?*
M. Kanazawa, *Algorithmic learning of grammars and the language aquisition device*
M. Moortgat, *Gramatical reasoning: logical constants and structural variation*

GENERAL PHILOSOPHY OF SCIENCE

Section 7: Methodology

Section Program Committee: D. Papineau (Chair), P. Achinstein, H. Mellor.

Invited Speakers:

J. Worrall, *New evidence for old*
A.F. Chalmers, *Experiment and the growth of experimental knowledge*
D. Mayo, *Theory testing, statistical methodology, and experimental knowledge*

Section 8: Probability, Induction, and Decision Theory

Section Program Committee: B. Skyrms (Chair), E. Eells, M.C. Galavotti.

Invited Speakers:

S. Zabell, *The evolution of 'the logic of chance'*
W. Rabinowicz, *Preference logic and radical interpretation*
A. Rubinstein, *Definable preferences*

PHILOSOPHICAL PROBLEMS CONCERNING THE SCIENCES

Section 9: Philosophy of Logic, Mathematics, and Computer Science

Section Program Committee: Ch. Parsons (Chair), R. Thomason, M. Urchs.

Invited Speakers:

W. Sieg, *Concepts and constructions in mathematics*
G. Sambin, *Steps towards a dynamical constructivism*
R. Murawski, *On the distinction proof-truth in matematics*

Section 10: Philosophy of the Physical Sciences

Section Program Committee: J. Faye (Chair), V. Rodriguez, N. Cartwright.

Invited Speakers:

M. Morrison, *Symmetries: physics or metaphysics?*
J. Butterfield, *Partial truth in physical theories*
M. Dorato, *Spacetime physics and the reality of time*

Section 11: Philosophy of the Biological Sciences

Section Program Committee: G. Wolters (Chair), J. Gayon, K. Sterelny.

Invited Speakers:

R.N. Brandon, *The modules of selection*
C. Debru, *Biology and the possible*
P. Godfrey-Smith, *On the theoretical role of 'genetic coding'*

Section 12: Philosophy of Cognitive Sciences and Artificial Intelligence

Section Program Committee: Ch. Habel (Chair), M. Boden, S. Tsuchiya.

Invited Speakers:

A. Sloman, *Architecture-based conceptions of mind*
M. Tomasello, *The cultural origins of human cognition*
P. Churchland, *Ten core problems in neuroscience*

Section 13: Philosophy of Linguistics

Section Program Committee: J. Hurford (Chair), L. Humberstone, H. Kamp.

Invited Speakers:

E. Itkonen, *Linguistics as a system of distinct types of ontology-cum-methodology*
F. Cowie, *The parameter-setting model of language acquisition: mechanism or metaphor?*
R. Botha, *Ontological problems in evolutionary linguistics*

Section 14: Philosophy of the Social Sciences

Section Program Committee: D. Sperber (Chair), S. Hurley, M. Salmon.

Invited Speakers:

B. Von Eckardt, *In defense of mental representation*
H. Kincaid, *Assessing game theoretic accounts in the social sciences*
M. Gilbert, *Philosophy and the social sciences*

ETHICAL, HISTORICAL, AND SOCIAL PERSPECTIVES ON PHILOSOPHY OF SCIENCE

Section 15: Ethics of Science and Technology

Section Program Committee: L. Bergstrom (Chair), T. Coady, P. Singer.

Invited Speakers:

H. Lacey, *The ways in which science is and is not value free*
H.E. Longino, *Scientific controversy and the public face of science*
E. Anderson, *Situated knowledge and the interplay of value judgements and evidence in scientific inquiry*

Section 16: History of Logic, Methodology, and Philosophy of Science

Section Program Committee: H. Ishiguro (Chair), R. Creath, S. McCall.

Invited Speakers:

M. Carrier, *Changing laws and shifting concepts: on the nature and impact of incommensurability*
H. Sinaceur, *Alfred Tarski: semantic shift, heuristic shift in metamathematics*
G. Sundholm, *A century of inference: 1837-1936*

Section 17: Philosophical Questions Raised by the History and Sociology of Science

Section Program Committee: S. Gaukroger (Chair), N. Roll-Hansen, D. Bloor.

Invited Speaker:

I. Hacking, *Historical ontology*

PLENARY SPEAKERS

Opening lecture

A.K. Wróblewski, *Development of science – order or arbitrariness*

Closing lecture

J. Maynard Smith, *The concept of information in biology*

Special lecture

S. Lem, *On the future of science*

SPECIAL SYMPOSIA

1. A Hundred Years of the Philosophy of Science

Invited Speakers:

A. Richardson, *Philosophy as science: the modernist agenda of philosophy of science 1900-1950*
E. McMullin, *The parting of the ways: philosophy of science 1950-2000*
G. Irzik, *Carnap and Kuhn: a belated encounter*

2. Cognitive Science Meets Philosophy of Science

Invited Speakers:

P. Churchland, *Cognitive science meets philosophy of science*
R.N. Giere, *Patterns of interaction between philosophers of science and cognitive scientists*
A. Gopnik, *Making causal maps: the evolutionary origins and computational character of the theory-formation system*

3. Special Round-Table Symposium on Ethics and Responsibility in Science

Invited Speakers:

W. Salmon, Opening Remarks
M. Kaiser, *Ethics and responsibility in science – presentation of the background paper for the world science conference of the standing committee of ethics and responsibility in science (SCRES)*
D. Mayo, *Uncertainty and values in the scientific assessment of risk*
H. Kincaid, *Do and should clinical trials involve informed consent*
M. Salmon, *Conflict of interest in the social sciences*
G. Wolters, *Eugenics in the Vatican*

Executive Committee of the Division of Logic, Methodology and Philosophy of Science, International Union of History and Philosophy of Science, 1995–1999:

W. Salmon (President, USA), P. Hájek (1st Vice-President, Czech Republic), M.L. Dalla Chiara (2nd Vice-President, Italy), E. Sober (Secretary General, USA), A. Preller (Treasurer, France), J.-E. Fenstad (Past President, Norway), J. Crossley (Assessor, Australia), P. Gärdenfors (Assessor, Sweden), K. Nomoto (Assessor, Japan), C. di Prisco (Assessor, Venezuela), M. Redhead (Assessor, UK), Q. Renzong (Assessor, China), B. Yudin (Assessor, Russia).

General Program Committee

P. Gärdenfors (Chair, Sweden), F. Jackson (Australia), L. Pacholski (Poland), E. Sober (USA), J. Stern (France), S. Uchii (Japan).

Local Organizing Committee

J. Woleński (Chair), M. Heller, J. Kabziński, A. Łomnicki, J. Misiek, A. Pelczar, Z. Piątek, T. Placek, A. Rojszczak, A. Staruszkiewicz, W. Suchoń, E. Szumakowicz, J. Szymura, E. Wojciechowski, A. Wroński, E. Żarnecka-Biały.

Supporting Institutions and Private Sponsors of the Congress:

The Ministry of National Education, the President of Cracow, the State Committee for Scientific Research, the Polish Academy of Science, the Polish Association of Logic and Philosophy of Science, the Polish Philosophical Association, the Jagiellonian University, the Institute of Philosophy (Jagiellonian University), the Institute of Polish Philology (Jagiellonian University), the Kasa im. J. Mianowskiego Foundation, Kluwer Academic Publishers, the computer firm "Optimus" s.a., the Polish Airlines LOT, the law firm "Studnicki, Pleszka, Cwiakalski, Gorski," (Cracow), Bank Przemysłowo Handlowy, Bank Współpracy Regionalnej, the construction firm "Nowator", the computer firm "Clico", Krzysztof Strykier "Auto-Hit" (Tychy) and Stanisław Bisztyga (Cracow).

Opening Lecture

DEVELOPMENT OF SCIENCE – DETERMINED OR ARBITRARY?

Andrzej K. Wróblewski
Warsaw University

Let me start with reminiscence. In 1974 I visited Stockholm to give some lectures at the university. I remember vividly one occasion when I have had the pleasure and honour to sit next to Professor Hannes Alfvén at dinner in the Royal Swedish Academy. Alfvén, the Nobel laureate in physics of 1970, was at that time occupied with the problem of antimatter in the universe. The apparent absence of antimatter in our part of the universe led him to invent a mechanism to explain the lack of symmetry between antimatter and ordinary matter (which he called koinomatter, after the Greek word: *koinos* – ordinary). Alfvén maintained that the universe since its origin had always been basically symmetric and contained equal amount of matter and antimatter but in the course of time the two became separated and now occupied different regions of space.

I remember that Alfvén was frustrated, or even infuriated, that very few people took interest in his papers and the book which he had published on the subject. "Let us imagine – he said – a planet permanently covered with a layer of clouds so thick that the stars can never be seen from the surface. Imagine an intelligent race living there. The scientists of that race had developed physics to a high degree of sophistication. They had studied the structure of matter and had discovered all elementary particles as we know them now. Then they began to explore their atmosphere. One day the first space rocket went above the clouds and the stars were seen for the first time. Then – Alfvén said – the very first question asked would have been whether the mysterious lights in the sky were made of matter or of antimatter! But our colleagues, the physicists of the earth, seem to care little for the solution of such a problem."

The encounter with Alfvén incited me to think about the ways science on earth had developed. Today I would like to discuss some problems of the history of science. As a physicist I shall try to stay as close as possible to my

P. Gärdenfors, J. Woleński and K. Kijania-Placek (eds.), In the Scope of Logic, Methodology and Philosophy of Science, Vol. I, 3-13.

area of expertise and thus I shall mainly discuss physics, but try to avoid being too technical.

The question in the title of my presentation may seem trivial for most of you. I presume also that most of you already know the answer to it. Let me, however, give two reasons why I consider the question worth discussing.

First reason – I may call it local – is that in this part of the world many of us still remember experience from a not too distant past when people were subject to indoctrination in dialectic and historic materialism which provided a ready answer to most problems. It also explained the development of science as determined by the strict laws discovered by the founders of that doctrine. Take a Soviet book on the history of physics and you find there a clear explanation of why physics developed the way it had, the only correct explanation supported by numerous quotations from Marx, Engels and Lenin (for some period also Stalin).

The second reason is that most scientists, and physicists in particular, have little enthusiasm for history. This is because the physicists try to discover new facts and develop new ideas. When new ideas prove effective they tend to make old ones obsolete. Thus, one may say that the physicists are interested in creating the future and destroying the past. Moreover, the description of the physics is usually logical and orderly, hence the impression is necessarily given that this was also the way this branch of science developed. Therefore, whatever pertains to the past is regarded to be not important and non interesting.

I think that it is a mistake to let preoccupation with the future close our eyes to history. To look ahead is to use past experience as a guide to the future.

As I said, many people have a distorted and mistaken picture of the progress of science. To this end let me quote my favourite excerpt from the speech by the eminent American astrophysicist Samuel Langley, delivered to the American Association for the Advancement of Science in 1889. Langley had this to say on the way the history of science is usually presented:

> We often hear [the progress of science] likened to the march of an army towards some definite end; but this, it has seemed to me, is not the way science usually does move, but only the way it seems to move in the retrospective view of the compiler, who probably knows almost nothing of the real confusion, diversity, and retrograde motion of the individuals comprising the body, and only shows us such parts of it as he, looking backward from his present standpoint, now sees to have been in the right direction.
>
> I believe this comparison of the progress of science to that of an army, which obeys an impulse from one head, has more error than truth in it; and, though all similes are more or less misleading, I would prefer to ask you to think rather of a moving crowd, where the direction of the whole comes somehow from the independent impulses of its individual members; not wholly unlike a pack of hounds, which in the long run perhaps catches its game, but where, nevertheless, when at fault, each individual goes his own way, by scent not by sight, some

> running back and some forward; where the louder voiced bring many to follow them nearly as often in a wrong path as in the right one; where the entire pack has even been known to move off bodily on a false scent; for this, if a less dignified illustration, would be one which had the merit of having a truth in it [...].

I believe that arbitrariness in the development of physics can be threefold, the first dictated by the initial natural conditions, the second resulting from social and economic matter, and the third arising from strictly personal choice.

On the first, the "natural" arbitrariness, you have just heard the story invented by Alfvén of a race which had developed elementary particle physics before the first observation of the stars. I can not help but tell you yet another beautiful example. It had been invented by Isaac Asimov, the well known author of science fiction stories.

One day in 1941 Asimov was shown a quotation from Ralph Emerson: "If the stars should appear one night in a thousand years, how would men believe and adore, and preserve for many generations the remembrance of the city of God?"

Asimov went home and in a stroke of genius wrote a short story entitled 'Nightfall'. I am convinced that many of you have read that masterpiece. I shall nevertheless summarize its essentials.

'Nightfall' is the story of a race living on a continent on one of the planets in a multiple star system. The planet, Lagash, can enjoy no less than six suns which provide permanent illumination so that the inhabitants do not know darkness, and do not know the stars.

But once every two thousand and forty nine years an extremely rare situation occurs when five of the suns are simultaneously below the horizon and at the same time a huge non luminous planetary body causes an eclipse of the only sun remaining above the horizon. Suddenly, for several hours, there is complete darkness on Lagash and the stars become visible in the dark sky. The inhabitants who know only illuminated world become scared to death by darkness. Trying to restore illumination they start fires and burn everything possible. In the madness that spreads over the whole inhabited continent the civilisation is destroyed and almost all people perish.

Afterwards, a new civilization is slowly built by the survivors and their offspring only to be destroyed in the next cataclysm 2049 years later.

Thus, the history of civilization on Lagash displays a cyclic character, which – when discovered by archaeologists – presents a mystery. In the slow process of rebuilding science certain laws of nature are being discovered again and again but the knowledge cannot grow beyond a certain limit because of a short span of each cycle.

Asimov's relation describes a situation shortly before a forthcoming eclipse when in the Lagash society there exists also the so-called cultist tradition developed over centuries. The cultists say that every two thousand fifty years

Lagash enters a huge cave so that all the suns disappear, and there comes total darkness all over the world. And then, things called stars appear which rob men of their souls and leave them unreasoning brutes, so that they destroy the civilisation they themselves have built up.

After reading 'Nightfall' it is easy to imagine what our science would have been if we had lived in a different planetary system. Just remember that the significance of the number seven had been established many millenia ago by our ancestors who had noticed that seven changes of day into night occured between the principal phases of the moon. If the earth had no moon our civilization would have been completely different. For one thing, we could not enjoy seven day weeks.

Or consider how physics would have developed if we lived instead on a planet in a multiple star system. Would there exist a Kepler to find order in the extremely complicated revolutions of the planets in such a system, and a Newton to discover the law of universal gravitation?

That social and economic conditions may decide on the development of science is so obvious – in particular in the present era of Big Science – that I do not need to spend a lot of time to discuss this problem.

I would like, nevertheless, to cite one famous example from the older history of science, that of the Danish astronomer Tycho Brahe, who had greatly impressed the king of Denmark, Frederick II, and received from him a substantial grant to pursue astronomy. The grant consisted not only of a large amount of cash and an annual allowance, but also of an island, Hven, together with its villages and farmers. Having such support Tycho was able to build the largest astronomical observatory of his time. Using larger, more stable and better calibrated instruments he recorded regularly the positions of the planets with accuracy almost an order of magnitude better than any previous observation. The data assembled by Tycho afterwards enabled Kepler to discover the laws of planetary motions, which led directly to the Newton's discovery of the law of universal gravitation.

What the course of events could have been if Frederick II was less generous to Tycho? Would there be 'Principia' written by Newton?

The history of classical physics can be summarized in the shortest way if we concentrate on the three great cases of unification which resulted in mechanics, electromagnetism, and statistical physics.

Firstly, the unified theory of gravitation by Newton (1687) provided consistent explanation of terrestrial movements and motions of celestial bodies. In the earlier Aristotelian system the two were believed to obey different laws.

Secondly, the electromagnetic theory by Maxwell (ca 1865) provided the unification of electricity, magnetism, and light. It has been the culmination of a series of investigations and discoveries by Volta, Oersted, Ampère, Faraday, and others.

Thirdly, statistical physics created in the third quarter of the 19th century by Boltzmann, Clausius, Kelvin, Maxwell, and others, provided the unification of atomic ideas with concepts concerning heat and mechanics.

These great theories not only sum up vast domain of experience but also predict unknown things and therefore expand knowledge. To most people today the order and timing of discoveries in physics appears to have been such an obvious development that things could have scarcely worked out otherwise. And yet, if we study this problem in more detail we find that they might well have done so. We discover that arbitrariness has played a very important role in the development of physics. In the following I shall present several examples to illustrate this point.

Let us consider first the electromagnetic theory. Why it has been constructed only 200 years after classical mechanics? Some electric and magnetic phenomena were known in the 17th century. Great Isaac Newton amused himself with several experiments but did not pursue his studies systematically enough to achieve important progress. In the 18th century investigations of electricity and magnetism were still rather scarce. In fact, most physicists of that time regarded electricity and magnetism as not sufficiently interesting to justify experimental effort – a clear example of arbitrariness.

The announcement in 1791 of the discovery of the so-called animal electricity by the Italian physician and physicist Luigi Galvani was received with great interest. Subsequent studies led quickly to the invention of the voltaic pile by Alessandro Volta and greatly accelerated development of electrical science.

But wait! Observations more or less similar to Galvani's were made and reported earlier, for example by the Dutch naturalist Jan Swammerdam in 1678, by the Swiss natural philosopher Johann Sulzer in 1752, and by the Italian anatomist Leopoldo Caldani in 1756, but aroused little interest.

Take another example. In April 1820 Hans Christian Oersted in Copenhagen discovered that an electric current acts on a magnetic needle. A few months later André Ampère discovered that wires carrying electric current attracted or repelled each other and he correctly attributed this phenomenon to magnetic properties of currents. This was the beginning of electromagnetism; for the first time connection has been found between electricity and magnetism.

And yet all this could have been established earlier. In 1801 none other than Jean Baptiste Biot and Pierre Simon de Laplace, both already famous physicists, reported their finding that the two wires which terminated the voltaic pile attracted one another and after contact adhered like two magnets. They noticed and announced that the attraction between the two wires, one of which received and the other lost the galvanic fluid, differed essentially from electrical attraction, since it was not followed by a repulsion after contact, but continued as long as the chain remained closed. Similar observation was simultaneously made by another French scientist Nicolas Gautherot. Thus three scientists have noticed and reported essentially the same what Ampère did in 1820. But in 1801 these reports were completely ignored.

False premises may strongly add to arbitrariness in the development of physics. For example, in the original paper by Oersted of 1820 we read that: "[...] by these experiments it seemed to be shown that the magnetic needle was moved from its position by the help of a galvanic apparatus, and that, when the galvanic circuit was closed, but not when open, as certain very celebrated physicists in vain attempted several years ago [...]."

Thus the relationship between electricity and magnetism was expected to be a static one. We know of numerous unsuccessful attempts to find such a relationship by experiment. The most famous example was that of the two French physicists Jean Nicolas Hachette and Charles Bernard Desormes who in 1805 experimented with a large voltaic pile. They tried to determine the direction which would be taken by the pile, whose poles were not joined, when freely suspended horizontally. Their pile, composed of 1480 plates of copper coated with zinc, was placed upon a boat floating on water in a large vat. But it assumed no fixed direction, although a magnetized steel bar, of a weight nearly equal to that of the pile, likewise placed on the boat, moved in a series of oscillations until it rested in the magnetic meridian.

According to all textbooks and encyclopedias electromagnetic waves were discovered in 1887 by Heinrich Hertz. But in 1879 and 1880 David Edward Hughes had already shown that signals sent from a spark transmitter could be detected at distances up to five hundred yards. Hughes demonstrated these experiments before Sir George Stokes and others and claimed that the signals were transmitted by electric waves in the air. Unfortunately for Hughes, Stokes was of the opinion that the effect might be explained by ordinary electromagnetic induction, and Hughes, discouraged, did not publish any account of his work until much later. Still earlier, in 1842, the eminent American physicist Joseph Henry had noticed that the "inductive" effect of the Leyden jar discharge could be observed at considerable distances, and even suggested a comparison with a spark from flint hit by steel which can be visible as light at a distance.

Oliver Lodge in the 1870's had worked experimentally on the flow of electricity in a metallic sheet, and had come very near to discovering the Hall effect. But he read in Maxwell's 'Treatise on Electricity and Magnetism' (vol. II, par. 501): "It must be carefully remembered, that the mechanical force which urges a conductor carrying a current across the lines of magnetic field acts, not on the electric current, but on the conductor which carries it"; and this deterred him from making the crucial test. But Edwin Hall either did not read the 'Treatise' or did not believe Maxwell and that is why we speak now of the Hall effect and not of Lodge effect.

Let us now change the field. The recent best-seller by Umberto Eco had added much to the popularity of the Foucault's pendulum which provides a convincing proof of the rotation of the earth. It was first demonstrated by Leon Foucault in Paris in 1851.

It is known, however, that the rotation of the oscillation plane of a pendulum had been observed long before that. Galileo may have had observed it first, but if so, he left no account. But one of Galileo's pupils and collaborators Vincenzo Viviani, and Giovanni Tozzetti, noticed the effect in 1661 in Florence. Eighty years later Giovanni Poleni observed it in Padua. Unfortunately, they did not investigate the phenomenon systematically enough to understand its meaning and their published accounts have been ignored by other scientists.

It is curious to speculate on the impression which would have been produced had Galileo presented the pendulum as a proof that the earth rotates around its axis. It is well known that his trial and persecution created a split between science and the Roman Catholic religion. The progress of science had been slowed down considerably, especially in the countries in which the Inquisition had much to say. The split continues until today in spite of the recent partial rehabilitation of Galileo by Pope John Paul II.

If Galileo's trial did not happen the history of science and of our civilisation would have been dramatically different.

I may remark in passing that the new evidence recently uncovered by Pietro Redondi seems to indicate that Galileo's accusation for Copernicanism may have been just a clever cover for the real thing, his atomism (which was treated as fatal for the Catholic doctrine of the Eucharist). This may be true, but the argument still holds.[1]

Hundred years ago, in August 1900, the first International Congress of Physics had taken place in Paris. An introductory lecture was given by famous mathematician and physicist Henri Poincaré, who discussed relations between experimental physics and theoretical physics. Scientists must order – he said. Science is build up with facts as a house is with stones, but a collection of facts is no more a science than a heap of stones is a house. He then compared experimental physics to a library. Theoretical physics arranges it and prepares the catalogue. It does not enrich it, but if it is well prepared it enables one to draw a greater profit from the former.

This picture is simple and appealing but unfortunately not true because real development of science has been much more complicated. Poincaré of course knew it so later in his talk he admitted that a given set of facts can be generalized in an infinite number of ways. This, of course, means that science develops in arbitrary way.

It is said sometimes that early physics (that is, physics in ancient and medieval times) was arbitrary only to a very small extent because our predecessors were simply discovering and investigating all the things in evidence without making any specific choice in collecting facts or interpreting them. Indeed, the notions such as position, extension, duration etc. which are the subject of study of mechanics are obvious to everyone. At first sight there can be little arbitrariness in comparing or measuring, say, the lengths of sticks, the periods of time or the speed of macroscopic objects. But then, of course, the development of mechanics required introduction of less obvious concepts such as instantaneous velocity, acceleration, force, momentum, and resistance of the medium. Twenty centuries of efforts, between Aristotle and Newton, to deduce correct relations between these quantities show conclusively the presence of arbitrariness even at this early stage of development of physical science.

It is true, however, that modern physics is largely arbitrary, because of the complex nature of so called "facts". These are no longer simple observations but readings of extremely complicated instruments which become "facts" only after an involved mathematical and statistical analysis. From such "facts" the physicists try to infer knowledge of things and processes inaccessible to our senses. It is not surprising that there is a lot of arbitrariness in theoretical explanation proposed by different scientists. It is enough to list here best known controversies in attempts to understand the nature of heat (mechanical theory versus caloric), the nature of light (corpuscular versus wave theory), and the constitution of matter (atomic versus continuum point of view). Thus, modern physics is said to be based largely on inventions rather than on discoveries. This is a fascinating but complicated subject.

Let me now change the subject and say a few words on standard models. Nowadays we, physicists, are proud of the Standard Model which has been with us for about twenty years. I would like simply to remind you that almost every generation of physicists had their standard model, and although they perhaps did not call it by that name, they were equally proud of it.

The standard model of the eighteenth century explained natural phenomena by the action of imponderable, subtle fluids of electricity, of magnetism, of light, and of heat (the so-called caloric). At that time the physicists knew already the concept of *vis viva* (which we now call kinetic energy), but restricted its use to mechanics. It was not yet known that phenomena of heat, electricity, magnetism, and light also involved energy which was as real as matter itself. Today we say that, for example, the difference between an electrified and a non-electrified body, or between a hot body and a cold body, results from differences in energy at the atomic and molecular scale. The answer provided by the eighteenth century standard model was different: everything was explained by the presence of the appropriate fluids. Thus an electrified body simply contained the electric fluid, a hot body contained more caloric than a cold body, etc.

The standard model of the end of the nineteenth century involved all of the classical physics. The physicists of that time had a feeling that they had discovered almost all what was really there. It appeared that the atomic point of view combined with the electromagnetic field concept could provide the ultimate explanation. Widely known is the story of young Max Planck who wanted to start working for his Ph.D. in theoretical physics and was discouraged by Philip von Jolly more or less in these words: "Young man, do not try to spoil your life but better make another choice. Physics is so advanced that it will be finished very soon and you will have no future." Peter Zeeman, who received Nobel prize in physics for 1902, enjoyed telling that he had been warned not to study physics: "Physics is no longer a promising subject; it is finished, there is no room for anything really new". And Albert Michelson in 1894 stated that: "The grand underlying principles have been firmly established [...] future truths of physics are to be looked for in the sixth place of decimals."

Today it is difficult for us to understand why the physicists of hundred years ago were so unprepared for the shock that came with the invention of the quantum theory and relativity. Had they restrained their curiosity about the peculiarities of thermal radiation and the discharge of electricity through gases at low pressure, they might have retained their confidence longer.

In 1901 Lord Kelvin wrote that the beauty and clearness of the dynamical theory, which asserted heat and light to be modes of motion, was only obscured by two clouds. The first involved the question, how could the earth move through an elastic solid, such as essentially was the luminiferous ether, and the second was the Maxwell-Boltzmann doctrine regarding the partition of energy. One must admit that Kelvin could certainly recognize the important clouds. One needed relativity, the other quantum theory, to blow it away.

What are the clouds obscuring the clearness and beauty of the present Standard Model? Perhaps the persistent absence of the Higgs particle is the one?

Thus far I have discussed physics developed by us, humans. But we know that our picture of the universe is very strongly influenced by our means of perception. Perhaps our sensory and psychological apparatus prejudices us in favour of some peculiar picture of the universe. Intelligent beings very different from us might have come up with a very different picture of the world.

In order to see how reasonable is the possibility that we are being fooled by our senses one could try to sketch how science might develop for a race of blind, deaf, but highly intelligent worms living in black, cold, sea-bottom muck, and possessing only senses of touch, temperature, and a kind of (chemical) taste. These hypothetical creatures are chosen because they seem to be about as different from human beings as they possibly can be and still be able to learn things about their environment. They have no eyes, ears, or hands. They cannot use the sun or the stars to give them notions of time. They do not have natural rulers because they have no feet and indeed their shape is very elastic. As a rule of the game, we impose the condition that the only concepts they can use are those that develop by intelligent and imaginative thinking about the experience available to them.

I have taken this example from the article 'Wiggleworm physics' by Jerome Rothstein, published in *Physics Today* in September 1962. You will have to read this fascinating article to learn how the hypothetical wiggleworms, or Wiggies, might have developed their physics. To go through all the details of their possible history would require much longer time than a part of this talk. Let me then only summarize some important points of Rothstein's analysis.

For example, one could expect chemistry and thermodynamics to be developed by wiggleworms rather early. Astronomy, along with many other sciences whose beginnings are primarily visual or mechanical, would develop much later. One would expect to find organic chemistry developed much earlier than inorganic chemistry. Light would probably be discovered by wiggleworms because of its heating effect (this was how we discovered the infrared region of the electromagnetic spectrum). Photovoltaic effects could give rise to the first electric cells, and electric currents would probably be first sensed either physiologically by direct contact or by chemical changes. Organic conductors and semiconductors would probably be highly developed before the discovery of metallic conductivity. Metals would probably be first produced electrolytically.

"Clearly the particular course of events would be strongly influenced by accidents and, as with us, serendipity would be an important and frequent factor in progress" – says Rothstein and decides that:

> It seems reasonable to conclude that though we shall always make errors or be mistaken to some extent in our ideas of the universe, we shall improve or correct our pictures sooner or later, and the same applies to Wiggy. Ultimately, when both he and we have covered the same accessible portions of the universe, we should find that the differences between our pictures would be only those of isomorphism. There seems to be no reason to believe, if an objectively "real" world exists, that Wiggy and we would end up with nonisomorphic pictures of the same body of experience.

An encounter of an alien intelligence in the universe, if it ever happens, will give humanity an unique opportunity to check these notions. Once more it may appear that Nature is richer than human imagination, that there "are more things in heaven and earth that are dreamt of in philosophy."

Notes

1. I could present many more examples of unnoticed or ignored discoveries. If you think that these things could not happen now it is enough to remind the paper of Yehuda Eisenberg of 1954 in which the first example of the triple strange hyperon Ω^- was reported. But, in spite of another example of the Ω^- reported few months later by Fry, Schneps, and Swami, no one treated it seriously until the "real" discovery of this particle in 1964. Not only experimental discoveries but also theoretical ideas have been often ignored. When in 1972 Leon van Hove wrote his thirty page summary of high energy physics, he did not mention partons and devoted only five lines to quarks (originally proposed in 1963), although the Stanford experiments revealing the existence of point-like constituents of nucleons were already much publicised. [see

Y. Eisenberg, *Physical Review* **96**, p. 541, 1954; F. Fry, J. Schneps and M.S. Swami, *Physical Review* **97**, p. 1189, 1955; *Il Nuovo Cimento* **2**, p. 346, 1955; L. Van Hove, *High Energy Physics 25 Years After the Discovery of the π Meson*, CERN Yellow Report, Geneva, 1972].

References

(and/or further reading)

Asimov, I.: (1941), *Nightfall, Astounding Science Fiction*, Street & Smith Publications, New York. (Reprinted since in many anthologies and translated into many languages).

Gliozzi, M.: (1965), *Storia della fisica*, Torino.

Harrison, H.: (1984), *West of Eden*, Bantam Books, Toronto-New York-London-Sydney-Auckland. (This science fiction novel describes a race of intelligent reptiles which "grow" necessary instruments and vehicles instead of constructing them).

Langley, S.P.: (1889), *American Journal of Science*, Ser. 3, 37, 1.

Lindsay, R.B.: (1967), 'Arbitrariness in Physics', *Physics Today* **20**, December, pp. 23–26.

Magie, W.F.: (1935), *A Source Book in Physics*, McGraw-Hill, New York.

Mottelay, P.F.: (1975), *Bibliographical History of Electricity and Magnetism*, Arno Press, New York.

Redondi, P.: (1983), *Galileo eretico*, Giulio Einaudi, Torino; Eng. trans. *Galileo Heretic*, 1987, Princeton University Press, Princeton, New Jersey.

Rothstein, J.: (1962), 'Wiggleworm Physics', *Physics Today* **15**, September, pp. 28–38.

Süsskind, Ch.: (1964), 'Observations of Electromagnetic Wave Radiation Before Hertz', *Isis* **55**, pp. 32–42.

Taton, R.: (1962), *Reason and Chance in Scientific Discovery*, Science Editions, New York.

Thomson, W. (Lord Kelvin): (1901), 'Nineteenth Century Clouds over the Dynamical Theory of Heat and Light', *Philosophical Magazine* **6** (2), pp. 1–40.

Thoren, E.: (1992), *The Lord of Uraniborg, A Biography of Tycho Brahe*, Cambridge University Press, Cambridge.

Whittaker, E.: (1960), *A History of the Theories of Aether and Electricity*, Harper & Brothers, New York.

I

PROOF THEORY

LOWER BOUNDS ON PROOF LENGTH IN AXIOMATIC THEORIES*

Vladimir Orevkov
Steklov Institute of Mathematics
St. Petersburg
orevkov@pdmi.ras.ru

Abstract The aim of this talk is to obtain lower bounds on proof length in axiomatic theories of algebraically closed and real-closed fields and in arithmetic without multiplication. We will also obtain upper bounds on the complexity of correct proofs in the theory of feasible numbers with induction scheme. The advantage of these bounds (over other author's bounds) is that they do not depend on the complexity of formulas in the proof.

1. Lower bounds on proof length in axiomatic theory of fields

We'll consider algebraically closed fields with characteristic 0 and real-closed ones. Let's fix an axiomatic Hilbert-type theory T (classical or intuitionistic) in language with $=, 0, 1, +$, and $\cdot$. Non-logical axioms of T are a finite list of closed formulas and open formulas

$$\exists x(x^n + t_{n-1} \cdot x^{n-1} + \cdots + t_0 = 0), \tag{1}$$

$$0 \neq \underbrace{1 + 1 + \cdots + 1}_{p \text{ times}}, \tag{2}$$

where $n > 0$, the variable x does not occur in terms $t_0, \ldots, t_{n-1}$; p is any prime number.

In case of real-closed fields the number n in (1) is odd and axioms (2) should be omitted.

The language of Presburger's arithmetic (or arithmetic without multiplication) contains constants 0 and 1, functional symbol +, predicates $=, <$. Non-logical axioms of this theory are a finite list of closed formulas and open for-

*Supported by INTAS (grant No. 96-0760) and RFBR (grants No. 94-01-01030 and 96-01-01612).

P. Gärdenfors, J. Woleński and K. Kijania-Placek (eds.), In the Scope of Logic, Methodology and Philosophy of Science, Vol. I, 17-26.

mulas

$$\exists yz(t = \underbrace{y+y+\cdots+y}_{n \text{ times}}+z \& z < \underbrace{1+1+\cdots+1}_{n \text{ times}}), \tag{3}$$

where $n > 0$, variables y and z do not occur in term t.

Let D be a proof in T. The *length* of D is the number of different formulas in D. The length of D will be denoted by $l[D]$. The expression

$$T \vdash_k A$$

means there exists a proof D of A in T in which

$$l[D] \leqslant k.$$

Theorems 1–3 following below are examples of generalization of proofs.

Theorem 1 *For any formula $A(x)$, any natural number m and any sufficiently large natural number n, if for any number i $(0 \leqslant i \leqslant n)$*

$$\mathbf{T} \vdash_{\frac{\log_2 n}{3\log_2 \log_2 n}} A(\mathbf{0}_{m+i}),$$

then

$$\mathbf{T} \vdash \forall x A(x),$$

where $\mathbf{T}$ *is the theory of algebraically closed fields with characteristic* 0 *or the theory of real-closed ones or arithmetic without multiplication. Here the expression* $\mathbf{0}_n$ *denotes the term*

$$\underbrace{0+0+\cdots+0}_{n \text{ times}}.$$

Below we will use the following notation:

$$\mathbf{1}_n \leftrightharpoons \begin{cases} \underbrace{1+1+\cdots+1}_{n \text{ times}}, & \text{if } n \geqslant 1, \\ 0, & \text{if } n = 0. \end{cases}$$

Theorem 2 *For any formula $A(x)$, any natural number m and any sufficiently large natural number n, if for any number i $(0 \leqslant i \leqslant n)$*

$$\mathbf{T} \vdash_{\frac{\log_2 n}{3\log_2 \log_2 n}} A(\mathbf{1}_{m+i}),$$

then there exist natural numbers $l_1, l_2, \ldots, l_k$ such that

$$\mathbf{T} \vdash \forall x \left(\left(\bigwedge_{i=1}^{k} (0 \neq \mathbf{1}_{l_i} + x) \right) \supset A(x) \right),$$

where **T** *is the theory of algebraically closed fields with characteristic 0.*

Theorem 3 *For any formula* $A(x)$, *any natural number* m *and any sufficiently large natural number* n, *if for any number* i $(0 \leqslant i \leqslant n)$

$$\mathbf{T} \vdash_{\frac{\log_2 n}{3\log_2 \log_2 n}} A(\mathbf{1}_{m+i}),$$

then there exists a natural number k *such that* $k \leqslant m+n$ *and*

$$\mathbf{T} \vdash \forall x((x > \mathbf{1}_k) \supset A(x)),$$

where **T** *is arithmetic without multiplication.*

In the proofs of theorems 1–3 we use some lemmas and estimates from Orevkov (1993) and the following section.

Theorems 1–3 can be used to obtain lower bounds on proof length.

The expression

$$\mathbf{T} \nvdash_k A$$

means the negation of the assertion $\mathbf{T} \vdash_k A$.

Theorem 4 *For infinitely many natural numbers* n, *the following conditions hold*

$$\mathbf{T} \nvdash_{\frac{\log_2 n}{3\log_2 \log_2 n}} 0 = \mathbf{0}_n,$$

$$\mathbf{T} \nvdash_{\frac{\log_2 n}{3\log_2 \log_2 n}} (\mathbf{1}_n + \mathbf{1}_n) \neq 1,$$

$$\mathbf{T} \nvdash_{\frac{\log_2 n}{3\log_2 \log_2 n}} (\mathbf{1}_n \cdot \mathbf{1}_n) \neq 1+1,$$

where **T** *is the theory of algebraically closed fields with characteristic 0 or the theory of real-closed ones or arithmetic without multiplication.*

We can also obtain upper bounds on proof length.

Theorem 5 *There is a natural number* c *such that for any* n

$$\mathbf{T} \vdash_{c\log_2 n} 0 = \mathbf{0}_n,$$

$$\mathbf{T} \vdash_{c\log_2 n} (\mathbf{1}_n + \mathbf{1}_n) \neq 1,$$

$$\mathbf{T} \vdash_{c\log_2 n} (\mathbf{1}_n \cdot \mathbf{1}_n) \neq 1+1,$$

where **T** *is the theory of algebraically closed fields with characteristic 0 or the theory of real-closed ones or arithmetic without multiplication.*

2. Logical deduction schemata in axiomatic theories

Let **T** be an axiomatic theory. We will assume that the set of non-logical axioms of **T** is a union a finite list Σ of *closed* formulas and recursively enumerable sets of *open* formulas $\mathfrak{A}_1, \mathfrak{A}_2, \ldots$ and $\mathfrak{A}_k$ $(k \geqslant 0)$.

We will supply applications of rules and axioms in proofs with analyses. The *analysis of a logical axiom* will indicate the schema of which the axiom is an instance; the *analysis of a non-logical axiom from* Σ will specify the number of this one; the *analysis of a non-logical axiom from* $\mathfrak{A}_i$ $(1 \leqslant i \leqslant k)$ will point out the code of this set; the *analysis of an application of a rule* will include the code of this rule, the indexes of the premises to which the rule is applied.

Let D be a proof. The *deduction schema (or skeleton) of proof D* will be the sequence of analyses of axioms and inference of rules from D.

Lemma 1 *For any sufficiently large natural number l the number of different deduction schema that contain at most l analyses does not exceed l^{3l}.*

Let A and B be two formulas. We will say that A is *transformable into* B if B can be obtained from A by simultaneous substitution of terms for free variables (where it is assumed, of course, that the terms are free for the variables). For example, formula

$$\forall x \exists y (P(x,u) \vee Q(z,y)) \tag{4}$$

is transformable into formula

$$\forall x \exists y (P(x,f(z)) \vee Q(u,y)). \tag{5}$$

An occurrence U of a term in a formula A will be called *free* if all occurrences of variables generated by U are free in A. We will say that A is a *q^*-type* of B if the following statements hold:

1. A is transformable into B;
2. A does not contain 0-place function symbols;
3. any object variable has at most one free occurrence in A;
4. if a term t occurs free in A, then t is a variable.

For example, the formula (4) is the q^*-type of (5). It is easy to show that q^*-types of the same formula coincide up to renaming of free variables.

We will say that B is a *$\exists$-matrix* of A if the following statements hold:

1. B does not begin with $\exists$;
2. A coincides with $\exists x_1 \ldots x_k B$, where $k \geqslant 0$.

q^*-type of the $\exists$-matrix of A will be called *$\exists q^*$-type* of A. For example, the formula

$$(t = \underbrace{y+y+\cdots+y}_{n \text{ times}}+z \& z < \underbrace{1+1+\cdots+1}_{n \text{ times}})$$

is the $\exists$-matrix of the formula (3). The $\exists q^*$-type of (3) is the formula

$$(a = b \& c < d).$$

The axiomatic theory **T** will be *the theory with a finite collection of $\exists q^*$-types of non-logical axioms* if for any i $(1 \leqslant i \leqslant k)$ $\exists q^*$-types of formulas from $\mathfrak{A}_i$ coincide up to renaming of free variables. For example, Presburger's arithmetic is a theory with a finite collection of $\exists q^*$-types of non-logical axioms. Other examples are axiomatic theories of algebraically closed with characteristic 0 and real-closed fields.

Let $\mathfrak{A}$ be the union of sets $\mathfrak{A}_1, \mathfrak{A}_2, \dots, \mathfrak{A}_k$ and the list Σ. We will consider the calculus **KGL**$(\mathfrak{A})$ from Orevkov (1993). This calculus contains the cut rule

$$\frac{\Gamma \to A, \Delta; \qquad \Gamma, A \to \Delta}{\Gamma \to \Delta}$$

and the $\mathfrak{A}$-cut rule for formulas in $\mathfrak{A}$

$$\frac{\Gamma, B \to \Delta}{\Gamma \to \Delta},$$

where $B \in \mathfrak{A}$.

Let B_i be the $\exists q^*$-type of formulas from $\mathfrak{A}_i$. Let $\mathfrak{A}_i^-$ denote the set of formulas A such that B_i is the q^*-type of A. Let $\mathfrak{A}^-$ be the union of sets $\mathfrak{A}_1^-$, $\mathfrak{A}_2^-, \dots, \mathfrak{A}_k^-$ and the list Σ.

Let D_1 and D_2 be proofs in **KGL**$(\mathfrak{A}^-)$. We will say that D_1 is *transformable into* D_2 if D_2 can be obtained from D_1 by simultaneous substitution of terms for free variables.

Lemma 2 *Let $A(x)$ be a formula, U be a deduction schema of a proof in* **KGL**$(\mathfrak{A}^-)$. *Suppose that U contains no applications of the cut rule and for a term t there is a proof of $A(t)$ in* **KGL**$(\mathfrak{A}^-)$ *whose deduction schema is U.*

Then there exists a term t_0 and a proof D_0 such that the following conditions hold:

1. *D_0 is a proof of $A(t_0)$ in* **KGL**$(\mathfrak{A}^-)$;
2. *the deduction schema of D_0 coincides with U;*
3. *for any term θ and any proof D of $A(\theta)$ in* **KGL**$(\mathfrak{A}^-)$ *if the deduction schema of D coincides with U, then D_0 is transformable into D;*

Theorem 6 *Let* **T** *be a theory with a finite collection of* $\exists q^*$*-types of non-logical axioms,* $A(x)$ *be a formula,* m *be a natural number,* n *be a sufficiently large natural number,* $\mathbf{c} = \mathbf{0}, \mathbf{1}$*. Suppose that for any number* i $(0 \leqslant i \leqslant n)$

$$\mathbf{T} \vdash_{\frac{\log_2 n}{3\log_2 \log_2 n}} A(\mathbf{c}_{m+i}).$$

Then there are a natural number k*, formulas* $B_1(x), B_2(x), \ldots, B_r(x)$ *and natural numbers* $l_1, l_2, \ldots, l_r$ *such that the following conditions hold:*

1. *formulas* $B_1(\mathbf{c}_{l_1}), B_2(\mathbf{c}_{l_2}), \ldots$ *and* $B_r(\mathbf{c}_{l_r})$ *are non-logical axioms of* **T**;
2. $k \leqslant n + m$;
3. $\mathbf{T} \vdash \forall x \left(\left(\bigwedge_{i=1}^{r} B_i(x) \right) \supset A((\cdots \underbrace{(x + \mathbf{c}) + \mathbf{c}) + \cdots + \mathbf{c}}_{k\ times}) \right).$

In the proofs of theorem 6 we use lemmas 1–2 and also some lemmas and estimates from Orevkov (1993). Theorems 1–3 are special cases of theorem 6.

3. Theory of feasible numbers with induction

Let **T** be an arbitrary consistent theory containing Peano's arithmetic **PA** (in the language with $=, <, 0, ', +$ and $\cdot$). Let θ be a term which contains no variables. The theory $\mathbf{T}_\theta$ is obtained by adding a new one-place predicate symbol F and the following non-logical axioms:

A. all non-logical axioms of **T** (in the language without predicate symbol F);

B. special axioms:

1. $F(0)$,
2. $\forall x(F(x) \supset F(x'))$,
3. $\forall xy(F(x) \supset (F(y) \supset F(x+y)))$,
4. $\forall xy(F(x) \supset (F(y) \supset F(x \cdot y)))$,
5. $\forall xy(x \leqslant y \supset (F(y) \supset F(x)))$,
6. $((A(0) \,\&\, \forall x(F(x) \supset (A(x) \supset A(x')))) \supset \forall x(F(x) \supset A(x)))$,
7. $\neg F(\theta)$.

Here formula A (in induction schema 6.) may contain predicate symbol F.

It is obvious that $\mathbf{T}_\theta$ is inconsistent. For the first time this theory (without scheme 6.) was given in Parikh (1971). The theory of feasible numbers with scheme 6 was considered in Bozhich (1987).

Let D be a proof in $\mathbf{T}_\theta$. Let's denote the number of different formulas containing predicate symbol F in D by $l_F(D)$.

Theorem 7 *There exists a function η such that the following conditions hold:*

1. *η defined by recursion up to ε_0;*
2. *for any formula B and any proof D of B in calculus $\mathbf{T}_\theta$, if B does not contain predicate symbol F and*

$$\mathbf{T} \vdash 0^{(\eta(l_F(D)))} < \theta,$$

then

$$\mathbf{T} \vdash B.$$

Here $0^{(\eta(l_F(D)))}$ denote 0 with $\eta(l_F(D))$ primes.

This theorem means that rather short proofs in $\mathbf{T}_\theta$ can give only true formulas. Our bounds depends only on the number of those members in a proof under consideration which contain predicate symbol F and *does not depend* on the complexity of formulas in this proofs.

R. Parikh proved the theorem about correctness of short proofs in theory with notions of feasibility without induction schema in Parikh (1971). Function η is a primitive recursive function in R. Parikh's theorem.

In the proof of theorem 7 we used the following lemma.

Lemma 3 *There exists a function φ such that the following conditions hold:*

1. *φ defined by recursion up to ε_0;*
2. *for any formula B and any proof D of B in calculus $\mathbf{T}_\theta$ if B does not contain predicate symbol F, then there is a proof D^* of B in which induction schema 6 is not applied and*

$$l[D^*] \leqslant \varphi(l[D]).$$

This lemma can also be used to obtain the following theorem.

Theorem 8 *Let A be a formula, θ and τ be terms. Suppose that the following conditions hold:*

1. *A does not contain predicate symbol F;*
2. *$\mathbf{T} \vdash \theta \neq \tau$;*
3. *there exists proof D_1 of A in $\mathbf{T}_\theta$ and proof D_2 of A in $\mathbf{T}_\tau$ such that deduction schema of D_1 and D_2 coincide.*

Then

$$\mathbf{T} \vdash A.$$

The condition 3 is main in this theorem.

4. Formula with high lower bound on the length of its proof

Let's consider a formula

$$\exists w((\forall y R(0,y,s(y)) \,\&\, \forall xz(R(x,s(0),z) \supset R(s(x),0,z)) \,\& \\ \&\, \forall xyzu((R(s(x),y,z) \,\&\, R(x,z,u)) \supset R(s(x),s(y),u))) \supset \\ \supset R(\underbrace{s(s(\ldots s}_{n \text{ times}}(0)\ldots)),\underbrace{s(s(\ldots s}_{m \text{ times}}(0)\ldots)),w)),$$

where s is an one-place function symbol.

This formula will be denoted by

$$\exists w A_{n,m}(w).$$

Let ψ be Ackermann's function defined by the recursion equation:

$$\begin{cases} \psi(0,n) = n+1, \\ \psi(m+1,0) = \psi(m,1), \\ \psi(m+1,n+1) = \psi(m,\psi(m+1,n)) \end{cases}$$

It is well known that Ackermann's function ψ grows faster than any primitive recursive function.

Let $\mathbf{C}$ be a Hilbert-style system for the classical predicate calculus.

Theorem 9

1. *For any term t*

$$\mathbf{C} \vdash A_{n,m}(t)$$

if and only if t is the term

$$\underbrace{s(s(\ldots s}_{\psi(n,m)\, times}(0)\ldots)).$$

2. *There is no primitive recursive function f with*

$$\mathbf{C} \vdash_{f(n)} \exists w A_{n,n}(w).$$

So $\exists w A_{n,n}(w)$ has very high lower bound on the length of its proof in predicate calculus.

Let F be a one-place predicate symbol. The theory $\mathbf{C}_F$ is obtained by adding the following non-logical axioms:

1. $F(0)$,

2. $\forall x(F(x) \supset F(s(x)))$,

3. $((A(0) \,\&\, \forall x(F(x) \supset (A(x) \supset A(s(x))))) \supset \forall x(F(x) \supset A(x)))$.

Here formula A (in induction schema 3.) may contain predicate symbol F.

Theorem 10 *For any formula B, if B does not contain predicate symbol F and*

$$\mathbf{C}_F \vdash B,$$

then

$$\mathbf{C} \vdash B.$$

So **C** is conservative extension of $\mathbf{C}_F$.

Let's compare lengths of direct and indirect proofs of existence theorems in predicate calculus with an induction schema.

Theorem 11

1. *There exists a proof D of* $\exists w A_{n,n}(w)$ *in* $\mathbf{C}_F$ *such that* $l[D]$ *does not exceed*

$$c_1 \log_2(n+1) + c_2.$$

2. *For any sufficiently large n and any term t the length of any proof of* $A_{n,n}(t)$ *in* $\mathbf{C}_F$ *exceeds*

$$c \cdot \sqrt{\psi(n,n)}.$$

3. *For infinitely many natural numbers n the length of any proof of* $\exists w A_{n,n}(w)$ *in* $\mathbf{C}_F$ *exceeds*

$$\frac{\log_2 n}{3 \log_2 \log_2 n}.$$

Here constants c, c_1 *and* c_2 *do not depend on the choice of a proof, t and n.*

5. Theory of feasible numbers with one-place functions

Let $\mathbf{PA}^1$ be a system of Peano's arithmetic whose language contains the constant 0, one-place function $'$, one-place function symbols for doubling and squaring, $=$ and three-place predicate for addition and multiplication. Let θ be a term which contains no variables. The theory $\mathbf{PA}^1{}_\theta$ is obtained by analogy with $\mathbf{T}_\theta$.

Let t be a term. Let's define the *height* $h[t]$ in a standard way:

1. if t is a variable or constant, then $h[t] = 0$;

2. $h[g(t_1, \ldots, t_k)] = 1 + \max\{h[t_1], \ldots, h[t_k]\}$.

The function $\mathbf{2}_i^n$ is defined by the recursion equations

$$\begin{cases} \mathbf{2}_0^n = n, \\ \mathbf{2}_{i+1}^n = 2^{(\mathbf{2}_i^n)}. \end{cases}$$

Theorem 12 *For any formula A and any proof D of A in* $\mathbf{PA^1}_\theta$, *if A does not contain predicate symbol F and*

$$\mathbf{PA^1} \vdash \mathbf{2}_4^m < \theta,$$

then

$$\mathbf{PA^1} \vdash A,$$

where

$$m = 12l + 2\rho + \log_2 \log_2(h+4),$$
$$l = l[D],$$

ρ *is the number of different subformulas in A and h is the maximal height of terms in A.*

In the proof of this theorem we used an upper bound on the exponent of periodicity of a solution of a system of word equations.

References

Bozhich, E.S.: (1987), 'On arithmetic with the notion of a feasible number', *Math. USSR-Izv.* **29**, no. 3, pp. 477–510. [Russian original in: Izv. Akad. Nauk SSSR, Ser. Mat., 50 (1986), no. 6, pp. 1123–1155].

Kraíček, J., Pudlák, P.: (1988), 'The number of proofs lines and size of proofs in first order logic', *Arch. Math. Logic* **27**, pp. 69–84.

Orevkov, V.P.: (1982), 'Lower bounds on the increase in complexity of deductions in cut elimination', *Journal of Soviet Mathematics* **20**, no. 4. [Russian original in: *Zapiski Nauchnykh Seminarov LOMI* **88** (1979), pp. 137–162].

Orevkov, V.P.: (1993), *Complexity of proofs and their transformations in axiomatic theories*, Amer. Math. Soc., Providence, Rhode Island.

Parikh, R.: (1971), 'Existence and Feasibility in Arithmetic', *J. Symb. Logic* **36**, pp. 494–508.

II

MODEL THEORY

FEW NON MINIMAL TYPES ON NON STRUCTURE*

SH603

Saharon Shelah
Institute of Mathematics, The Hebrew University
Jerusalem, Israel
Rutgers University, Department of Mathematics
New Brunswick, NJ USA

Abstract We deal with abstract elementary classes $\mathfrak{K}$ which has amalgamation in λ. Our main result is that if ($2^\lambda < 2^{\lambda^+}$ and) the minimal types on members of K_λ are not dense (among non algebraic (complete) types over models in K_λ, extending our given model), then the number of models in K of cardinality λ^+ or λ^{++} is maximal. For this we deal with some claims in pcf. This improves a result in Shelah (2001), but the amount of relying is small, mostly of a "black box" character.

Keywords: model theory, abstract elementary classes, classification theory, categoricity, nonstructure theory, pcf theory

Annotated Content

0. Introduction [We explain our aim and define our framework.]

1. Non minimal types and nonstructure [We define unique amalgamation, UQ, and try to use it for building many models in λ^- when $2^\lambda < 2^{\lambda^+}$ (so the weak diamond holds). If this approach fails we still get the many models in λ^{++} by the "easy" criterion of Shelah (2001, §3) but it works only if the weak diamond ideal on λ^+ is not λ^{++}-saturated.]

2. Remarks on pcf [We prove some pcf observations needed here.]

*I thank Alice Leonhardt for the beautiful typing
Split from Sh600; work done Spring of 1995
Latest version - 2000/Apr/6

P. Gärdenfors, J. Woleński and K. Kijania-Placek (eds.), In the Scope of Logic, Methodology and Philosophy of Science, Vol. I, 29-53.

3. Finishing the many models [We prove the result of Section 1 without the extra assumption on the saturation of the weak diamond ideal.]

4. A minor debt [There was one point in Shelah (2001) where we use $\lambda > \aleph_0$, though our aim there was to generalize theorem known for $\lambda = \aleph_0$. We eliminate this use.]

0. Introductions

In Shelah (2001) there was an important point where we used as assumption $I(\lambda^{+3}, K) = 0$. This was fine for the purpose there, but is unsuitable in other frameworks, like Shelah (200x): we want to analyze what occurs in higher cardinals, so our main aim here is to eliminate its use and add to our knowledge on non-structure.

The point was "the minimal triples in K^3_λ are dense" (Shelah 2001, 3.17t). For this we assume we have a counterexample, and try to build many nonisomorphic models. Hence we get cases of amalgamation which are necessarily unique. Those "unique amalgamations" are normally too strong (even for first order superstable theories), but here they help us to prove positive theorems, controlling omitting types. So we try to build many models in λ^+ by omitting "types" over models of size λ, in a specific way where unique amalgamation holds. If this argument fails, we prove $\boldsymbol{C}^1_{\mathfrak{K},\lambda}$ has weak λ^+-coding (see Shelah 2001, §3) and by it get $2^{\lambda^{++}}$ non-isomorphic models except when the weak diamond ideal on λ^+ is λ^{++}-saturated; this is done in Section 1. In Section 3 we work harder and by partition to cases relying on pcf theory we succeed to get the full result. We work also to get large IE (many models no one $\leqslant_{\mathfrak{K}}$-embedding to another). The pcf lemmas (which are pure infinite combinatorics) are dealt with in Section 2.

There was also another point left in Shelah (2001, 4.2t), for the case $\lambda = \aleph_0$ only, this is filled in Section 4.

* * *

Definition 0.1 We say $\mathfrak{K} = (K, \leqslant_{\mathfrak{K}})$ is an abstract elementary class, aec or a.e.c. in short, if ($\tau = \tau_K$ is a fixed vocabulary, K a class of τ-models (and $Ax\ 0$ holds and)) $Ax\ I - VI$ hold where:

$Ax\ 0$: The holding of $M \in K, N \leqslant_{\mathfrak{K}} M$ depends on N, M only up to isomorphism i.e. $[M \in K, M \cong N \Rightarrow N \in K]$, and [if $N \leqslant_{\mathfrak{K}} M$ and f is an isomorphism from M onto the τ-model M' mapping N onto N' <u>then</u> $N' \leqslant_{\mathfrak{K}} M'$].

$Ax\ I$: If $M \leqslant_{\mathfrak{K}} N$ then $M \subseteq N$ (i.e. M is a submodel of N).

$Ax\ II$: $M_0 \leqslant_{\mathfrak{K}} M_1 \leqslant_{\mathfrak{K}} M_2$ implies $M_0 \leqslant_{\mathfrak{K}} M_2$ and $M \leqslant_{\mathfrak{K}} M$ for $M \in K$.

Ax III: If λ is a regular cardinal, $M_i\,(i<\lambda)$ is $\leqslant_{\mathfrak{K}}$-increasing (i.e. $i<j<\lambda$ implies $M_i \leqslant_{\mathfrak{K}} M_j$) and continuous (i.e. for limit ordinal $\delta<\lambda$ we have $M_\delta = \bigcup_{i<\delta} M_i$) then $M_0 \leqslant_{\mathfrak{K}} \bigcup_{i<\lambda} M_i$.

Ax IV: If λ is a regular cardinal, $M_i\,(i<\lambda)$ is $\leqslant_{\mathfrak{K}}$-increasing continuous, $M_i \leqslant_{\mathfrak{K}} N$ then $\bigcup_{i<\lambda} M_i \leqslant_{\mathfrak{K}} N$.

Ax V: If $M_0 \subseteq M_1$ and $M_\ell \leqslant_{\mathfrak{K}} N$ for $\ell = 0,1$, then $M_0 \leqslant_{\mathfrak{K}} M_1$.

Ax VI: $LS(\mathfrak{K})$ exists[1], where $LS(\mathfrak{K})$ is the minimal cardinal λ such that: if $A \subseteq N$ and $|A| \leqslant \lambda$ then for some $M \leqslant_{\mathfrak{K}} N$ we have $A \subseteq |M \leqslant \lambda$ and we demand for simplicity $|\tau| \leqslant \lambda$.

Notation 0.2 1) $K_\lambda = \{M \in K : \|M\| = \lambda\}$ and $K_{<\lambda} = \bigcup_{\mu<\lambda} K_\mu$.
See more in Shelah (2001, §0).

Definition 0.3

1) For $\mu \geqslant LS(\mathfrak{K})$ and $M \in K_\mu$ we define $\mathscr{S}(M)$ as $\{\mathrm{tp}(a,M,N) : M \leqslant_{\mathfrak{K}} N \in K_\mu \text{ and } a \in N\}$ where $\mathrm{tp}(a,M,N) = (M,N,a)/E_M$ where E_M is the transitive closure of E^{at}_M, and the two-place relation E^{at}_M is defined by:

$$(M,N_1,a_1)E^{at}_M(M,N_2,a_2) \text{ \underline{iff} there is } N \in K_\mu \text{ and } \leqslant_{\mathfrak{K}}\text{-embeddings}$$
$$f_\ell : N_\ell \to N \text{ for } \ell = 1,2 \text{ such that:}$$
$$f_1 \restriction M = \mathrm{id}_M = f_2 \restriction M \text{ and } f_1(a_1) = f_2(a_2).$$
$$(\text{of course } M \leqslant_{\mathfrak{K}} N_1, M \leqslant_{\mathfrak{K}} N_2 \text{ and } a_1 \in N_1, a_2 \in N_2)$$

2) We say "a realizes p in N" if $a \in N, p \in \mathscr{S}(M)$ and for some $N' \in K_\mu$ we have $M \leqslant_{\mathfrak{K}} N' \leqslant_{\mathfrak{K}} N$ and $a \in N'$ and $p = \mathrm{tp}(a,M,N')$; so $M, N' \in K_\mu$ but possibly $N \notin K_\mu$.

3) We say "a_2 strongly realizes $(M,N^1,a^1)/E^{at}_M$ in N" if for some N^2, a^2 we have $M \leqslant_{\mathfrak{K}} N^2 \leqslant_{\mathfrak{K}} N$ and $a_2 \in N^2$ and $(M,N^1,a^1)\,E^{at}_M\,(M,N^2,a^2)$. (Note: if M_0 is an amalgamation base, see below, then the difference between realize and strongly realize disappears).

4) We say $M_0 \in \mathfrak{K}_\lambda$ is an amalgamation base if: for every $M_1, M_2 \in \mathfrak{K}_\lambda$ and $\leqslant_{\mathfrak{K}}$-embeddings $f_\ell : M_0 \to M_\ell$ (for $\ell = 1,2$) there is $M_3 \in \mathfrak{K}_\lambda$ and $\leqslant_{\mathfrak{K}}$-embeddings $g_\ell : M_\ell \to M_3$ (for $\ell = 1,2$) such that $g_1 \circ f_1 = g_2 \circ f_2$.

5) We say $\mathfrak{K}$ is stable in λ if $LS(\mathfrak{K}) \leqslant \lambda$ and $M \in K_\lambda \Rightarrow |\mathscr{S}(M)| \leqslant \lambda$.

6) We say N is λ-universal over M if for every $M', M \leqslant_{\mathfrak{K}} M' \in K_\lambda$, there is a $\leqslant_{\mathfrak{K}}$-embedding of M' into N over M. If we omit λ we mean $\|N\|$.

7) We say N is (λ,κ)-saturated over M if there is as $\leqslant_{\mathfrak{K}}$-increasing continuous sequence $\langle M_i : i \leqslant \kappa\rangle$, such that $M = M_0, M_\kappa = N, M_i \in K_\lambda, M_{i+1}$ is an amalgamation base, M_{i+1} is universal over M_i. Replacing κ by $*$ means "for some $\kappa = \operatorname{cf}(\kappa) \leqslant \lambda$". We omit "over M" to mean "for some $M \in K_\lambda$".

8) $K^3_\lambda = \{(M,N,a) : M \leqslant_{\mathfrak{K}} N, a \in N \backslash M$ and $M,N \in \mathfrak{K}_\lambda\}$, with the partial order $\leqslant$ defined by $(M,N,a) \leqslant (M',N',a')$ iff $a = a', M \leqslant_{\mathfrak{K}} M'$ and $N \leqslant_{\mathfrak{K}} N'$.

9) We say (M,N,a) is minimal if $(M,N,a) \leqslant (M',N_\ell,a) \in K^3_\lambda$ for $\ell = 1,2$ implies $\operatorname{tp}(a,M',N_1) = \operatorname{tp}(a,M',N_2)$.

10) We say $\mathfrak{K}$ has amalgamation in λ if every $M \in \mathfrak{K}_\lambda$ is an amalgamation base.

1. Non-minimal triples and non-structure

We assume here some knowledge of Shelah (2001, §2).

Context 1.1

(a) $\mathfrak{K}$ abstract elementary class with $LS(\mathfrak{K}) \leqslant \lambda$

(b) $\mathfrak{K}$ has amalgamation in λ.

Remark. Alternatively we can use a weaker context.

Definition 1.2

1) For $x \in \{a,d\}$ we say $UQ^x_\lambda(M_0,M_1,M_2,M_3)$ if:

   (a) $M_\ell \in K_\lambda$ for $\ell \leqslant 3$

   (b) $M_0 \leqslant_{\mathfrak{K}} M_\ell$ for $\ell = 1,2$

   (c) if for $i \in \{1,2\}$ we have $M^i_\ell \in K_\lambda$, for $\ell < 4$ and $M^i_0 \leqslant_{\mathfrak{K}} M^i_\ell \leqslant_{\mathfrak{K}} M^i_3$ for $i = 1,2, \ell = 1,2$ and $[x{=}d \Rightarrow M^i_1 \cap M^i_2 = M^i_0]$ and f^i_ℓ is an isomorphism from M_ℓ onto M^i_ℓ for $\ell < 3$ and $f^i_0 \subseteq f^i_1, f^i_0 \subseteq f^i_2$ <u>then</u> there are M'_3, f_3 such that $M^2_3 \leqslant_{\mathfrak{K}} M'_3$ and f_3 is a $\leqslant_{\mathfrak{K}}$-embedding of M^1_3 into M'_3 extending $(f^2_1 \circ (f^1_1)^{-1}) \cup (f^2_2 \circ (f^1_2)^{-1})$ i.e. $f_3 \circ f^1_1 = f^2_1 \,\&\, f_3 \circ f^1_2 = f^2_2$

   (d) $M_0 \leqslant_{\mathfrak{K}} M_\ell \leqslant_{\mathfrak{K}} M_3 \in K_\lambda$ for $\ell = 1,2$

   (e) $x = d \Rightarrow M_1 \cap M_2 = M_0$.

2) We say $UQ^x_\lambda(M_0,M_1,M_2)$ if $UQ^x_\lambda(M_0,M'_1,M'_2,M_3)$ for some M_3 and M'_1,M'_2 isomorphic over M_0 to M_1,M_2 respectively.

3) If we omit x, we mean $x = a$.

4) $K_\lambda^{3,*}$ is the family of triple $(M,N,a) \in K_\lambda^3$ such that there is no minimal triple above it.

5) $K_\lambda^{2,*}$ is the family $\{(M,N) :$ for some $a, (M,N,a) \in K_\lambda^{3,*}\}$.

6) $\mathscr{S}_*(M) = \{p \in \mathscr{S}(M)$: for some $(M,N,a) \in K_\lambda^{3,*}$ we have $p = \text{tp}(a,M,N)\}$.

Claim 1.3

1) Symmetry: assuming $x \in \{a,d\}$ we have $UQ_\lambda^x(M_0,M_1,M_2,M_3) \Rightarrow UQ_\lambda^x(M_0,M_2,M_1,M_3)$; we can also omit M_3.

2) $UQ_\lambda^a(M_0,M_1,M_2) \Rightarrow UQ_\lambda^d(M_0,M_1,M_2)$ recalling M_0 is an amalgamation base (in $\mathfrak{K}_\lambda$) by clause (b) of 1.1.

3) $UQ_\lambda^a(M_0,M_1,M_2,M_3)$ iff clauses (a), (b), (d), (e) of Definition 1.2(1),(2) hold and also (c)$^-$, i.e. clause (c) restricted to the case $M_\ell^1 = M_\ell$ for $\ell \leqslant 3$.

4) If $UQ_\lambda^x(M_0,M_1,M_2,M_3), M_3 \leqslant_{\mathfrak{K}} M_3' \in K_\lambda$ then $UQ_\lambda^x(M_0,M_1,M_2,M_3')$; and also the inverse: if $UQ_\lambda^x(M_0,M_1,M_2,M_3')$ and $M_1 \cup M_2 \subseteq M_3 \leqslant_{\mathfrak{K}} M_3'$ then $UQ_\lambda^x(M_0,M_1,M_2,M_3)$.

5) Assume $(M,N,a) \in K_\lambda^3$ and it is not minimal then $\neg UQ(M,N,N)$.

Proof. 1), 2) Trivial.
3) Chasing arrows, we should prove clause (c) of Definition 1.2(1). Assume we are given $\langle M_\ell^1 : \ell < 4\rangle, \langle M_\ell^2 : \ell < 4\rangle, \langle f_\ell^i : \ell < 3\rangle$ as there for $i = 1,2$. First for $i = 1,2$ apply clause $(c)^-$ to $\langle M_\ell^i : \ell < 4\rangle, \langle f_\ell^i : \ell < 3\rangle$. So there are N_3^i, f_3^i such that: $M_3^i \leqslant_{\mathfrak{K}} N_3^i \in K_\lambda$, and f_3^i a $\leqslant_{\mathfrak{K}}$-embedding of M_3 into N_3^i extending $f_1^i \cup f_2^i$. As $\mathfrak{K}$ has amalgamation in λ (by 1.1(b)) there are $N \in K_\lambda$ and $\leqslant_{\mathfrak{K}}$-embeddings $g^i : N^i \to N$ such that $g^1 \circ f_3^1 = g^2 \circ f_3^2$, so we are done.
4) Again by the amalgamation i.e. 1.1(b).
5) Easy; and essentially by Shelah (2001, 2.6t(1)). $\square_{1.3}$

Claim 1.4

1) transitivity: If $UQ_\lambda(M_\ell,N_\ell,M_{\ell+1},N_{\ell+1})$ for $\ell = 0,1$ then $UQ_\lambda(M_0,N_0,M_2,N_2)$.

2) If $\theta = cf(\theta) < \lambda^+$, and $\langle M_i : i \leqslant \theta\rangle$ is $\leqslant_{\mathfrak{K}}$-increasing continuous and $\langle N_i : i \leqslant \theta\rangle$ is $\leqslant_{\mathfrak{K}}$-increasing and $UQ_\lambda(M_i,N_i,M_{i+1},N_{i+1})$ for each $i < \theta$ then $UQ_\lambda(M_0,N_0,M_\theta,N_\theta)$.

3) *Assume:*

(a) $\alpha, \beta < \lambda^+$

(b) $M_{i,j} \in K_\lambda$ *for* $i \leqslant \alpha, j \leqslant \alpha$

(c) $i_1 \leqslant i_2 \leqslant \alpha \,\&\, j_1 \leqslant j_2 \leqslant \beta \Rightarrow M_{i_1,j_1} \leqslant_{\mathfrak{K}} M_{i_2,j_2}$

(d) $\langle M_{i,j} : i \leqslant \alpha \rangle$ *is* $\leqslant_{\mathfrak{K}}$*-increasing continuous for each* $j \leqslant \beta$

(e) $\langle M_{i,j} : j \leqslant \beta \rangle$ *is* $\leqslant_{\mathfrak{K}}$*-increasing continuous for each* $i \leqslant \alpha$

(f) $UQ_\lambda(M_{i,j}, M_{i+1,j}, M_{i,j+1}, M_{i+1,j+1})$.

Then $UQ_\lambda(M_{0,0}, M_{\alpha,0}, M_{0,\beta}, M_{\alpha,\beta})$.

4) *If* $UQ^x_\lambda(M_0, M_1, M_2)$ *and* $M_0 \leqslant_{\mathfrak{K}} M'_1 \leqslant_{\mathfrak{K}} M_1$ *and* $M_0 \leqslant_{\mathfrak{K}} M'_2 \leqslant_{\mathfrak{K}} M_2$ *then* $UQ_\lambda(M_0, M'_1, M'_2)$.

Proof. Chasing arrows (note: $UQ = UQ^a$ is easier than UQ^d, for UQ^d the parallel claim is not clear at this point, e.g. seemingly transitivity fails). $\square_{1.4}$

Definition 1.5

1) $\mathscr{T}_\lambda[\mathfrak{K}] = \{(M, \Gamma) : M \in K_\lambda, \Gamma \subseteq K_\lambda, |\Gamma| \leqslant \lambda \text{ and } N \in \Gamma \Rightarrow M \leqslant_{\mathfrak{K}} N\}$.

2) We define relations on $\mathscr{T}_\lambda[\mathfrak{K}]$:

(α) $(M_1, \Gamma_1) \leqslant_h (M_2, \Gamma_2)$ iff

(a) $M_1 \leqslant_{\mathfrak{K}} M_2$

(b) h is a partial function from Γ_2 onto Γ_1

(c) if $h(N_2) = N_1$ then $UQ_\lambda(M_1, N_1, M_2, N_2)$.

(β) $(M_1, \Gamma_1) \leqslant^h (M_2, \Gamma_2)$ iff h is a one to one function from Γ_1 to Γ_2 such that $(M_1, \Gamma_2) \leqslant_{h^{-1}} (M_2, \Gamma_2)$

(γ) $(M_1, \Gamma_1) \leqslant^{h_2}_{h_1} (M_2, \Gamma_2)$ iff $(M_1, \Gamma_1) \leqslant_{h_1} (M_2, \Gamma_2), (M_1, \Gamma_2) \leqslant^{h_2} (M_2, \Gamma_2)$ and $h_1 \circ h_2 = \mathrm{id}_{\Gamma_1}$.

3) We write $<_h$ (or $<^h$ or $<^{h_2}_{h_1}$) if in addition $M_1 \neq M_2$, moreover[2] $(M_1, M_2) \in K^{2,*}_\lambda$.

4) $(M_1, \Gamma_1) \leqslant (M_2, \Gamma_2)$ means:

(a) $\langle N \backslash M_\ell : N \in \Gamma_\ell \rangle$ are pairwise disjoint non-empty (and $(M_\ell, \Gamma_\ell) \in \mathscr{T}_\lambda[\mathfrak{K}]$) for $\ell = 1, 2$

(b) for every $N_1 \in \Gamma_1$ there is exactly one $N_2 \in \Gamma_2$ such that $(N_1 \backslash M_1) \cap (N_2 \backslash M_2) \neq \emptyset$

(c) $(M_1, \Gamma_1) \leqslant^h (M_2, \Gamma_2)$ when $h : \Gamma_1 \to \Gamma_2$ is defined by: $h(N_1) = N_2 \leftrightarrow (N_1 \backslash M_1) \cap (N_2 \backslash M_2) \neq \emptyset$.

(So $(M_1,\Gamma_1) \leqslant (M_1,\Gamma_1)$ just means clause (a) and we call (M_1,Γ_1) disjoint).

5) Let $\mathscr{T}_\lambda^{dis}[\mathfrak{K}] = \{(M,\Gamma) \in \mathscr{T}_\lambda[\mathfrak{K}] : (M,\Gamma) \leqslant (M,\Gamma)\}$.

Remark.

1) Of course, except for bookkeeping only the isomorphism type of the $N \in \Gamma$ over M matters.

2) In 1.6(2) below, if the continuity is not demanded, the result may be that $h_{i,\delta}$ is not onto Γ_i. So we may use $\leqslant_{h_1}^{h_2}$.

3) Note that we can use 1.6(3) and not 1.6(2).

Claim 1.6

1) If $(M_1,\Gamma_1) \leqslant_{h_1} (M_2,\Gamma_2) \leqslant_{h_2} (M_3,\Gamma_3)$ then $(M_1,\Gamma_1) \leqslant_{h_1 \circ h_2} (M_3,\Gamma_3)$.

2) If $\delta < \lambda^+$ is a limit ordinal, for $i < j < \delta, (M_i,\Gamma_i) \leqslant^{h_{i,j}} (M_j,\Gamma_j)$ and the $h_{i,j}$ commute, and for limit $\alpha < \delta, (\Gamma_\alpha, \langle h_{i,\alpha} : i < \alpha\rangle)$ is the (direct) limit of $\langle \Gamma_i, h_{i,j} : i < j < \alpha\rangle$ <u>then</u> we can find (M_δ,Γ_δ) and $\langle h_{i,\delta} : i < \delta\rangle$ such that $(M_i,\Gamma_i) \leqslant_{h_{i,\delta}} (M_\delta,\Gamma_\delta)$ and $M_\delta = \bigcup_{i<\delta} M_i$ and $(\Gamma_\delta, \langle h_{i,\delta} : i < \delta\rangle)$ is the limit of $\langle \Gamma_i, h_{i,j} : i < j < \delta\rangle$.

3) $(\mathscr{T}_\lambda[\mathfrak{K}], \leqslant)$ is a partial order in which any increasing sequence of length $< \lambda^+$ has a lub (= the natural limit). (Actually, only $\mathscr{T}_\lambda^{dis}[\mathfrak{K}]$ matters.)

Proof. 1) The main point is the use of 1.4(1).
2) In addition to chasing arrows we use 1.4(2) (or 1.4(3) for $\beta = 1$).
3) Reformulation of 1), 2). $\square_{1.6}$

Claim 1.7 *Assume $2^\lambda < 2^{\lambda^+}$ or at least the definitional*[3] *weak diamond; i.e. DfWD$^+(\lambda^+)$, see Shelah (2001, 1.7t). If $(*)_\lambda$ or at least $(*)'_\lambda$ below holds (hence above some triple from K_λ^3 there is no minimal one), <u>then</u>*

$$I(\lambda^+, K) \geqslant \mu_{wd}(\lambda^+) \text{ where}$$

$(*)_\lambda$ *for every $(M,\Gamma) \in \mathscr{T}_\lambda[\mathfrak{K}]$ for some h, M', Γ' we have $(M,\Gamma) <^h (M',\Gamma')$ (without loss of generality h is onto Γ') or just*

$(*)'_\lambda$ *for some $(M_0,\Gamma_0) \in \mathscr{T}_\lambda[\mathfrak{K}]$, if $(M_0,\Gamma_0) \leqslant^{h_0} (M,\Gamma)$ then for some h, M', Γ' we have $(M,\Gamma) <^h (M',\Gamma')$.*

Proof. We choose by induction on $\alpha < \lambda$, $\langle (M_\eta, \Gamma_\eta, \Gamma_\eta^+) : \eta \in {}^\alpha 2\rangle$ such that:

(a) $M_\eta \in K_\lambda$ has universe $\gamma_\eta < \lambda^+$

(b) $(M_\eta, \Gamma_\eta) \in \mathscr{T}_\lambda[\mathfrak{K}]$

(c) (M_η, Γ_η) is disjoint (see 1.5(4)(b)) and $N \in \Gamma_\eta \to (N \backslash M_\eta) \cap \lambda^+ = \emptyset$

(d) $\nu \triangleleft \eta \Rightarrow (M_\nu, \Gamma_\nu) < (M_\eta, \Gamma_\eta)$

(e) $(M_\eta, \Gamma_\eta^+) \in \mathscr{T}_\lambda[\mathfrak{K}]$ is disjoint (see 1.5(4)(b)) and $N \in \Gamma_\eta^+ \Rightarrow (N \backslash M_\eta) \cap \lambda^+ = \emptyset$

(f) $\Gamma_\eta \subseteq \Gamma_\eta^+$

(g) $(M_\eta, \Gamma_\eta^+) \leqslant (M_{\eta\hat{}\langle 0\rangle}, \Gamma_{\eta\hat{}\langle 0\rangle})$

(h) for some $N \in \Gamma_\eta^+$ we have $N \cong_{M_\eta} M_{\eta\hat{}\langle 1\rangle}$

(i) $(M_\eta, M_{\eta\hat{}\langle 1\rangle}) \in K_\lambda^{2,*}$, that is for some $a \in M_{\eta\hat{}\langle 1\rangle}$ we have $(M_\eta, M_{\eta\hat{}\langle 1\rangle}, a) \in K_\lambda^3$ and above it there is no minimal triple from K_λ^3.

There is no problem to carry the induction with Γ_η^+ ($\eta \in {}^\alpha 2$) chosen in the $(\alpha+1)$-th step. For $\alpha = 0$ let $(M_{<>}, \Gamma_{<>})$ be the (M_0, Γ_0) from $(*)'_\lambda$ except that we rename the elements to make the relevant parts of clauses (a), (c) true. For α limit use 1.6(3) (part on lub). For $\alpha = \beta+1, \eta \in {}^\beta 2$, by $(*)'_\lambda$ we can find $(M_{\eta\hat{}\langle 1\rangle}, \Gamma_{\eta\hat{}\langle 1\rangle})$ such that $(M_{\eta\hat{}\langle 1\rangle}, \Gamma_{\eta\hat{}\langle 1\rangle}) \in \mathscr{T}_\lambda[\mathfrak{K}], (M_\eta, \Gamma_\eta) < (M_{\eta\hat{}\langle 1\rangle}, \Gamma_{\eta\hat{}\langle 1\rangle})$. So by the Definition 1.5(3) of $<$, for some $a \in M_{\eta\hat{}\langle 1\rangle} \backslash M_\eta$, no triple in K_λ^3 above $(M_\eta, M_{\eta\hat{}\langle 1\rangle}, a)$ is minimal. By renaming without loss of generality the universe of $M_{\eta\hat{}\langle 1\rangle}$ is some $\gamma_{\eta\hat{}\langle 1\rangle} (\in (\gamma_\eta, \lambda^+))$ and clause (c) holds. Let N_η be isomorphic to $M_{\eta\hat{}\langle 1\rangle}$ over M_η with universe disjoint to $\lambda^+ \cup \bigcup_{N \in \Gamma_\eta} |N| \backslash \gamma_\eta$ and let $\Gamma_\eta^+ = \Gamma_\eta \cup \{N_\eta\}$, so $(M_\eta, \Gamma_\eta^+) \in \mathscr{T}_\lambda[\mathfrak{K}]$ is disjoint, now apply to it $(*)'_\lambda$ to get $(M_{\eta\hat{}\langle 0\rangle}, \Gamma_{\eta\hat{}\langle 0\rangle})$. Why does clause (h) hold? By the choice of N_η. So $M_\eta, \Gamma_\eta, \Gamma_\eta^+$ ($\eta \in {}^{\lambda^+}2$) are defined.

Note: if $\eta\hat{}\langle 0\rangle \triangleleft \nu \in {}^{\lambda^+ >}2$, <u>then</u> $M_{\eta\hat{}\langle 1\rangle}$ is not $\leqslant_{\mathfrak{K}}$-embeddable into M_ν over M_η (by clause (g) + (h) because by 1.3(5) and clause (i) we have

$$\neg UQ(M_\eta, M_{\eta\hat{}\langle 1\rangle}, N_\eta)).$$

By Shelah (2001, 1.4t) we get the desired conclusion (really, usually also on IE). $\square_{1.7}$

Claim 1.8 *[Categoricity in λ] An equivalent condition for $(*)_\lambda$ or just $(*)'_\lambda$ of 1.7 is (respectively):*

$(**)_\lambda$ *for every $M \leqslant_{\mathfrak{K}} N$ from K_λ for some M' we have $M <_{\mathfrak{K}} M' \in K_\lambda$ and $UQ_\lambda(M, M', N) \& (M, M') \in K_\lambda^{2,*}$*

or just

$(**)'_\lambda$ *for some* $M_0 \in K_\lambda$ *if* $M_0 \leqslant_{\mathfrak{K}} M \leqslant_{\mathfrak{K}} N \in K_\lambda$ *then for some* M', $M <_{\mathfrak{K}} M' \in K_\lambda$ *and* $UQ_\lambda(M, M', N) \& (M, M') \in K^{2,*}_\lambda$.

Proof. For any $(M, \Gamma) \in \mathscr{T}_\lambda[\mathfrak{K}]$, by "$\mathfrak{K}$ has amalgamation in λ" (and properties of abstract elementary classes) there are $N^*, \langle f_N : N \in \Gamma\rangle$ such that:

(a) $M \leqslant_{\mathfrak{K}} N^* \in K_\lambda$

(b) for $N \in \Gamma, f_N$ is a $\leqslant_{\mathfrak{K}}$-embedding of N into N^* over M.

This shows $(**)_\lambda \Rightarrow (*)_\lambda$ and also $(**)'_\lambda \Rightarrow (*)'_\lambda$.
The other direction are by applying $(*)_\lambda$ (or $(*)'_\lambda$) to $(M, \{N\})$. $\square_{1.8}$

Claim 1.9 *Assume*

(a) $(**)_\lambda$ *of 1.8 fails*

(b) $M \in K_\lambda \Rightarrow |\mathscr{S}_*(M)| > \lambda^+$ *(follows from "above* $(M, N, a) \in K^3_\lambda$ *there is no minimal triple"* $+2^\lambda > \lambda^+$*)*

(c) K is categorical in λ

(d) K is categorical in λ^+

(e) $2^\lambda < 2^{\lambda^+} < 2^{\lambda^{++}}$.

Then $I(\lambda^{++}, K) = 2^{\lambda^{++}}$ *except possibly when*

$(*)_\lambda$ *WDmId*(λ^+) *is* λ^{++}*-saturated (normal ideal on* λ^+*).*

Proof. We rely on Shelah (2001, §3). By (a) (and (c)) for every $M \in K_\lambda$ for some $N = N_{[M]}$ we have $M <_{\mathfrak{K}} N \in K_\lambda$ and without loss of generality even $(M, N) \in K^{2,*}_\lambda$ and $(M, M') \in K^{2,*}_\lambda \Rightarrow \neg UQ(M, M', N)$, of course, using N' where $N \leqslant_{\mathfrak{K}} N' \in K_\lambda$ will do too as $\mathfrak{K}$ has amalgamation in λ.

Let $\boldsymbol{C} = \boldsymbol{C}^1_{\lambda^+}[\mathfrak{K}]$, see Shelah (2001, Definition 3.3t). So there is $\langle M^*_i : i < \lambda^+\rangle$ such that $(M^*_i, M^*_{i+1}) \cong (M^*_i, N_{[M^*_i]})$.

Now for $\langle M_i : i < \lambda^+\rangle \in \boldsymbol{Seq}_{\lambda^+}[\boldsymbol{C}]$, by clause (b) there is $p^* \in \mathscr{S}_*(M_0)$ not realized in $\bigcup_{i<\lambda^+} M_i$ hence $M_i <_{\mathfrak{K}} M' \in K_\lambda \&$ (some $a \in M'$ realizes p^*) $\Rightarrow$ $(M, M') \in K^{2,*}_\lambda$; for a club of i's, for some $j \in (i, \lambda^+)$ we have, by 1.4(4) as K is categorical in λ^+ and in λ:

(α) for some $\leqslant_{\mathfrak{K}}$-embedding f_i of $N_{[M^*_i]}$ into M_j, f_i maps M^*_i onto M_i

so by assumption (a)

(β) $M_i <_{\mathfrak{K}} M' \in K_\lambda$ & some $a \in M'$ realizes $p^* \Rightarrow \neg UQ_\lambda(M_i, M_j, M')$.

From this it follows that for some λ^+-amalgamation function F, F has the weak λ^+-coding property for $\boldsymbol{C}$ (see Shelah 2001, 3.8t(2)) hence the weaker version, too. Hence by Shelah (2001, 3.12t), if WDmId(λ^+) is not λ^{++}-saturated then $I(\lambda^{++}, K) = 2^{\lambda^{++}}$. $\square_{1.9}$

Remark 1.10

1) We can get more abstract results.

2) Note $\neg(*)_\lambda$ of 1.9 is a "light" assumption, in fact, e.g. its negation has high consistency strength.

2. Remarks in pcf

The following will provide us a useful division into cases (it is from pcf theory; on $\mu_{\mathrm{wd}}(\lambda)$ see Shelah (2001, 1.1t)), we can replace λ^+ by regular λ such that $2^\theta = 2^{<\lambda} < 2^\lambda$ for some θ).

Fact 2.1 Assume $2^\lambda < 2^{\lambda^+}$. Then one of the following cases occurs:

$(A)_\lambda$ we can find μ such that letting $\chi^* = 2^{\lambda^+}$

- (α) $\lambda^+ < \mu \leqslant 2^\lambda$ and $\mathrm{cf}(\mu) = \lambda^+$
- (β) $\mathrm{pp}(\mu) = \chi^*$, moreover $\mathrm{pp}(\mu) =^+ \chi^*$ and $\chi^* > 2^\lambda$
- (γ) $(\forall \mu')(\mathrm{cf}(\mu') \leqslant \lambda^+ < \mu' < \mu \to \mathrm{pp}(\mu') < \mu)$ hence $\mathrm{cf}(\mu') \leqslant \lambda^+ < \mu' < \mu \Rightarrow \mathrm{pp}_{\lambda^+}(\mu') < \mu$
- (δ) for every regular cardinal χ in the interval $(\mu, \chi^*]$ there is an increasing sequence $\langle \lambda_i : i < \lambda^+ \rangle$ of regular cardinals $> \lambda^+$ with limit μ such that $\chi = \mathrm{tcf}\left(\prod_{i<\lambda^+} \lambda_i / J^{\mathrm{bd}}_{\lambda^+}\right)$
- (ε) for some regular $\kappa \leqslant \lambda$, for any $\mu' < \mu$ there is a tree T with $\leqslant \lambda$ nodes, κ levels and $|\mathrm{Lim}_\kappa(T)| \geqslant \mu'$ (in fact e.g. $\kappa = \mathrm{Min}\{\kappa : 2^\kappa \geqslant \mu\}$ is appropriate; without loss of generality $T \subseteq {}^{\kappa >}\lambda$; we can get, of course, a tree T with $\mathrm{cf}(\kappa)$ levels, $\leqslant \lambda$ nodes and $|\mathrm{Lim}_{\mathrm{cf}(\kappa)}(T)| \geqslant \mu'$).

$(B)_\lambda$ for some μ, χ^* we have: clauses (α) – (ε) from above (so $2^\lambda < \chi^*$) and

- (ζ) there is $\langle T_\zeta : \zeta < \chi^* \rangle$ such that: $T_\zeta \subseteq {}^{\lambda^+ >}2$ a tree, of cardinality $\leqslant \lambda^+$ and ${}^{\lambda^+}2 = \bigcup_{\zeta < \chi^*} \mathrm{Lim}_{\lambda^+}(T_\zeta)$ and $\chi^* < 2^{\lambda^+}$
- (η) $2^\lambda < \chi^* < \mu_{\mathrm{wd}}(\lambda^+, 2^\lambda)$ (but $< \mu_{\mathrm{wd}}(\lambda^+, 2^\lambda)$ is not used here, see Shelah (2001, Definition 1.1t(5)))
- (θ) for some $\zeta < \chi^*$ we have $\mathrm{Lim}_{\lambda^+}(T_\zeta) \notin \mathrm{WDmTId}(\lambda^+)$, not used

(ι) if there is a normal λ^{++}-saturated ideal on λ^+, e.g. the ideal $\mathrm{WDmId}(\lambda^+)$ is, then $2^{\lambda^+} = \lambda^{++}$ (so as $2^\lambda < 2^{\lambda^+}$, necessarily $2^\lambda = \lambda^+$)

(κ) $\mathrm{cov}(\chi^*, \lambda^{++}, \lambda^{++}, \aleph_1) = \chi^*$, equivalently $\chi^* = \sup\{\mathrm{pp}(\chi) : \chi \leqslant 2^\lambda, \aleph_1 \leqslant \mathrm{cf}(\chi) \leqslant \lambda^+ < \chi\}$ by Shelah (1994, Ch. II, 5.4)

$(C)_\lambda$ letting $\chi^* = 2^\lambda$ we have $(\zeta), (\eta), (\theta), (\iota), (\kappa)$ of clause (B) and

(λ) for no $\mu \in (\lambda^+, 2^\lambda]$ do we have $\mathrm{cf}(\mu) \leqslant \lambda^+$, $\mathrm{pp}(\mu) > 2^\lambda$ equivalently $2^\lambda > \lambda^+ \Rightarrow \mathrm{cf}([2^\lambda]^{\lambda^+}, \subseteq) = 2^\lambda$ hence (see the proof) $\mu_{\mathrm{wd}}(\lambda^+, 2^\lambda) = 2^{\lambda^+}$ except (maybe) when $\lambda < \beth_\omega$ and there is $\mathscr{A} \subseteq [\mu_{\mathrm{wd}}(\lambda^+, 2^\lambda)]^\lambda$ such that $A \neq B \in \mathscr{A} \Rightarrow |A \cap B| = \aleph_0$.

Remark. Remember that

$$\mathrm{cov}(\chi, \mu, \theta, \sigma) = \chi + \mathrm{Min}\{|\mathscr{P}| : \mathscr{P} \subseteq [\chi]^{<\mu} \text{ and every member of } [\chi]^{<\theta} \text{ is included in the union of } < \sigma \text{ members of } \mathscr{P}\}.$$

Proof. This is related to Shelah (1994, II, 5.11); we assume basic knowledge of pcf (or a readiness to believe). Note that if $2^\lambda > \lambda^+$ then $\mathrm{cf}([2^\lambda]^{\leqslant\lambda^+}, \subseteq) = 2^\lambda \leftrightarrow \mathrm{cov}(2^\lambda, \lambda^{++}, \lambda^{++}, \aleph_0) = 2^\lambda$ and $\mathrm{cov}(2^\lambda, \lambda^{++}, \lambda^{++}, \aleph_0) \geqslant \mathrm{cov}(2^\lambda, \lambda^{++}, \lambda^{++}, \theta) = 2^\lambda$ for $\theta \in [\aleph_0, \lambda]$.

<u>Possibility 1</u>: $\chi^* =: \mathrm{cov}(2^\lambda, \lambda^{++}, \lambda^{++}, \aleph_1) = 2^\lambda$.

We shall show that case (C) holds.

Now by the definition of cov, clause (ζ) is obvious, as well as (κ). As on the one hand by Shelah (1998, AP, 1.16 + 1.19) we have $(\mu_{\mathrm{wd}}(\lambda^+, 2^\lambda))^{\aleph_0} = 2^{\lambda^+} > 2^\lambda = \chi^*$ and on the other hand $(\chi^*)^{\aleph_0} = (2^\lambda)^{\aleph_0} = 2^\lambda = \chi^*$ necessarily $\chi^* < \mu_{\mathrm{wd}}(\lambda^+, 2^\lambda)$ so clause (η) follows; now clause (θ) follows from clause (ζ) as $\mathrm{WDmTId}(\lambda^+)$ is $(2^\lambda)^+$-complete by Shelah (2001, 1.2t(5)) and we have chosen $\chi^* = 2^\lambda$. Now if $2^{\lambda^+} > \lambda^{++}$, (so $2^{\lambda^+} \geqslant \lambda^{+3}$), then for some $\zeta < \chi^*$, T_ζ is (a tree with $\leqslant \lambda^+$ nodes, λ^+ levels and) at least λ^{+3} λ^+-branches which is well known (see e.g. [J]) to imply "no normal ideal on λ^+ is λ^{++}-saturated"; so we got clause (ι). As for (λ) the definition of χ^* and the assumption $\chi^* = 2^\lambda$ we have the first two phrases, as for $\mu_{\mathrm{wd}}(\lambda^+, 2^\lambda) = 2^{\lambda^+}$ by Shelah (1998, AP 1.14 + 1.16, pp. 956–958) there is $\mathscr{A}$ as mentioned in (λ) and by Shelah (2000) we get $\lambda < \beth_\omega$. The "equivalently" holds as $(2^\lambda)^{\aleph_0} = 2^\lambda$.

<u>Possibility 2</u>: $\chi^* =: \mathrm{cov}(2^\lambda, \lambda^{++}, \lambda^{++}, \aleph_1) > 2^\lambda$.

Let $\mu = \mathrm{Min}\{\mu : \mathrm{cf}(\mu) \leqslant \lambda^+, \lambda^+ < \mu \leqslant 2^\lambda \text{ and } \mathrm{pp}(\mu) = \chi^*\}$; clearly we may replace $\mathrm{pp}(\mu) = \chi^*$ by $\mathrm{pp}(\mu) =^+ \chi^*$. We know by Shelah (1994, II, 5.4) that μ exists and (by Shelah 1994, II, 2.3(2)) clause (γ) holds, also $2^\lambda < \mathrm{pp}(\mu) \leqslant$

$\mu^{\mathrm{cf}(\mu)}$ hence $\mathrm{cf}(\mu) = \lambda^+$. So clauses $(\alpha), (\beta), (\gamma)$ hold (of course, for clause (β) use Shelah (1994, Ch. II, 5.4(2))), and by (γ) + Shelah (1994, VIII, §1) also clause (δ) holds.

For clause (ε) let $\Upsilon = \mathrm{Min}\{\Upsilon : 2^\Upsilon \geqslant \mu\}$, clearly $\alpha < \Upsilon \Rightarrow 2^{|\alpha|} < \mu$ and $\Upsilon \leqslant \lambda$ (as $2^\lambda \geqslant \mu$) hence $\mathrm{cf}(\Upsilon) \leqslant \Upsilon \leqslant \lambda < \lambda^+ = \mathrm{cf}(\mu)$ hence $2^{<\Upsilon} < \mu$. Now we shall first prove

$(*)$ there[4] is a tree with λ^+ nodes, $\mathrm{cf}(\Upsilon)$ levels and $\geqslant \mu$ Υ-branches.

Why $(*)$ holds? Otherwise we shall get contradiction to the claim 2.3 below with $\sigma, \kappa, \theta_0, \theta_1, \mu, \chi$ there standing for $\mathrm{cf}(\Upsilon), \lambda^+, \lambda^+, 2^{<\Upsilon}, \mu, (2^\lambda)^+$ here and T^* defined below; let us check the conditions there:

Clause (a): It says $\mathrm{cf}(\Upsilon) < \lambda^+ = \mathrm{cf}(\mu) \leqslant \lambda^+ \leqslant 2^{<\Upsilon} < \mu$ which is readily checked except the inequality $\lambda^+ \leqslant 2^{<\Upsilon}$ but if it fails we immediately get more than required.

Clause (b): This is clause (γ) of (A) which we have proved.

Clause (c): The tree T^* is $({}^{\Upsilon >}2, \triangleleft)$ restricted to an unbounded set of levels of order type $\mathrm{cf}(\Upsilon)$.

Clause (d): Let $\theta_2 =: \mathrm{cov}(2^{<\Upsilon}, \lambda^{++}, \lambda^{++}, \mathrm{cf}(\Upsilon)^+)$.
So the statement we have to prove is $\mathrm{pp}(\mu) \geqslant \chi = \mathrm{cf}(\chi) > \theta_2^{\mathrm{cf}(\Upsilon)}$. Now $\mathrm{pp}(\mu) \geqslant \chi$ holds by the choice of μ and $\chi = \mathrm{cf}(\chi)$ as $\chi = (2^\lambda)^+$. For the last inequality, by Shelah (1994, Ch. II, 5.4) and the choice of μ, as we have shown $2^{<\Upsilon} < \mu$ we know $\theta_2 < \mu$, but $\mu \leqslant 2^\lambda$ so $\theta_2^{\mathrm{cf}(\Upsilon)} \leqslant (2^\lambda)^{\mathrm{cf}(\Upsilon)} \leqslant (2^\lambda)^\Upsilon \leqslant (2^\lambda)^\lambda = 2^\lambda < \chi$.

So we have verified clauses $(a) - (d)$ of 2.3 hence its conclusion holds, but this gives $(*)$, i.e. the desired conclusion in clause (ε) of Case A in 2.1; well not exactly, it gives only $|T^*| \leqslant \lambda^+$, so $T^* = \bigcup_{i<\lambda^+} T_i, T_i$ increases continuously with each T_i of cardinality $\leqslant \lambda$, so for every $\mu' < \mu$ for some i we have $|\mathrm{Lim}_{\mathrm{cf}(\Upsilon)}(T_i)| \geqslant \mu'$, so we have clauses $(\alpha) - (\varepsilon)$ there.

Subpossibility 2a: $\chi^* < 2^{\lambda^+}$.

We shall prove $(B)_\lambda$, so we are left with proving $(\zeta) - (\kappa)$ when $\chi^* < 2^{\lambda^+}$. By the choice of χ^*, easily clause (ζ) (in Case B of 2.1) holds. In clause (η), "$2^\lambda < \chi^*$" holds as we are in possibility 2.

Also as $\mathrm{pp}(\mu) = \chi^*$ by the choice of μ necessarily (by transitivity of pcf, i.e. Shelah 1994, Ch. II, 2.3(2)) $\mathrm{cf}(\chi^*) > \lambda^+$ but $\mu > \lambda^+$. Easily $\chi \leqslant \chi^* \wedge \mathrm{cf}(\chi) \leqslant \lambda^+ \Rightarrow \mathrm{pp}(\chi) \leqslant \chi^*$ hence $\mathrm{cov}(\chi^*, \lambda^{++}, \lambda^{++}, \aleph_1) = \chi^*$ by Shelah (1994, Ch. II, 5.4), which gives clause (κ), so as $(\lambda^+)^{\aleph_0} \leqslant 2^\lambda < \chi^*$ certainly there is no family of $> \chi^*$ subsets of χ^* each of cardinality λ^+ with pairwise finite intersections, hence (by Shelah (1982, Ch. XIV, §1) or see Shelah (2001, 1.2(1)) or Shelah (1998, AP, 1.16)) we have $\chi^* < \mu_{wd}(\lambda^+, 2^\lambda)$ thus completing the proof of (η).

Now clause (θ) follows by $(\zeta) + (\eta)$ by Shelah (2001, 1.2t(5)). Also if $2^{\lambda^+} \neq \lambda^{++}$ then $2^{\lambda^+} \geqslant \lambda^{+3}$ so by clause (ζ) (as $\chi^* < 2^{\lambda^+}$), for some ζ, $|\mathrm{Lim}_{\lambda^+}(T_\zeta)| \geqslant \lambda^{+3}$, which is well known to imply no normal ideal on λ^+ is λ^{++}-saturated; i.e. clause (ι). So we have proved that case $(B)_\lambda$ holds.

Subpossibility 2b: $\chi^* = 2^{\lambda^+}$.

We have proved that case $(A)_\lambda$ holds, as we already defined μ and χ^* and proved $(\alpha), (\beta), (\gamma), (\delta), (\varepsilon)$ we are done.

Still we depend on 2.3 below but first we prove 2.2.

Claim 2.2 *Assume*

(a) $\sigma < \kappa = \mathrm{cf}(\mu) \leqslant \theta_0 \leqslant \theta_1 < \mu \leqslant \theta_1^\sigma$

(b) $(\forall \mu')[\theta_0 < \mu' < \mu \;\&\; \mathrm{cf}(\mu') \leqslant \kappa \Rightarrow \mathrm{pp}(\mu') < \mu]$

(c) $\theta_2 = \theta_1 + \mathit{cov}(\theta_1, \theta_0^+, \kappa^+, \sigma^+)$ *(by clause (b) and Shelah (1994, Ch. II, 5.4) we know that it is* $< \mu$*)*

(d) $\mathrm{pp}(\mu) \geqslant \chi = \mathit{cf}(\chi) > \theta_2^\sigma$ $(\geqslant \theta_1^\sigma \geqslant \mu)$.

Then $\theta_0^\sigma \geqslant \mu$.

Remark. In fact $\theta_2^\sigma \geqslant \mathrm{cov}(\theta_1, \theta_0^+, \kappa^+, 2)$.

Proof. Assume toward contradiction $\theta_0^\sigma < \mu$. By Shelah (1994, Ch. II, 2.3(2)) and clause (b) of the assumption we have $\sup\{\mathrm{pp}(\mu') : \theta_C^- \leqslant \mu' \leqslant \theta_0^\sigma$ and $\mathrm{cf}(\mu') \leqslant \kappa\} < \mu$ hence by Shelah (1994, Ch. II, 5.4) it follows that $\kappa^* =: \mathrm{cov}(\theta_0^\sigma, \theta_0^+, \kappa^+, \sigma^+) < \mu$. We can by assumptions (b) + (d) and Shelah (1994, Ch. II, 3.5) + Shelah (1994, Ch. VIII, §1) find $T \subseteq {}^{\kappa>}\mu$, a tree with $\leqslant \mu$ nodes, $|\mathrm{Lim}_\kappa(T)| \geqslant \chi$, (if $\chi = \mathrm{pp}(\mu)$, the supremum in the definition of $\mathrm{pp}(\mu)$ is obtained by Shelah 1994, II, 5.4(2)). Moreover, by the construction there is $\Xi \subseteq \mathrm{Lim}_\kappa(T), |\Xi| = \chi$ such that $\Xi' \subseteq \Xi \;\&\; |\Xi'| \geqslant \chi \Rightarrow |\{\eta \restriction \alpha : \alpha < \kappa, \eta \in \Xi'\}| = \mu$. By renaming (and also by the construction), without loss of generality

$\otimes$ if $\eta_0{}^\frown\langle\alpha_0\rangle \neq \eta_1{}^\frown\langle\alpha_1\rangle$ belongs to T then $\alpha_0 \neq \alpha_1$.

So let $\eta_i \in \mathrm{Lim}_\kappa(T)$ for $i < \chi$ be pairwise distinct, listing Ξ.
As $\mu \leqslant \theta_1^\sigma$ there is a sequence $\bar{F} = \langle F_\varepsilon : \varepsilon < \sigma\rangle$ satisfying: F_ε a function from μ to θ_1 such that $\alpha < \beta < \mu \Rightarrow (\exists \varepsilon < \sigma) F_\varepsilon(\alpha) \neq F_\varepsilon(\beta)$.

Let $w_{i,\varepsilon} = \{F_\varepsilon(\eta_i(\alpha)) : \alpha < \kappa\}$, so $w_{i,\varepsilon} \in [\theta_1]^\kappa$. By assumption (c) we have $\theta_2 = \theta_1 + \mathrm{cov}(\theta_1, \theta_0^+, \kappa^+, \sigma^+)$ so there is $\mathscr{P} \subseteq [\theta_1]^{\theta_0}, \theta_2 = |\mathscr{P}|$ such that: any

$w \in [\theta_1]^\kappa$ is included in a union of $\leqslant \sigma$ members of $\mathscr{P}$. So we can find $X_{i,\varepsilon,\zeta} \in \mathscr{P}$ for $\zeta < \sigma$ such that $w_{i,\varepsilon} \subseteq \bigcup_{\zeta<\sigma} X_{i,\varepsilon,\zeta}$. So $\bigcup_{\varepsilon<\sigma} w_{i,\varepsilon} \subseteq Y_i =: \bigcup_{\zeta,\varepsilon<\sigma} X_{i,\varepsilon,\zeta}$. Let $\mathscr{P}^* = \{\bigcup_{\varepsilon<\sigma} X_\varepsilon : X_\varepsilon \in \mathscr{P}$ for $\varepsilon < \sigma\}$, so $\mathscr{P}^*$ is a family of $\leqslant |\mathscr{P}|^\sigma \leqslant \theta_2^\sigma$ sets and $i < \chi \Rightarrow Y_i \in \mathscr{P}^*$.

For each $Y \in \mathscr{P}^*$ let

$$Z_Y = \{\alpha < \mu : (\forall \varepsilon < \sigma)(F_\varepsilon(\alpha) \in Y)\}$$

clearly $Y = Y_i \Rightarrow \operatorname{Rang}(\eta_i) \subseteq Z_Y$, also $|Y| \leqslant \theta_0$ hence $|Z_Y| \leqslant \theta_0^\sigma < \mu$ hence there is a family Q_Y of cardinality $\kappa^* =: \operatorname{cov}(\theta_0^\sigma, \theta_0^+, \kappa^+, \sigma^+) < \mu$ whose members are subsets of Z_Y each of cardinality $\leqslant \theta_0$ such that any $X \in [Z_Y]^{\leqslant \kappa}$ is included in the union of $\leqslant \sigma$ of them. For each $Y \in \mathscr{P}^*$ and $W \in Q_Y$ let $T'_W = \{\eta \in T :$ for some $\alpha < \kappa$ we have: $\alpha + 1 = \ell g(\eta)$ and $\eta(\alpha) \in W\}$ and $T_W = \{\eta \in T : (\exists \nu)(\eta \lhd \nu \in T_W)\}$.

So by $\otimes$ above we have: T'_W, hence T_W is a set of $\leqslant |W| + \kappa \leqslant \theta_0$ nodes in T, $\lhd$-downward closed. Also

$$(*)\quad \left|\bigcup_{Y \in \mathscr{P}^*} Q_Y\right| \leqslant |\mathscr{P}^*| \times \sup_{Y \in \mathscr{P}^*} |Q_Y| \leqslant \theta_2^\sigma \times \kappa^* \leqslant \theta_2^\sigma + \operatorname{cov}(\theta_0^\sigma, \theta_0^+, \kappa^+, \sigma^+) < \mu.$$

However, for every $i < \chi, Y_i \in \mathscr{P}^*$ and $\operatorname{Rang}(\eta_i) \in [Y_i]^{\leqslant \kappa}$ so for some $W \in Q_{Y_i}, (\exists^\kappa \alpha < \kappa)[\eta_i(\alpha) \in W]$ hence $\eta_i \in \operatorname{Lim}_\kappa(T_W)$.

By assumption (d) and $(*)$ above for some $W \in \bigcup_{Y \in \mathscr{P}^*} Q_Y$ we have

$$|\{i < \chi : \eta_i \in \operatorname{Lim}_\kappa(T_W)\}| = \chi.$$

T_W is (essentially) a tree with $\leqslant \theta_0$ nodes and contradict the choice of $\Xi = \{\eta_i : i < \chi\}$.
(We could have instead using κ^*, Q_Y to fix $Y_i = Y$ as $|\mathscr{P}^*| < \chi = \operatorname{cf}(\chi)$.) $\square_{2.2}$

Claim 2.3 *Assume*

(a) $\sigma < \kappa = \operatorname{cf}(\mu) \leqslant \theta_0 \leqslant \theta_1 < \mu$

(b) $(\forall \mu')[\theta_0 < \mu' < \mu \;\&\; \operatorname{cf}(\mu') \leqslant \kappa \rightarrow \operatorname{pp}(\mu') < \mu]$

(c) T^* *is a tree with* $\leqslant \theta_1$ *nodes,* σ *levels and* $\geqslant \mu$ σ*-branches*

(d) $\operatorname{pp}(\mu) \geqslant \chi = \operatorname{cf}(\chi) > \theta_2^\sigma$ *where* $\theta_2 = \operatorname{cov}(\theta_1, \theta_0^+, \kappa^+, \sigma^+)$.

Then for some subtree $Y \subseteq T^*, |Y| \leqslant \theta_0$ *and* $|Lim_\sigma(Y)| \geqslant \mu$ *(for 2.1 enough* $\geqslant \mu'$ *for any given* $\mu' < \mu$*).*

Proof. Let $T, \Xi = \{\eta_i : i < \chi\}$ be as in the proof of the previous claim. Let $\{\nu_\zeta : \zeta < \mu\}$ list μ distinct σ-branches of T^* (see clause (c)). Without loss of generality the set of nodes of T^* is θ_1. Choose for each $\varepsilon < \sigma$ the function $F_\varepsilon : \mu \to \theta_1$ by $F_\varepsilon(\gamma) = \nu_\gamma(\varepsilon)$. Define $w_{i,\varepsilon}, \mathscr{P}, X_{i,\varepsilon,\zeta}, Y_i, \mathscr{P}^*$ as in the proof of 2.2. But for $Y \in \mathscr{P}^*$ we change the choice of Z_Y, first

$$Y' = \{\beta < \theta_1 : \text{for some } \alpha \in Y, \text{ we have } \beta <_{T^*} \alpha\}$$

So $|Y'| \leqslant \sigma + |Y|$ and let

$$Z_Y = \{\alpha < \mu : (\forall \varepsilon < \sigma)(F_\varepsilon(\alpha) \in Y')\}.$$

We continue as in the proof of 2.2. $\square_{2.3}, \square_{2.1}$

Remark 2.4

1) We could have used in 2.2, 2.3, $\theta_2 = \text{cov}(\theta_0^\sigma, \theta_0^+, \kappa^+, \kappa^+)$ instead $\text{cov}(\theta_0^\sigma, \theta_0^+, \kappa^+, \sigma^+)$ and similarly in the proof of 2.1.

2) We can also play with assumption (b) as 2.2, 2.3.

It may be useful to note (actually $\lambda^{\langle \text{tr},\kappa \rangle} = \lambda$ suffices)

Fact 2.5 If $T \subseteq {}^{\lambda^+>}2$ is a tree, $|T| \leqslant \lambda^+$ and $\lambda \geqslant \beth_\omega$ <u>then</u> for every regular $\kappa < \beth_\omega$ large enough, we can find $\langle Y_\delta : \delta < \lambda^+, \text{cf}(\delta) = \kappa \rangle, |Y_\delta| \leqslant \lambda$ such that: for every $\eta \in \text{Lim}_{\lambda^+}(T)$ for a club of $\delta < \lambda^+$ we have $\text{cf}(\delta) = \kappa \Rightarrow \eta \restriction \delta \in Y_\delta$.

Fact 2.6 Assume $\mathfrak{K}$ is an abstract elementary class with amalgamation in λ, and above $(M, N, a) \in K_\lambda^3$ there is no minimal pair.

1) Assume T is a tree with $\delta < \lambda^+$ levels and $\leqslant \lambda$ nodes. <u>Then</u> we can find $(M^*, N_\eta, a) \in K_\lambda^3$ above (M, N, a) for $\eta \in \text{Lim}_\delta(T)$ such that $\text{tp}(a, M^*, N_\eta)$ for $\eta \in \text{Lim}_\delta(T)$ are pairwise distinct. We can add "(M^*, N_η, a) is reduced" (see Shelah (2001)).

2) If $M \in K_\lambda$ is universal then $\mathscr{S}(M) \geqslant \sup\{\text{Lim}_\delta(T) : T \text{ a tree with} \leqslant \lambda \text{ nodes and } \delta \text{ levels}\}$.

Proof. 1) Straight (or see the proof of 3.3(1)).
2) As for any $N \in K_\lambda$ there is a model $N' \leqslant_{\mathfrak{K}} M$ isomorphic to N, now $p \mapsto p \restriction N'$ is a function from $\mathscr{S}(M)$ onto $\mathscr{S}(N')$ by Shelah (2001, §2) hence $|\mathscr{S}(M)| \geqslant |\mathscr{S}(N')| = |\mathscr{S}(N)|$. Now use part (1). $\square_{2.6}$

The following is needed when we like to get in the model theory not just many models but many models no one $\leqslant_{\mathfrak{K}}$-embeddable into another.

Fact 2.7 Assume:

(a) $\mathrm{cf}(\mu) \leqslant \kappa < \mu$, $\mathrm{pp}_\kappa(\mu) =^+ \chi^*$, moreover $\mathrm{pp}_\kappa(\mu) =^+ \chi^*$ and $\kappa^+ < \theta < \chi^*$

(b) $\boldsymbol{F}$ is a function, with domain $[\mu]^\kappa$, such that: for $a \in [\mu]^\kappa, \boldsymbol{F}(a)$ is a family of $< \theta$ members of $[\mu]^\kappa$

(c) F is a function with domain $[\mu]^\kappa$ such that
$$a \in [\mu]^\kappa \Rightarrow a \subseteq F(a) \in \boldsymbol{F}(a).$$

<u>Then</u> we can find pairwise distinct $a_i \in [\mu]^\kappa$ for $i < \chi^*$ such that $\mathscr{I} = \{a_i : i < \chi\}$ is $(F, \boldsymbol{F})$-independent which means

$(*)_{F,\boldsymbol{F},\mathscr{I}}$ if $a \neq b$ & $a \in \mathscr{I}$ & $b \in \mathscr{I}$ & $c \in \boldsymbol{F}(a) \Rightarrow \neg(F(b) \subseteq c)$.

Remark.

1) Clearly this claim is similar to Hajnal's free subset theorem (Hajnal 1961).

2) Note that we can let $F(a) = a$.

3) Note that if $\lambda = \mathrm{cf}([\mu]^\kappa, \subseteq)$ then for some $\boldsymbol{F}, F$ as in the Fact

 $(*)$ if $a_i \in [\mu]^\kappa$ for $i < \lambda^+$ are pairwise distinct then not every pair $\{a_i, a_j\}$ is $(\boldsymbol{F}, F)$-independent
 [why? let $\mathscr{P} \subseteq [\mu]^\kappa$ be cofinal (under $\subseteq$) of cardinality λ, and let $\boldsymbol{F}$ be such that
 $\boldsymbol{F}(a) \subseteq \{b \in [\mu]^\kappa : a \subseteq b$ and $b \in \mathscr{P}\}$ has a $\subseteq$-maximal member $F(a)$; clearly there is such $\boldsymbol{F}$. Now clearly

 $(*)_1$ if $a \neq b$ are from $[\mu]^\kappa$ and $\boldsymbol{F}(a) \cap \boldsymbol{F}(b) \neq \emptyset$ then $\{a, b\}$ is not $(\boldsymbol{F}, F)$-independent.
 Also if $\mu_1 \leqslant \mu$, $\mathrm{cf}(\mu_1) \leqslant \kappa \leqslant \kappa + \theta < \mu_1$ and $\mathrm{pp}_\kappa(\mu_1) \geqslant \mu$ then by Shelah (1994, Ch. II, 2.3) the Fact for μ_1 implies the one for μ.]

Proof. We can define $g : [\mu]^{\leqslant\kappa} \to [\mu]^\kappa, F', \boldsymbol{F}'$ functions with domain $[\mu]^{\leqslant\kappa}$ as follows:

$$g(a) = \{\kappa + \alpha : \alpha \in a\} \cup \{\alpha : \alpha < \kappa\}$$
$$\boldsymbol{F}'(a) = \{\{\alpha : \kappa + \alpha \in b\} : b \in \boldsymbol{F}(g(a))\}$$
$$F'(a) = \{\alpha : \kappa + \alpha \in F(g(a))\}.$$

Now $\boldsymbol{F}', F'$ are as above only replacing everywhere $[\mu]^\kappa$ by $[\mu]^{\leqslant\kappa}$, and if $\mathscr{I} = \{a_i : i < \chi\} \subseteq [\mu]^{\leqslant\kappa}$ with no repetitions satisfying $(*)_{F',\boldsymbol{F}',\mathscr{I}}$ then $\mathscr{I}' = \{g(a_i) : i < \chi\}$ is with no repetitions and $(*)_{F,\boldsymbol{F},\mathscr{I}'}$.
So we conclude that we can replace $[\mu]^\kappa$ by $[\mu]^{\leqslant\kappa}$. Without loss of generality $\kappa^+ < \theta$.

Assume $\theta < \chi = \mathrm{cf}(\Pi\mathfrak{a}/J)$ where $\mathfrak{a} \subseteq \mu \cap \mathrm{Reg}\backslash\kappa^+, |\mathfrak{c}| \leqslant \kappa, \sup(\mathfrak{a}) = \mu$, $J^{bd}_{\mathfrak{a}} \subseteq J$ and for simplicity $\chi = \max \mathrm{pcf}(\mathfrak{a})$ so as explained above without loss of generality $|\mathfrak{a}| \leqslant \kappa$ and let $f = \langle f_\alpha : \alpha < \chi\rangle$ be a sequence of members of $\Pi\mathfrak{a}, <_J$-increasing, and cofinal in $(\Pi\mathfrak{a}, <_J)$, so, of course, $\chi \leqslant \chi^*$. Without loss of generality $f_\alpha(\tau) > \sup(\mathfrak{a} \cap \tau)$ for $\tau \in \mathfrak{a}$. Also for every $a \in [\mu]^\kappa$, define $f_a \in \Pi\mathfrak{a}$ by $f_a(\tau) = \sup(a \cap \tau)$ for $\tau \in \mathfrak{a}$ so for some $\zeta(a) < \chi$ we have $f_a <_J f_{\zeta(a)}$ (as $\langle f_\alpha : \alpha < \chi\rangle$ is cofinal in $(\Pi\mathfrak{a}, <_J)$). So for each $a \in [\mu]^\kappa$, as $|\boldsymbol{F}(a)| < \theta < \chi = \mathrm{cf}(\chi)$ clearly $\xi(a) = \sup\{\zeta(b) : b \in \boldsymbol{F}(a)\}$ is $< \chi$, and clearly $(\forall b \in \boldsymbol{F}(a))[f_b <_J f_{\xi(a)}]$. So
$C = \{\gamma < \chi : \text{for every } \beta < \gamma, \xi(\kappa \cup \mathrm{Rang}\, f_\beta) < \gamma\}$ is a club of χ.

For each $\alpha < \chi$, $\mathrm{Rang}(f_\alpha) \cup \kappa \in [\mu]^\kappa$, hence $\boldsymbol{F}(\mathrm{Rang}(f_\alpha) \cup \kappa)$ has cardinality $< \theta$, but $\theta < \chi = \mathrm{cf}(\chi)$ hence for some $\theta_1 < \theta$; we have $\theta_1 > \kappa^{++}$ and $\chi = \sup\{\alpha < \chi : |\boldsymbol{F}(\mathrm{Rang}(f_\alpha) \cup \kappa)| \leqslant \theta_1\}$, so without loss of generality $\alpha < \chi \Rightarrow \theta_1 \geqslant |\boldsymbol{F}(\mathrm{Rang}(f_\alpha) \cup \kappa)|$.

As $\kappa^+ < \theta_1$, there is a stationary $S \subseteq \{\delta < \theta_1^+ : \mathrm{cf}(\delta) = \kappa^+\}$ which is in $I[\theta_1^+]$, by Shelah (1993, §1) and let $\langle d_i : i < \theta_1^+\rangle$ witness it, so $\mathrm{otp}(d_i) \leqslant \kappa^+, d_i \subseteq i$, $[j \in d_i \Rightarrow d_j = d_i \cap i]$ and $i \in S \Rightarrow i = \sup(d_i)$, and for simplicity: for every club E of θ_1^+ for stationarily many $\delta \in S$ we have $(\forall \alpha \in d_\delta)(\exists \beta \in E)(\sup(\alpha \cap d_\delta) < \beta < \alpha)]$, exists by Shelah (1993, §1). Now try to choose by induction on $i < \theta_1^+$, a triple (g_i, α_i, w_i) such that:

(a) $g_i \in \Pi\mathfrak{a}$

(b) $j < i \Rightarrow g_j <_J g_i$

(c) $(\forall \tau \in \mathfrak{a})(\sup\limits_{j \in d_i} g_j(\tau) < g_i(\tau))$

(d) $\alpha_i < \chi, \alpha_i > \sup(\bigcup_{j<i} w_j)$

(e) $j < i \Rightarrow \alpha_j < \alpha_i$

(f) $g_i <_J f_{\alpha_i}$

(g) $\beta \in \bigcup_{j<i} w_j \Rightarrow \xi(\beta) < \alpha_i \;\&\; f_\beta <_J g_i$

(h) w_i is a maximal subset of (α_i, χ) satisfying

$(*)$ $\beta \in w_i \& \gamma \in w_i \& \beta \neq \gamma \& a \in \boldsymbol{F}(\kappa \cup \mathrm{Rang}(f_\beta)) \Rightarrow \neg(F(\mathrm{Rang}\, f_\gamma) \subseteq a)$

or just

$(*)^+$ $\beta \in w_i \;\&\; \gamma \in w_i \;\&\; \beta \neq \gamma \;\&\; a \in \boldsymbol{F}(\mathrm{Rang}(f_\beta)) \Rightarrow$
$\{\tau \in \mathfrak{a} : f_\gamma(\tau) \in a\} \in J.$

[note that really

$\otimes$ if $w \subseteq (\alpha_i, \chi)$ satisfies $(*)^+$ then it satisfies $(*)$

why? let us check $(*)$, so let $\beta \in w, \gamma \in w, \beta \neq \gamma$ and $a \in \boldsymbol{F}(\mathrm{Rang}(f_\beta))$; by $(*)^+$ we know that $\mathfrak{a}' = \{\tau \in \mathfrak{a} : f_\gamma(\tau) \in a\} \in J, J$ is a proper ideal on $\mathfrak{a}$ clearly for some $\tau \in \mathfrak{a}$ we have $\tau \notin \mathfrak{a}'$, hence $f_\gamma(\tau) \notin a$ but $f_\gamma(\tau) \in \mathrm{Rang}(f_\gamma) \subseteq F(\mathrm{Rang}(f_\gamma))$ hence $f_\gamma(\tau) \in F(\mathrm{Rang}(f_\gamma)) \backslash a$ so $\neg(F(\mathrm{Rang}(f_\gamma) \subseteq a)$, as required.]

We claim that we cannot carry the induction because if we succeed, then as $\mathrm{cf}(\chi) = \chi > \theta \geqslant \theta_1^+$ there is α such that $\bigcup_{i<\theta_1^+} \alpha_i < \alpha < \chi$ and let $\boldsymbol{F}(\kappa \cup \mathrm{Rang}(f_\alpha)) = \{a^\alpha_\zeta : \zeta < \theta_1\}$ (possible as $1 \leqslant |\boldsymbol{F}(\kappa \cup \mathrm{Rang}(f_\alpha))| \leqslant \theta_1$). Now for each $i < \theta_1^+$, by the choice of w_i clearly $w_i \cup \{\alpha\}$ does not satisfy the demand in clause (h), so as $\beta \in w_i \Rightarrow \xi(\beta) < \alpha_{i+1} < \alpha$, necessarily for some $\beta_i \in w_i$ and $\zeta_i < \theta_1$ we have

$$\mathfrak{a}_i = \{\tau \in \mathfrak{a} : f_{\beta_i}(\tau) \in a^\alpha_{\zeta_i}\} \notin J.$$

[why use the ideal? In order to show that $\mathfrak{b}_\varepsilon \neq \emptyset$.] Now $\mathrm{cf}(\theta_1^+) = \theta_1^+ > \theta_1$, for some $\zeta(*) < \theta_1^+$ we have $A = \{i : \zeta_i = \zeta(*)\}$ is unbounded in θ_1^+. Hence $E = \{\alpha < \theta_1^+ : \alpha$ a limit ordinal and $A \cap \alpha$ is unbounded in $\alpha\}$ is a club of θ_1^+. So for some $\delta \in S$ we have $\delta = \sup(A \cap \delta)$, moreover if $d_\delta = \{\alpha_\varepsilon : \varepsilon < \kappa^+\}$ (increasing) then $(\forall \varepsilon)[E \cap (\sup_{\zeta<\varepsilon} \alpha_\zeta, \alpha_\varepsilon) \neq \emptyset]$ hence we can find $i(\delta,\varepsilon) \in (\sup_{\zeta<\varepsilon} \alpha_\zeta, \alpha_\varepsilon) \cap A$ for each $\varepsilon < \kappa^+$.

Clearly for each $\varepsilon < \kappa^+$

$$\mathfrak{b}_\varepsilon = \{\tau \in \mathfrak{a} : g_{i(\delta,\varepsilon)}(\tau) < f_{\alpha_{i(\delta,\varepsilon)}}(\tau) < f_{\beta_{i(\delta,\varepsilon)}}(\tau) < g_{i(\delta,\varepsilon)+1} < f_{\alpha_{i(\delta,\varepsilon)+1}}(\tau) < f_\alpha(\tau)\} = \mathfrak{a} \bmod J$$

hence $\mathfrak{b}_\varepsilon \cap \mathfrak{a}_{i(\delta,\varepsilon)} \notin \emptyset$. Moreover, $\mathfrak{b}_\varepsilon \cap \mathfrak{a}_{i(\delta,\varepsilon)} \notin J$. Now for each $\tau \in \mathfrak{a}$ let $\varepsilon(\tau)$ be $\sup\{\varepsilon < \kappa^+ : \tau \in \mathfrak{b}_\varepsilon \cap \mathfrak{a}_{i(\delta,\varepsilon)}\}$ and let $\varepsilon(*) = \sup\{\varepsilon(\tau) : \tau \in \mathfrak{a}$ and $\varepsilon(\tau) < \kappa^+\}$ so as $|\mathfrak{a}| \leqslant \kappa$ clearly $\varepsilon(*) < \kappa^+$. Let $\tau^* \in \mathfrak{b}_{\varepsilon(*)+1} \cap \mathfrak{a}_{i(\delta,\varepsilon(*)+1)}$, so $B = \{\varepsilon < \kappa^+ : \tau^* \in \mathfrak{b}_\varepsilon \cap \mathfrak{a}_{i(\delta,\varepsilon)}\}$ is unbounded in κ^+, $\langle f_{\beta_{i(\delta,\varepsilon)}}(\tau^*) : \varepsilon \in B\rangle$ is strictly increasing (see clause (c) above and the choice of $\mathfrak{b}_\varepsilon$) and $\varepsilon \in B \Rightarrow f_{\beta_{i(\delta,\varepsilon)}}(\tau^*) \in a_{\zeta(*)}$ (by the definition of $\mathfrak{a}_{i(\delta,\varepsilon)}$, and $\zeta(*)$ as $\zeta_{i(\delta,\varepsilon)} = \zeta(*)$). We get contradiction to $a \in \boldsymbol{F}(\kappa \cup \mathrm{Rang}(f_\alpha)) \Rightarrow |a| \leqslant \kappa$.

So really we cannot carry the induction so we are stuck at some i. If $i = 0$, or i limit, or $i = j+1$ & $\sup w_j < \chi$ we can find g_i and then α_i and then w_i as required. So necessarily $i = j+1, \sup(w_j) = \chi$. Now if $\chi = \chi^*$, then this w_j is as required in the fact. As $\mathrm{pp}^+(\mu) = (\chi^*)^+$, the only case we cannot have is when χ^* is singular. Let $\chi^* = \sup_{\varepsilon < \mathrm{cf}(\chi^*)} \chi_\varepsilon$ and $\chi_\varepsilon \in (\mu, \chi^*) \cap \mathrm{Reg}$ is (strictly) increasing with ε. By Shelah (1994, Ch. II, §1) we can find, for each $\varepsilon < \mathrm{cf}(\chi^*)$, $\mathfrak{a}_\varepsilon, J_\varepsilon, \bar{f}^\varepsilon = \langle f^\varepsilon_\alpha : \alpha < \chi_\varepsilon\rangle$ as above, but in addition

(∗) $\bar{f}^\varepsilon$ is μ^+-free i.e. for every $u \in [\chi_\varepsilon]^\mu$ there is $\langle \mathfrak{b}_\alpha : \alpha \in u\rangle$ such that $\mathfrak{b}_\alpha \in J_\varepsilon$ and for each $\tau \in \mathfrak{a}_\varepsilon$, $\langle f^\varepsilon_\alpha(\tau) : \alpha$ satisfies : $\tau \notin \mathfrak{b}_\alpha\rangle$ is strictly increasing.

So for every $a \in [\mu]^\kappa$ and $\varepsilon < \text{cf}(\chi^*)$ we have

$$\big\{\alpha < \chi_\varepsilon : \{\tau \in \mathfrak{a}_\varepsilon : f_\alpha(\tau) \in a\} \notin J_\varepsilon\big\} \text{ has cardinality } \leqslant \kappa.$$

Hence for each $a \in [\mu]^\kappa$

$$\big\{(\varepsilon,\alpha) : \varepsilon < \text{cf}(\chi^*) \text{ and } \alpha < \chi_\varepsilon \text{ and } \{\tau \in \mathfrak{a}_\varepsilon : f_\alpha(\tau) \in a\} \notin J_\varepsilon\big\}$$

has cardinality $\leqslant \kappa + \text{cf}(\chi^*) = \text{cf}(\chi^*)$ as for singular $\mu > \kappa \geqslant \text{cf}(\mu)$ we have $\text{cf}(\text{pp}_\kappa(\mu)) > \kappa$.

Define: $X = \{(\varepsilon,\alpha) : \varepsilon < \text{cf}(\chi^*), \alpha < \chi_\varepsilon\}$

$$F'\big((\varepsilon,\alpha)\big) = \big\{(\varepsilon',\alpha') : (\varepsilon',\alpha') \in X\backslash\{(\varepsilon,\alpha)\} \text{ and for some } d \in \boldsymbol{F}(\text{Rang}(f^\varepsilon_\alpha)) \text{ we have } \{\tau \in \mathfrak{a}_\varepsilon : f^{\varepsilon'}_{\alpha'}(\tau) \in d\} \notin J_{\varepsilon'}\big\}$$

so $F'\big((\varepsilon,\alpha)\big)$ is a subset of X of cardinality $< \text{cf}(\chi^*)^+ + \theta < \chi^*$.

So by Hajnal's free subset theorem (Hajnal 1961) we finish (we could alternatively, for χ^* singular, have imitated his proof). $\square_{2.7}$

3. Finishing the many models

Recall from Shelah (2001, 3.13t(3)).

Claim 3.1

1) Assume

(a) $2^\lambda < 2^{\lambda^+}$ *and Case A or B of Fact 2.1 holds for* μ, χ^* *(or just the conclusion there)*

(b) $\mathfrak{K}$ *is an abstract elementary class with* $LS(\mathfrak{K}) \leqslant \lambda$

(c) $K_{\lambda^+} \neq 0$

(d) $\mathfrak{K}$ *has amalgamation in* λ

(e) in K^3_λ, *the minimal triples are not dense.*

<u>Then</u>

$(*)_1$ *for any regular* $\chi \leqslant \mu$ *we have:*

$(*)^1_\chi$ *there is* $M \in K_\lambda, |\mathscr{S}(M)| \geqslant \chi$.

2) If in part (1) we strengthen clause (d) *to* $(d)^+$, *<u>then</u> we get* $(*)^+_1$ *where:*

$(d)^+$ $\mathfrak{K}$ *has amalgamation in* λ *and a universal member in* λ

$(*)_1^+$ *for some* $M \in K_\lambda$ *we have* $|\mathscr{S}(M)| \geqslant \mu$.

3) *Assume (a), (b), (c), (e) of part (1) and* $(d)^+$ *of part (2)* *then:*

$(*)_2$ $I(\lambda^+, K) \geqslant \chi^*$ *and if* $(2^\lambda)^+ < \chi^*$ *then* $IE(\lambda^+, \mathfrak{K}) \geqslant \chi^*$.

4) *If in clause* (a) *of part (1) we restrict ourselves to Case A of 2.1,* *then* $\chi^* = 2^{\lambda^+}$ *so in part (3) we get*

$(*)_2^+$ $I(\lambda^+, K) = 2^{\lambda^+}$ *and* $(2^\lambda)^+ < 2^{\lambda^+} \Rightarrow IE(\lambda^+, K) \geqslant 2^{\lambda^+}$.

Remark 3.2

1) We can restrict clause (b) to $\mathfrak{K}_\lambda$, interpreting in (c) + (e), K_{λ^+} as $\{\bigcup_{i<\lambda^+} M_i : M_i \in K_\lambda$ is $<_\mathfrak{K}$-increasing (strictly and) continuous$\}$, but see Shelah (2001, §0), mainly 0.31t.

2) Part (3) of 3.1 is like Shelah (1994, Ch. II, 4.10E), Kojman and Shelah (1992, §2).

3) We can apply this to λ^+ standing for λ.

4) We can state the part of (A) of 2.1 used (and can replace 2^{λ^+} by smaller cardinals).

5) We can replace λ^+ by a weakly inaccessible cardinal with suitable changes.

Proof. 1) Note that μ is singular (as by clause (α) of (A) of 2.1, $\mathrm{cf}(\mu) = \lambda^+ < \mu$). By 2.6(1) it suffices for each $\mu' < \mu$ to have $\delta < \lambda^+$ and a tree with $\leqslant \lambda$ nodes and $\geqslant \mu'$ δ-branches. They exist by clause (ε) of (A) of 2.1.
2) Similarly using 2.6(2).
3) So assume $M^* \in K_\lambda$ and $\mathscr{S}(M^*)$ has cardinality $\geqslant \mu$, let $p_\eta \in \mathscr{S}(M^*)$ for $\eta \in Z$ be pairwise distinct, $|Z| \geqslant \mu$ and let $M^* \leqslant_\mathfrak{K} N_\eta \in K_\lambda, p_\eta = \mathrm{tp}(a_\eta, M^*, N_\eta)$. Now for every $X \in [Z]^{\lambda^+}$, as $\mathfrak{K}$ has amalgamation in λ there is $M_X \in K_{\lambda^+}$ such that $M^* \leqslant_\mathfrak{K} M_X$ and $\eta \in X \Rightarrow N_\eta^*$ is embeddable into M_X over M^* (hence p_η is realized in M_X). Let $Y[X] = \{\eta \in Z : p_\eta$ is realized in $M_X\}$. So $X \subseteq Y[X] \in [Z]^{\lambda^+}$, so $\{Y[X] : X \in [Z]^{\lambda^+}\}$ is a cofinal subset of $[Z]^{\lambda^+}$, hence (see clause (β) of case (A) of fact 2.1)

$$\begin{aligned} &|\{(M_X, c)_{c \in M^*} / \cong : X \in [Z]^{\lambda^+}\}| \geqslant \\ &\qquad |\{Y[X] : X \in [Z]^{\lambda^+}\}| \geqslant \mathrm{cf}([Z]^{\lambda^+}, \subseteq) \geqslant \\ &\qquad \mathrm{cf}([\mu]^{\lambda^+}, \subseteq) \geqslant pp(\mu) = \chi^*. \end{aligned}$$

As $2^\lambda < \chi^*$ also $|\{M_X / \cong : X \in [Z]^{\lambda^+}\}| \geqslant \chi^*$ (clear or see Shelah (1978, Ch. VIII, 1.2) because $\|M_X\| = \lambda^+, \|M^*\| = \lambda$ and $(\lambda^+)^\lambda < \mu$) but $I(\lambda^+, K)$ is $\geqslant$ than the former.

Lastly we shall prove $(2^\lambda)^+ < 2^{\lambda^+} \Rightarrow IE(\lambda^+, K) \geqslant \chi^*$ (so the reader may skip this, sufficing himself with the estimate on $I(\lambda^+, K)$).
For each $X \in [\mu]^{\lambda^+}$, let $\boldsymbol{F}_X = \{f : f \text{ a } \leqslant_{\mathfrak{K}}\text{-embedding of } M^* \text{ into } M_X\}$, and for $f \in F_X$ let

$$\mathscr{Z}_{X,f} = \{X_1 \in [Z]^{\lambda^+} : \text{there is a } \leqslant_{\mathfrak{K}}\text{-embedding of } M_{X_1} \text{ into } M_X \text{ extending } f\},$$

and let $\mathscr{S}_{X,f} = \{p \in \mathscr{S}(M^*) : f(p) \text{ is realized in } M_X\}$, so $\mathscr{Z}_{X,f} \subseteq \{X_1 : X_1 \subseteq \mathscr{S}_{X,f}\}$ and $|\mathscr{S}_{X,f}| \leqslant \lambda^+$.

Now the result follows from the the fact 2.7 above.

4) Should be clear from the proof of part (3).

Claim 3.3

1) Assume

(a) $2^\lambda < 2^{\lambda^+} < 2^{\lambda^{+2}}$ *and case B or C of Fact 2.1 for* λ *occurs (so* χ^*, T_ζ *are determined)*

(b) $\mathfrak{K}$ *is an abstract elementary class* $LS(\mathfrak{K}) \leqslant \lambda$

(c) $K_{\lambda^{++}} \neq \emptyset$,

(d) $\mathfrak{K}$ *has amalgamation in* λ *and in* λ^+

(e) in K^3_λ, *the minimal triples are not dense.*

Then

(α) *for each* $\zeta < \chi^*$ *for some* $M \in K_{\lambda^+}$ *we have* $|\mathscr{S}_*(M)| \geqslant |Lim_{\lambda^+}(T_\zeta)|$ *(the tree from clause* (ζ) *of 2.1).*

2) If K satisfies (a)-(e) and is categorical in λ^+ *or just has a universal member in* λ^+, *then for some* $M \in K_{\lambda^+}$ *we have* $|\mathscr{S}_*(M)| = 2^{\lambda^+}$.

3) If $(f)^+$, *then* $I(\lambda^{++}, K) \geqslant 2^{\lambda^{++}}$ *where*

$(f)^+$ $\mathfrak{K}$ *is categorical in* λ *and* λ^+.

4) Assume (a)-(e) of part (1) then $\boldsymbol{C}^1_{\mathfrak{K},\lambda}$ *has weaker* λ*-coding (if we have restricted to* (M,N,a) *in Definition of* $\boldsymbol{C}$*).*

Remark. Note that for 3.4 we do not use 3.3.

Proof. 1) Let $\zeta < \chi^*$. Let
$T_\zeta = \bigcup_{\alpha<\lambda^+} T^\zeta_\alpha$ where T^ζ_α are pairwise disjoint in α, each T^ζ_α has cardinality $\leqslant \lambda$,
$T^\zeta_0 = \{<>\}$ and $\eta \in T^\zeta_\alpha \;\&\; \beta < \ell g(\eta) \Rightarrow \eta \restriction \beta \in \bigcup_{\gamma<\alpha} T^\zeta_\gamma$, and

$\eta \in T^\zeta_\alpha \Rightarrow \bigwedge_{\ell<2} \eta\hat{\ }\langle \ell \rangle \in T^\zeta_{\zeta,\alpha+1}$ so $T_{\zeta,\alpha+1} = \{\eta\hat{\ }\langle \ell \rangle : \eta \in T^\zeta_\alpha$ and $\ell < 2\}$. For $\eta \in T^\zeta_\delta, \delta$ a limit ordinal, necessarily both $\ell g(\eta)$ and $\alpha(\eta) = \sup\{\gamma :$ for some $\varepsilon < \ell g(\eta), \eta \restriction \varepsilon \in T^\zeta_\gamma\}$ are limit ordinals $\leqslant \delta$.

Let $(M,N,a) \in K^3_\lambda$ be such that there is no minimal triple above it. We now choose by induction on $\alpha < \lambda^+, \langle M^\zeta_\alpha, M^\zeta_\eta, N^\zeta_\eta : \eta \in T^\zeta_\alpha \rangle$ such that:

(a) $(M^\zeta_\alpha, N^\zeta_\eta, a) \in K^3_\lambda$ and is reduced if $\eta \in T^\zeta_\alpha, \alpha$ non-limit

(b) $(M^\zeta_0, N^\zeta_{<>}, a) = (M,N,a)$

(c) $\nu \in T^\zeta_\beta, \eta \in T^\zeta_\alpha, \nu \triangleleft \eta, \beta < \alpha \Rightarrow (M^\zeta_\beta, N^\zeta_\nu, a) \leqslant (M^\zeta_\alpha, N^\zeta_\eta, a)$ in the order of K^3_λ

(d) if δ is a limit ordinal then: $M^\zeta_\delta = \bigcup_{\beta<\delta} M^\zeta_\beta$

(e) if δ is a limit ordinal and $\eta \in T^\zeta_\delta$ then
$N^\zeta_{\ell g(\eta)} = \bigcup_{\beta<\delta} N^\zeta_{\eta\restriction\beta}$ hence $(M_{\alpha(\eta)}, N^\zeta_\eta, a) \in K^3_\lambda$

(f) if $\eta \in T^\zeta_\alpha$ then $\mathrm{tp}(a, M_{\alpha+1}, N_{\eta\hat{\ }\langle 0 \rangle}) \neq \mathrm{tp}(a, M_{\alpha+1}, N_{\eta\hat{\ }\langle 1 \rangle})$

(g) $M^\zeta_\alpha \neq M^\zeta_{\alpha+1}$.

There is no problem to carry the definition. Let $M_\zeta = \bigcup_{\alpha<\lambda^+} M^\zeta_\alpha \in K_{\lambda^+}$, and for each $\nu \in \mathrm{Lim}_{\lambda^+}(T_\zeta)$ let $N^\zeta_\nu = \bigcup_{\alpha<\lambda^+} N^\zeta_{\nu\restriction\alpha}$, clearly $M_\zeta \leqslant_{\mathfrak{K}} N^\zeta_\nu$ and $a \in N^*_\nu$ and $\langle \mathrm{tp}(a, M_\zeta, N^\zeta_\nu) : \nu \in \mathrm{Lim}_{\lambda^+}(T_\zeta) \rangle$ are pairwise distinct members of $\mathscr{S}(M_\zeta)$. This proves clause (α) of part (1).
2) This part follows by 2.6(2).
3) Now 2.1 also applies to $(\lambda^+, \lambda^{++})$ in place of (λ, λ^+), so if for this case, clause $(A)_{\lambda^+}$ holds then by $(f)^+$ we get that 3.1(4) applies (with $(f)^+$ providing assumption $(\mathrm{d})^+$ there) hence we get $I(\lambda^{++}, K) = 2^{\lambda^{++}}$, so we can assume $(C)_{\lambda^+} \vee (B)_{\lambda^+}$ occurs.

Now if $(2^\lambda > \lambda^+) +$ WDmId(λ^+) is not λ^{++}-saturated we get the desired result as follows: by 1.9 if $\neg(**)$ of 1.8 holds and by 1.7 we get a contradiction to $I(\lambda^+, K) = 1$ if $(*)_\lambda$ (or $(*)'_\lambda$) of 1.7 holds, but by 1.8 one of the cases applies. Now $2^\lambda = \lambda^+$, as otherwise we are in case $(B)_\lambda$ or $(C)_\lambda$ (see (a) of 3.3(1)) by clause (ι) of 2.1 we have $2^\lambda = \lambda^+, 2^{\lambda^+} = \lambda^{++}$. However, once we know $2^\lambda = \lambda^+$ we deduce that there is a model in λ^+ saturated over λ so Claim Shelah (2001, 3.16At) applies.
4) Left to the reader as not used. $\square_{3.3}$

Claim 3.4 *Assume*

(a) $2^\lambda < 2^{\lambda^+} < 2^{\lambda^{++}}$

(b) K *is categorical in* λ, λ^+

(c) $1 \leqslant I(\lambda^{++}, K) < 2^{\lambda^{++}}$.

Then above every $(M^*, N^*, a) \in K^3_\lambda$ *there is a minimal triple.*

Proof. Assume toward contradiction that above $(M^*, N^*, a) \in K^3_\lambda$ there is no minimal type. If $2^\lambda = \lambda^+$, then there is a $M \in K_{\lambda^+}$ saturated over λ hence we finish by Shelah (2001, 3.16At), possibility $(*)_2$. So we can assume $2^\lambda > \lambda^+$, hence (Shelah (1996, 6.3) with λ^+ here for λ there so μ there is $\leqslant \lambda$ so $\delta < \lambda^+$ hence $|T| \leqslant |\delta| \leqslant \lambda$ and let $\kappa = \text{cf}(\delta)$) there are $\kappa \leqslant \lambda$ and[5] tree T with $\leqslant \lambda$ nodes and κ levels with $|\text{Lim}_\kappa(T)| > \lambda^+$ hence for some $M \in K_\lambda, |\mathscr{S}_*(M)| > \lambda^+$ (e.g. by the proof of 3.3(1)). If WDmId(λ^+) is not λ^{++}-saturated then in 1.9 assumption (b) holds, and assumptions (c) + (d) + (e) holds by the assumptions of the present claim but not the conclusion, so (a) fails, that is $(**)_\lambda$ of 1.8 holds hence by 1.8, $(*)_\lambda$ of 1.7 holds. But now 1.7 contradicts clause (b) of the assumption, so we have to assume that WDmId(λ^+) is λ^{++}-saturated. Hence clause (ι) of 2.1, Case B does not occur, hence Cases B, C of 2.1 do not occur and hence Case A occurs. So by 3.1(3) we get a contradiction to categoricity in λ^+. $\square_{3.4}$

4. A minor debt

Claim 4.1 *We can prove (Shelah 2001, 4.2t) also for* $\lambda = \aleph_0$.

Proof. We ask:

Question 1: are there $M <_{\mathfrak{K}} N$ in K_λ such that for no $a \in N \backslash M$ is tp(a, M, N) minimal?

If the answer is yes, we can find $\langle M^1_i : i < \lambda^+ \rangle$ a representation of a model $M^1 \in K_{\lambda^+}$ such that: $a \in M^1_{i+1} \backslash M^1_i \Rightarrow$ tp(a, M^1_i, M^1_{i+1}) is not minimal. This implies $a \in M^1 \backslash M^1_i \Rightarrow$ tp(a, M^1_i, M^1) is not minimal (as for some $j \in [i, \lambda^+)$ we have $a \in M^1_{j+1} \backslash M^1_j$ so $(M^1_i, M^1_{j+1}, a) \leqslant (M^1_j, M^1_{j+1}, a)$ and the latter is not minimal). But we can build another representation $\langle M^2_i : i < \lambda^+ \rangle$ of $M^2 \in K_{\lambda^+}$ such that for each $i < \lambda^+$ for some $a \in M^1_{i+1} \backslash M^1_i$, tp$(a, M^1_i, M^1_{i+1})$ is minimal (as there is a minimal triple). So $M^1 \not\cong M^2$.

So we assume the answer is no.

Question 2: If $M \in K_\lambda, \Gamma \subseteq \Gamma^*_M =: \{p \in \mathscr{S}(M) : p \text{ minimal}\}$ and $|\Gamma| \leqslant \lambda$, is there N such that: $M <_{\mathfrak{K}} N \in K_\lambda$ and N omit every $p \in \Gamma$?

If the answer to question 2 is yes, we can build $\langle M_\eta : \eta \in {}^{\lambda^+>}2 \rangle$ as in the proof of 1.7 (more exactly $\eta \triangleleft \nu \Rightarrow M_\eta \leqslant_{\mathfrak{K}} M_\nu, M_\eta \in K_\lambda$) and we also have $\Gamma_\eta \subseteq$

$\{p : \text{for some } N \leqslant_{\mathfrak{K}} M_\eta, N \in K_\lambda \text{ and } p \in \mathscr{S}(N) \text{ is minimal not realized in } M_\eta\}$ have cardinality $\leqslant \lambda, \eta \triangleleft \nu \Rightarrow \Gamma_\eta \subseteq \Gamma_\eta$ and there is $p \in \Gamma_{\eta^\frown\langle 1\rangle}$ realized in $M_{\eta^\frown\langle 0\rangle}$ (and if you like also $p' \in \Gamma_{\eta^\frown\langle 0\rangle}$ realized in $M_{\eta^\frown\langle 1\rangle}$). So by Shelah (2001, 1.6t) we get $I(\lambda^+, K) = 2^\lambda$. So assume the answer is no and for every $M \in K_\lambda$ let Γ_M be a counterexample. Let $\langle M_i^1 : i < \lambda^+\rangle$, representing a model $M^1 \in K_{\lambda^+}$ be such that $i < \lambda^+ \;\&\; p \in \Gamma_{M_i} \Rightarrow p$ realizes in M. Now as in the proof of saturated = model homogeneous (see Shelah (2001, 0.21t)) we can prove M^1 is saturated. But this proves more than required: $|\mathscr{S}(M_\ell^1)| \leqslant \lambda^+$. $\square_{4.1}$

Notes

1. We normally assume $M \in \mathfrak{K} \Rightarrow \|M\| \geqslant LS(\mathfrak{K})$, here there is no loss in it. It is also natural to assume $|\tau(\mathfrak{K})| \leqslant LS(\mathfrak{K})$ which means just increasing $LS(\mathfrak{K})$, but no real need.
2. Note that: if there is no minimal triple in K_λ^3 then every $a \in M_2 \backslash M_1$ is as required so the moreover is not necessary.
3. We could make the relation on $(M_{\eta^\frown\langle 0\rangle}, \Gamma_{\eta^\frown\langle 0\rangle}), (M_{\eta^\frown\langle 1\rangle}, \Gamma_{\eta^\frown\langle 1\rangle})$ symmetric.
4. The less easy point is when $\mathrm{cf}(\Upsilon) = \aleph_0$, otherwise we can get the conclusion differently (by Shelah (1994, II, 5.4)), so 2.1(A) suffice.
5. Let $\kappa = \mathrm{Min}\{\sigma : 2^\sigma > \lambda^+\}$, so if $2^{<\sigma} \leqslant \lambda$ then $({}^{\sigma>}2, \triangleleft)$ is okay, otherwise ${}^{\sigma>}2 = \bigcup_{i<\lambda^+} T_i, |T_i| \leqslant \lambda$, T_i increasing with i so for some $i, |\{\eta \in {}^\sigma 2 : \bigwedge_{\alpha<\sigma} \eta \restriction \alpha \in T_i\}| > \lambda^+$.

References

[References of the form `math.LO/ ...` refer to `xxx.lanl.gov` archive.]

Hajnal, A.: (1961), 'Proof of a conjecture of S. Ruziewicz', *Fundamenta Mathematicae* **50**, pp. 123–128. [Ha61]

Kojman, M., Shelah, S.: (1992), 'Non-existence of Universal Orders in Many Cardinals', *The Journal of Symbolic Logic* **57**, pp. 875–891. `math.LO/9209201`. [KjSh 409]

Shelah, S.: (1978), *Classification theory and the number of nonisomorphic models*, Vol. 92 of *Studies in logic and the Foundations of Mathematics*, North-Holland Publishing Company, Amsterdam. [Sh:a]

Shelah, S.: (1982), *Proper forcing*, Vol. 940 of *Lecture Notes in Mathematics*, Springer Verlag, New York, Berlin, Heildelberg, Tokyo. [Sh:b]

Shelah, S.: (1993), 'Advances in Cardinal Arithmetic', in N.W. Sauer, et al (ed.), *Finite and Infinite Combinatorics in Sets and Logic*, Kluwer Academic Publishers, Dordrecht, pp. 355–383. [Sh 420]

Shelah, S.: (1994), *Cardinal Arithmetic*, Vol. 29 of *Oxford Logic Guides*, Oxford University Press, Oxford. [Sh:g]

Shelah, S.: (1996), 'Further cardinal arithmetic', *Israel Journal of Mathematics* **95**, pp. 61–114. `math.LO/9610226`. [Sh 430]

Shelah, S.: (1998), *Proper and impropoer forcing*, Perspectives in Mathematical Logic, Springer Verlag, New York, Berlin, Heildelberg, Tokyo. [Sh:f]

Shelah, S.: (2000), 'The Generalized Continuum Hypothesis revisited', *Israel Journal of Mathematics* **116**, pp. 285–321. `math.LO/9809200`. [Sh 460]

Shelah, S.: (2001), 'Categoricity of an abstract elementary class in two successive cardinals', *Israel Journal of Mathematics* **126**, pp. 29–128. math.LO/9805146. [Sh 576]

Shelah, S.: (200x), 'Categoricity in abstract elementary classes: going up inductive step', in preparation. [Sh 600]

III

RECURSION THEORY

MINIMISATION VS. RECURSION ON THE PARTIAL CONTINUOUS FUNCTIONALS

Ulrich Berger
Department of Computer Science
University of Wales Swansea

Abstract We study the relationship between partial continuous higher type functionals defined by minimisation on the one hand and recursion on the other hand. We prove that already at type level two minimisation is weaker than recursion.

1. Introduction

There are two well-known ways of extending the schemata for the primitive recursive functions such that all partial recursive functions are obtained. One is recursion the other is minimisation. In Kleene (1959a) both are extended to higher types: recursion via the schemata (S1-S9) and minimisation via μ-recursion. Kleene showed that at type two (S1-S9) and μ-recursion coincide, but already at type three the latter is weaker than the former. In Bergstra (1976) it is proved that also on the total continuous functionals (Kleene 1959b, Kreisel 1959) the schemata (S1-S9) are weaker than recursive continuity (having a recursive associate). These results were based on an interpretation of computations on *total* (continuous) functionals, whereas it is now common to interpret them on the *partial* continuous functionals. Normann showed that this makes a difference by proving that under the new interpretation (S1-S9) is as powerful as recursive continuity when restricted to the total continuous functional (Normann 2000). To be precise Normann works with the functional language PCF (Plotkin 1977), which however, as essentially shown in Platek (1966), is equivalent to (S1-S9) on the partial continuous functionals.

In this paper it is shown that the relationship between minimisation and recursion is affected by the choice of the underlying domain, too. We prove that already at type level two minimisation is weaker than recursion. From our result we deduce that at those types minimisation is not only denotationally but also operationally weaker than PCF.

P. Gärdenfors, J. Woleński and K. Kijania-Placek (eds.), In the Scope of Logic, Methodology and Philosophy of Science, Vol. I, 57-64.

After giving in section 2 the basic definitions we present in 3 as a starter a new short proof that every computable monotone functions of type level one is μ-recursive in the parallel OR, a result first proved in Trakhtenbrot (1976). Section 4 contains the main result already discussed above, and in section 5 we conclude by discussing some open problems arising at higher types.

2. Partial continuous functionals

The hierarchy of *partial continuous functionals* is a family of effective Scott-Ershov-domains (Scott 1982, Griffor, Lindström and Stoltenberg-Hansen 1993) defined by

$$D_\iota := \mathbf{N}_\perp, \quad D_o := \mathbf{B}_\perp, \quad D_{\rho\to\sigma} := D_\rho \to D_\sigma,$$

where $\mathbf{N}_\perp := \mathbf{N}\cup\{\perp\}$ and $\mathbf{B}_\perp := \mathbf{B}\cup\{\perp\}$ ($\mathbf{B} = \{\mathrm{tt},\mathrm{ff}\}$) are the flat domains of partial integers and boolean values respectively. $D \to E$ denotes the domain of continuous functions from D to E, i.e. the exponential in the cartesian closed category of Scott-Ershov domains and continuous functions. We let $\mathcal{D}$ be the union of the D_ρ. By a *functional* (of type ρ) we mean an element of $\mathcal{D}$ (D_ρ). When writing $f\colon \rho$ we mean that f is a functional of type ρ. A functional of type ρ is called *computable* if it is a computable element of the effective Scott-Ershov domain D_ρ, i.e. its compact approximations are recursively enumerable.

We let τ range over the base types ι and o. A type $\rho_1 \to (\rho_2 \to \ldots \to (\rho_n \to \sigma)\ldots)$ will often be written $\rho_1 \to \rho_2 \to \ldots \to \rho_n \to \sigma$ or simply $\vec{\rho} \to \sigma$. An iterated application $f(x_1)\ldots(x_n)$ will also be written $fx_1\ldots x_n$ or $f(x_1,\ldots,x_n)$. For $\rho = \vec{\rho} \to \tau$ we define $\perp_\rho\colon \rho$ by $\perp_\rho(\vec{x}) := \perp$. The level of a type ρ is defined as usual by $\mathrm{lev}(\iota) = \mathrm{lev}(o) := 0$, $\mathrm{lev}(\rho \to \sigma) := \max(\mathrm{lev}(\rho)+1, \mathrm{lev}(\sigma))$. The level of a functional $f \in D_\rho$ is the level of ρ. A functional $g \in \mathcal{D}$ is *explicitly definable* from a set of functionals $\mathcal{F} \subseteq \mathcal{D}$ if g can be defined from elements of $\mathcal{F}$ by application and λ-abstraction.

In the following we will frequently omit type information as long as it can be recovered from the context.

A functional of type level $\leqslant 1$ will be called a *monotone function*. The following monotone functions will be used frequently (see also Plotkin (1977)). We let i, j, m, n, k range over natural numbers.

$\mathrm{if}_\tau\colon o \to \tau \to \tau \to \tau$	$\mathrm{if}_\tau(\mathrm{tt},x,y) = x$, $\mathrm{if}_\tau(\mathrm{ff},x,y) = y$, $\mathrm{if}_\tau(\perp,x,y) = \perp$
$\mathrm{Z}\colon \iota \to o$	$\mathrm{Z}(0) := \mathrm{tt}$, $\mathrm{Z}(n+1) := \mathrm{ff}$, $\mathrm{Z}(\perp) := \perp$
$(+1)\colon \iota \to \iota$	$(+1)(n) := n+1$, $(+1)(\perp) := \perp$
$(-1)\colon \iota \to \iota$	$(-1)(0) := 0$, $(-1)(n+1) := n$, $(-1)(\perp) := \perp$

The set $\{0, \mathrm{if}, \mathrm{Z}, (+1), (-1)\}$ is called the set of *basic functions*.

Recursive definitions in $\mathcal{D}$ are modelled by the *fixed point operators* $Y_\rho\colon (\rho \to \rho) \to \rho$,

$$Y(f) := \text{ the least fixed point of } f = \bigsqcup\{f^n(\bot_\rho) \mid n \in \mathbf{N}\}.$$

A functional is definable in the functional programming language PCF (Plotkin 1977) iff it is explicitly definable from the basic functions and the fixed point operators.

A functional is *μ-recursive* if it is explicitly definable from the basic functions, the *primitive recursor* $R\colon \iota \to (\iota \to \iota \to \iota) \to \iota \to \iota$,

$$R(x,h,0) := x, \quad R(x,h,n+1) := h(n, R(x,h,n)), \quad R(x,h,\bot) := \bot,$$

and the *minimisation functional* $\mu\colon (\iota \to o) \to \iota$,

$$\mu(f) := n \text{ if } f(n) = \mathrm{tt} \text{ and } f(i) = \mathrm{ff} \text{ for } i < n, \quad \mu(f) := \bot \text{ otherwise}$$

A functional g is *μ-recursive in a set of functionals* $\mathcal{F}$ if there are $f_1, \ldots, f_k \in \mathcal{F}$ and a μ-recursive functional h such that $g = h(f_1, \ldots, f_n)$. The notions '*PCF-definable in*' and '*computable in*' are defined similarly.

The following implications are well-known:

$$\mu\text{-recursive} \quad \Rightarrow \quad \text{PCF-definable} \quad \Rightarrow \quad \text{computable}$$

Of course these implications also hold when relativised to a set of functionals.

A *strict function* is a monotone function f such that $f(\vec{x}) = \bot$ whenever $x_i = \bot$ for some i. It follows from ordinary recursion theory that a strict function is computable iff it is μ-recursive.

3. Monotone functions

In Sazonov (1975) it is proved that a monotone function is PCF-computable iff it is an *effectively sequential function* (Vuillemin 1975). In Trakhtenbrot (1976) it is shown that the effectively sequential functions are precisely those definable from computable strict functions and if_τ by composition. Since computable strict functions are μ-recursive we have:

Theorem 1 (Sazonov, Trakhtenbrot) *A monotone function is PCF-definable iff it is μ-recursive.*

The monotone function *parallel or* $\mathrm{OR}\colon o \to o \to o$ is defined by

$$\mathrm{OR}(\mathrm{tt}, x) = \mathrm{OR}(x, \mathrm{tt}) = \mathrm{tt},\ \mathrm{OR}(\mathrm{ff}, \mathrm{ff}) = \mathrm{ff},\ \mathrm{OR}(x,y) = \bot \text{ otherwise}.$$

Note that $\mathrm{OR}(\mathrm{tt}, \bot) = \mathrm{OR}(\bot, \mathrm{tt}) = \mathrm{tt}$, but $\mathrm{OR}(\bot, \bot) = \bot$. Hence OR is not sequential and therefore not PCF-definable (although computable, trivially).

Trakhtenbrot (Trakhtenbrot 1976) showed that all computable monotone functions are μ-recursive in OR. Below we give a new short proof of this result.

Theorem 2 (Trakhtenbrot) *A monotone function is computable iff it is μ-recursive in OR.*

Proof. We have to show that every computable monotone function f is μ-recursive in OR. We prove this first for boolean-valued functions, i.e. f of a type $\tau_1 \to \ldots \to \tau_n \to o$. We use the vote functional $\mathrm{MAJ}\colon o \to o \to o \to o$ defined by $\mathrm{MAJ}(u,v,w) := b \in \mathbf{B}$ if at least two of the arguments are equal b. Otherwise $\mathrm{MAJ}(u,v,w) := \bot$ (Trakhtenbrot's Γ_d in Trakhtenbrot (1976)). MAJ can be defined using OR and the dual parallel AND by (using infix notation)

$$\mathrm{MAJ}(u,v,w) = (u\,\mathrm{AND}\,v)\,\mathrm{OR}\,(u\,\mathrm{AND}\,w)\,\mathrm{OR}\,(v\,\mathrm{AND}\,w).$$

Obviously it suffices to prove the following statement.

> (+) Given base types $\vec{\tau} = \tau_1, \ldots, \tau_n$ then every computable monotone function $f\colon \vec{\tau} \to \vec{\tau}' \to o$ such that $f(\vec{x}) \in D_{\vec{\tau}' \to o}$ is strict for all $\vec{x} \in D_{\vec{\tau}}$ is explicitly definable from MAJ and computable strict functions.

We prove (+) by induction on n (the length of $\vec{\tau}$). If $n = 0$ then f is strict and hence the assertion holds trivially. Now let $f\colon \vec{\tau} \to \tau \to \vec{\tau}' \to o$ be given such that $f(\vec{x},y)\colon \vec{\tau}' \to o$ is strict for all $\vec{x}\colon \vec{\tau}$, $y\colon \tau$. For $b \in \mathbf{B}$ we define a computable monotone function $g_b\colon \vec{\tau} \to \vec{\tau}' \to o$ by $g_b(\vec{x},\vec{z}) := b$, if $f(\vec{x},y,\vec{z}) := b$ for some $y\colon \tau$, $g_b(\vec{x},\vec{z}) := \neg b$, if $f(\vec{x},\bot,\vec{z}) := \neg b$. Otherwise $g_b(\vec{x},\vec{z}) := \bot$. Furthermore define a computable monotone function $h\colon \vec{\tau} \to \tau \to \vec{\tau}' \to o$ by $h(\vec{x},y,\vec{z}) := f(\vec{x},y,\vec{z})$, if $y \neq \bot$, $h(\vec{x},\bot,\vec{z}) := \bot$ otherwise. Clearly $g_b(\vec{x})$ and $h(\vec{x})$ are strict for all $\vec{x}\colon \vec{\tau}$, and hence, by induction hypothesis, are explicitly definable from MAJ and computable strict functions. Now one easily verifies that

$$f(\vec{x},y,\vec{z}) = \mathrm{MAJ}(g_{\mathrm{tt}}(\vec{x},\vec{z}), g_{\mathrm{ff}}(\vec{x},\vec{z}), h(\vec{x},y,\vec{z})).$$

So far we have shown that every boolean-valued computable monotone function is μ-recursive in OR. Now let $f\colon \tau \to \tau$ be an arbitrary computable monotone function. Define the graph of f, $g\colon \tau \to \tau \to o$ by $g(\vec{x},y) := \mathrm{eq}_\tau(f(\vec{x}),y)$, where $\mathrm{eq}_\tau\colon \tau \to \tau \to o$ is strict equality. g is a boolean-valued computable monotone function and hence μ-recursive in OR. No we can define f by $f(\vec{x}) = \mu(\lambda y.g(\vec{x},y))$. □

4. μ-recursion at type level two

While at type level 1 μ-recursion is as strong as PCF this is no longer true at type level 2 in the following strong sense: there is a PCF-definable functional H

of level 2 such that there is no μ-recursive functional F coinciding with H on all PCF-definable arguments. This still holds when we allow F to be μ-recursive in arbitrary monotone functions.

Let D_1 be the set of all monotone functions, i.e. $D_1 = \bigcup\{D_\rho \mid \mathrm{lev}(\rho) \leqslant 1\}$.

Lemma Let $F\colon (\iota \to \iota \to \iota) \to \vec{\iota} \to \tau$ be μ-recursive in D_1 and let $\vec{x}\colon \vec{\iota}$ be such that $F(\bot,\vec{x}) = \bot$. Then there is a natural number k such that for every $g\colon \iota \to \iota \to \iota$ such that $F(g,\vec{x}) \neq \bot$ we have $g(m_1,m_2) \neq \bot$ for some $m_1,m_2 \leqslant k$.

Proof. Let $\mathcal{F}$ be the set of functionals of a type $(\iota \to \iota \to \iota) \to \vec{\iota} \to \tau$ (the length of $\vec{\iota}$ and the base type τ may vary) which are μ-recursive in D_1. Every $F \in \mathcal{F}$ is the denotation of a simply typed λ-term with constants for R, μ and elements of D_1. By the well-known normalisation theorem for the simply typed λ-calculus this term may be assumed to be in long β-normal form, i.e. of the form $\lambda g \lambda \vec{x}.M$ where $M\colon \tau$ is in long β-normal form with free variables among $g,\vec{x}$. Analysing these terms M one arrives at the following inductive generation of the set $\mathcal{F}$.

$$
\begin{aligned}
F(g,\vec{x}) &= x_i && (1)\\
F(g,\vec{x}) &= f(F_1(g,\vec{x}),\ldots,F_n(g,\vec{x})) && (2)\\
F(g,\vec{x}) &= g(F_1(g,\vec{x}),F_2(g,\vec{x})) && (3)\\
F(g,\vec{x}) &= \mathrm{R}(F_0(g,\vec{x}),\lambda y \lambda z.F_1(g,\vec{x},y,z),F_2(g,\vec{x})) && (4)\\
F(g,\vec{x}) &= \mu(\lambda y.F_1(g,\vec{x},y)) && (5)
\end{aligned}
$$

In (2) $f \in D_1$ is arbitrary. Of course these equations have to be read as rules of an inductive definition: if in an equation the F_i are in $\mathcal{F}$ then F defined by the equation is in $\mathcal{F}$. Now the assertion of the lemma can be proved by induction along this inductive definition of $\mathcal{F}$. We only treat the rules (3), (4), and (5), since (1) is trivial and (2) is similar to (3) but simpler.

(3): Assume $F(\bot,\vec{x}) = \bot$. Set $y_i := F_i(g,\vec{x})$. For any $i \in \{1,2\}$ such that $y_i = \bot$ we have by induction hypothesis k_i for F_i and $\vec{x}$. Let k be such that for all $i \in \{1,2\}$ if $y_i = \bot$ then $k_i \leqslant k$ and if $y_i \in \mathbf{N}$ then $y_i \leqslant k$. Now assume $F(g,\vec{x}) \neq \bot$. Set $z_i := F_i(g,\vec{x})$. If $(y_1,y_2) = (z_1,z_2)$ then $g(y_1,y_2) \neq \bot$. In this case set $m_i := y_i$ if $y_i \in \mathbf{N}$, $m_i := 0$ otherwise. Then $m_1,m_2 < k$ and $g(m_1,m_2) \neq \bot$. If $(y_1,y_2) \neq (z_1,z_2)$ then w.l.o.g. $y_1 = \bot$ and $z_1 \neq \bot$, i.e. $F_1(\bot,\vec{x}) = \bot$ and $F_1(g,\vec{x}) \neq \bot$. Hence by the choice of k_1 there are $m_1,m_2 < k_1 < k$ such that $g(m_1,m_2) \neq \bot$.

(4): Assume $F(\bot,\vec{x}) = \bot$. If $F_2(\bot,\vec{x}) = \bot$ we have by induction hypothesis k for F_2 and $\vec{x}$, and this k does the job. Now assume $F_2(\bot,\vec{x}) = n \in \mathbf{N}$. Set $y_0 := F_0(\bot,\vec{x})$, $y_{i+1} := F_1(\bot,\vec{x},i,y_i)$. By induction hypothesis if $y_0 = \bot$ we have k_0 for F_0 and $\vec{x}$, and for every $i < n$ such that $y_{i+1} = \bot$ we have k_{i+1} for F_1 and $\vec{x},y_i$. Let k be greater or equal all these k_i. Now assume $F(g,\vec{x}) \neq \bot$. Since $F_2(g,\vec{x}) = F_2(\bot,\vec{x}) = n$ we have $F(g,\vec{x}) = z_n$ where $z_0 := F_0(g,\vec{x})$, $z_{i+1} :=$

$F_1(g,\vec{x},i,z_i)$. Let $j \leqslant n$ be minimal such that $y_j \neq z_j$ ($y_n \neq z_n$!). Hence $y_j = \bot$ and $z_j \neq \bot$. If $j = 0$ then $F_0(\bot,\vec{x}) = \bot$ and $F_0(g,\vec{x}) \neq \bot$. Hence there are $m_1, m_2 \leqslant k_i$ such that $g(m_1,m_2) \neq \bot$ and we are done since $k_0 \leqslant k$. If $j = i+1$ then $F_1(\bot,\vec{x},i,y_i) = y_j = \bot$ and $F_1(g,\vec{x},i,y_i) = F_1(g,\vec{x},i,z_i) = z_j \neq \bot$. Hence we are done since $k_{i+1} \leqslant k$.

(5): Assume $F(\bot,\vec{x}) = \bot$. Set $z_n := F_1(\bot,\vec{x},n)$. If $z_n = \text{ff}$ for all n then we may take $k := 0$ since then $F(g,\vec{x}) = \bot$ for all g. Otherwise there is some minimal n such that $z_n = \bot$ and $z_i = \text{ff}$ for all $i < n$. The k for F_1 and $\vec{x}, n$ will do. □

Theorem 3 *There is a PCF-definable functional* $H\colon (\iota \to \iota \to \iota) \to \iota$ *such that for every functional* F *of the same type that is* μ*-recursive in* D_1 *there is a PCF-definable* $g\colon \iota \to \iota \to \iota$ *such that* $H(g) \neq F(g)$.

Proof. Define $G\colon (\iota \to \iota \to \iota) \to \iota \to \iota$ recursively by

$$G(g,x) = g(x, G(x+1))$$

and from that

$H\colon (\iota \to \iota \to \iota) \to \iota$ by $H(g) := G(g,0)$, i.e. $H(g) = g(0, g(1, g(2, \ldots$
Clearly H is PCF-definable. Now let $F\colon (\iota \to \iota \to \iota) \to \iota$ be μ-recursive in D_1. If $F(\bot) \neq \bot$ then F differs from H at $\bot$ since $H(\bot) = \bot$. If $F(\bot) = \bot$ then according to the lemma there is some k such that $F(g) = \bot$ whenever $g(m_1,m_2) = \bot$ for all $m_1, m_2 \leqslant k$. Now define $g\colon \iota \to \iota \to \iota$ by $g(x,y) := \text{if}(x \geqslant k, k, \text{if}(y \geqslant k, k, \bot))$ where $\geqslant$ is strict. Clearly g is PCF-definable and $g(m_1,m_2) = \bot$ for all $m_1, m_2 < k$. Hence $F(g) = \bot$. But $H(g) = k$. □

In the following corollary we view μ-recursion in an obvious way as a subsystem of PCF. The notion of operational equivalence is taken from Plotkin (1977).

Corollary There is a closed PCF-term M of type level 2 which is not operationally equivalent to any μ-recursive term.

Proof. Let M be a PCF-term defining the functional H from theorem 3. If some μ-recursive term K were operationally eqivalent to M then it would follow from the adequacy result in Plotkin (1977) that the value F of K defines a functional that agrees with H on all PCF-definable arguments. But according to theorem 3 this is impossible.

Rounding off this section we briefly review the well-known results on computability of total functionals of type level two. A functional $f\colon \vec{\rho} \to \tau$ is *total* if $f(\vec{x}) \neq \bot$ for all total $\vec{x}$ (this is a definition by induction on the type level). It is well-known that every total computable functional F of a type level $\leqslant 2$

is μ-recursive in the sense that there is a μ-recursive functional agreeing with F on all total arguments. In fact a slightly more general statement holds.

Lemma To every computable functional F of type level $\leqslant 2$ there is a μ-recursive functional agreeing with F on all strict total arguments.

Proof. W.l.o.g. assume that $F\colon (\iota \to \iota) \to \iota$ and $F \neq \bot$. Let $(g_i, n_i)_{i\in\mathbf{N}}$ some primitive recursive enumeration of the pairs (g, n) such that g is a finite strict function and $F(g) = n$. For $f\colon \iota \to \iota$ define $G(f) := n_i$ if i is such that $f(m) = g_i(m)$ for all m in the domain of g_i and for all $j < i$ and all m' in the domain of g_j $f(m')$ is defined but for at least one such m' $f(m') \neq g_j(m')$. If such an i does not exist let $G(f) := \bot$. Clearly G is μ-recursive and coincides with F on all strict total arguments.

Since every total function is an extension of a strict total function this immediately implies that every total computable functional F of a type level $\leqslant 2$ is μ-recursive.

5. Higher types and open problems

We conclude by discussing the relationship between minimisation and recursion at higher types and propose some open problems to be studied.

1. In Bergstra (1976) is shown that at type three μ-recursion is weaker than (S1-S9) where both are interpreted over the total continuous functionals. It seems that this carries over to the interpretation over the partial continuous functionals, which would mean that under the interpretation of μ-recursion and PCF given in this paper, PCF defines more total continuous functionals than μ-recursion does. However to confirm this, details still have to be checked.

2. We defined μ-recursion as the explicit closure of the basic functions, the minimisation operator μ, and the primitive recursor R. One may ask what happens when we replace R by primitive recursors of higher types $\mathrm{R}\colon \rho \to (\iota \to \rho \to \rho) \to \iota \to \rho$ (Gödel 1958). It is well-known that without μ the computational power of the R_ρ strictly increases. Is this still the case when μ is present?

3. Similarly one may consider subsystems PCF_n instead of full PCF where only fixed point operators Y_ρ with ρ of type level $\leqslant n$ are allowed. Do the PCF_n form a proper hierarchy?

References

Bergstra, J.: (1976), *Computability and continuity in finite types*, Ph.D.-Thesis, University of Utrecht.

Gödel, K.: (1958), 'Über eine bisher noch nicht benützte Erweiterung des finiten Standpunktes', *Dialectica* **12**, pp. 280–287.

Griffor, E., Lindström, I., Stoltenberg-Hansen, V.: (1993), *Mathematical Theory of Domains*, Cambridge University Press, Cambridge.

Kleene, S.C.: (1959a), 'Recursive functionals and quantifiers of finite type I', *T.A.M.S.* **91**, pp. 1–52.

Kleene, S.C.: (1959b), 'Countable functionals', in A. Heyting (ed.), *Constructivity in Mathematics*, North-Holland, Amsterdam, pp. 81–100.

Kreisel, G.: (1959), 'Interpretation of analysis by means of constructive functionals of finite types', in A. Heyting (ed.), *Constructivity in Mathematics*, North-Holland, Amsterdam, pp. 101–128

Normann, D.: (2000), 'Computability over the partial continuous functionals', *Journal of Symbolic Logic* **5(3)**, pp. 1133–1142.

Platek, R.A.: (1966), *Foundations of recursion theory*, PhD-thesis, Department of Mathematics, Stanford University.

Plotkin, G.: (1977) 'LCF considered as a programming language', *Theoretical Computer Science* **5**, pp. 223–255.

Sazonov, V.: (1975), 'Sequentially and parallely computable functionals', Symposium on Lambda-Calculus and Computer Science Theory, *LNCS* **37**, pp. 312–318.

Scott, D.: (1982), 'Domains for denotational semantics', in E. Nielsen, E.M. Schmidt (eds.), *Automata, Languages and Programming*, *LNCS* **140**, pp. 577–613.

Trakhtenbrot, M.B.: (1976), 'Relationships between classes of monotonic functions', *Theoretical Computer Science* **2**, pp. 225–247.

Vuillemin, J.: (1975), 'Correct and optimal implementations of recursion in a simple programming language', *Journal of Computer and System Science* **9**, pp. 332–354.

GLOBAL PROPERTIES OF DEGREE STRUCTURES*

André Nies
Department of Mathematics, The University of Chicago, USA
nies@math.uchicago.edu

Abstract We investigate degree structures induced by many-one reducibility and Turing reducibility on the computably enumerable (c.e.), the arithmetical and all subsets of $\mathbb{N}$. We ask which subsets of the degree structure are automorphism bases: For instance, the minimal degrees form an automorphism base of the c.e. many one degrees, but not for the other two degree structures based on $\leqslant_m$. We develop a method to show that a subset is an automorphism base and apply it to the c.e. m-degrees. Moreover, we use it to give a modified proof of Ambos-Spies' result that nontrivial initial intervals of the c.e. Turing degrees are automorphism bases. This new proof may be useful in showing that $\mathcal{R}_T$ is biinterpretable with $\mathbb{N}$ in parameters. Finally, we show that the arithmetical m-degrees form a prime model.

The study of sets of natural numbers under a notion of relative computability is a central topic of computability theory. Specifying a reducibility and an appropriate class of sets gives rise to a degree structure, which may have interesting "global" properties. This is the case for the degree structures induced by many-one reducibility (denoted $\leqslant_m$) and Turing reducibility ($\leqslant_T$) on all sets, the arithmetical and the computably enumerable (c.e.) sets. For $r \in \{m, T\}$, these structures are denoted by $\mathcal{D}_r$, $\mathcal{A}_r$ and $\mathcal{R}_r$. The least element of a degree structure is denotes by **0**, and the greatest (if any) by **1**. In analyzing a degree structure $\boldsymbol{A}$, frequently the following program is carried out:

1. *Understanding the basic algebraic properties.* In other words, one determines whether $\boldsymbol{A} \models \varphi$, for certain φ of algebraic significance. Examples are: $\mathcal{D}_T, \mathcal{R}_m$ have minimal elements above **0**, and $\mathcal{R}_T$ is dense.

2. *Studying more general algebraic properties of $\boldsymbol{A}$.* This includes embedding theorems and extensions of embeddings theorems for partial orders, embeddings of finite lattices (which is most interesting for $\mathcal{R}_T$) and characterizing initial segments.

*This research was partially supported under NSF grant DMS–9803482.

P. Gärdenfors, J. Woleński and K. Kijania-Placek (eds.), In the Scope of Logic, Methodology and Philosophy of Science, Vol. I, 65-80.

3. *Analyzing properties of Th(**A**)*. First, one proves that Th(**A**) is undecidable. This has been done for most degree structures. Next, one turns to fragments of the theories. All the Π_2-theories of the degree structures based on $\leqslant_m$, in the language of u.s.l., are decidable (Degtev 1979), while, even in the language of p.o., their Π_3-theories are undecidable (Nies 1996). The Π_3-theory of the p.o. $\mathcal{R}_T$ is undecidable by Lempp et al. (1998). A further question of interest is whether Th(**A**) is atomic.

4. *Model theoretic properties of **A***. Now one considers the structure itself, not only its theory. Interesting topics are the interaction of subsets of **A** with the whole (esp. which subsets are automorphism bases), defining a copy of $\mathbb{N}$ without parameters, and whether (in the countable case) **A** is a prime model of its theory, or even biinterpretable with $\mathbb{N}$ in parameters (i.e. if there is a parameter defined copy M of $(\mathbb{N},+,\times)$ and a parameter definable 1-1 map $f : \boldsymbol{A} \mapsto M$).

This article contributes mainly to Part 4 of the program. The plan is as follows. In the preparatory Section 1 we state the extension property for the structures based on $\leqslant_m$, use it to recall the characterizations of $\mathcal{D}_m$ and $\mathcal{R}_m$ and to give a characterization of $\mathcal{A}_m$. A subset B of a structure **A** is an *automorphism base* if the only automorphism of **A** fixing B pointwise is the identity. In Section 2 we explore automorphism bases of $\mathcal{D}_m$ and $\mathcal{A}_m$. In Section 3 we consider codings of copies of $(\mathbb{N},+,\times)$ in these structures and use them to show that $\mathcal{A}_m$ is a prime model and, under the assumption that $V = L$, Th$(\mathcal{D}_m)$ is atomic. Finally, in Section 4 we develop a general method to show that a subset B of a structure **A** is an automorphism base. This method is applied firstly to show that each definable $D \subseteq \mathcal{R}_m$, $D \not\subseteq \{\mathbf{0},\mathbf{1}\}$, is an automorphism base. Secondly, we improve the proof of Ambos-Spies' result that each interval $[\mathbf{0},\boldsymbol{c}]$ of $\mathcal{R}_T$, $\boldsymbol{c} \neq \mathbf{0}$, is an automorphism base. As explained in Nies (ta), the different language used here promises to be useful in showing that $\mathcal{R}_T$ is biinterpretable with $\mathbb{N}$ in parameters.

Notation. Let (ψ_e) be an effective list of partial computable functions defined on initial segments of $\mathbb{N}$ which includes all the total computable functions. Given a degree $\boldsymbol{x}$ in a degree structure, $\widehat{\boldsymbol{x}}$ denotes $[\mathbf{0},\boldsymbol{x}]$. For a subset X of an uppersemilattice (u.s.l.) U, $[X]_{id}$ denotes the ideal of U generated by X.

1. The extension property for $\mathcal{D}_m, \mathcal{A}_m$ and $\mathcal{R}_m$

All three structures $\mathcal{D}_m, \mathcal{A}_m$ and $\mathcal{R}_m$ can be characterized as distributive upper semilattices (d.u.s.l.) with certain basic properties and an extension property (EP). The EP for the degree structure **A** states that the diagram below can be completed by a map c such that $g = c \circ h$, where I, J are ideals of $\mathcal{D}$ and the maps g, h, c are embeddings as ideals, and I, J, g, h, c satisfy the restrictions given in Table 1.

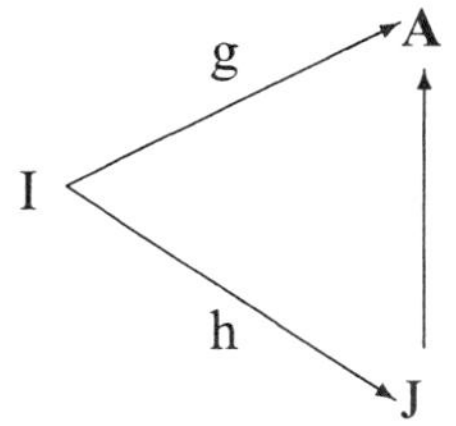

Table 1.

	$\mathcal{D}_m$	$\mathcal{A}_m$	$\mathcal{R}_m$
I,J	D.u.s.l. of size $< 2^\omega$	Arithmetical d.u.s.l. with **1**	Lachlan semilattices
Maps	any embeddings as ideals	arithmetical embeddings	effective embeddings

$\mathcal{D}_m$ can be characterized as follows. It is the unique distributive u.s.l. of size 2^ω such that all topped initial segments are countable and the extension property holds for the embeddings described by Table 1 (see Ershov 1975).

A structure $\boldsymbol{A}$ over a finite symbol set is called *arithmetical* if there is an onto map $\beta : \mathbb{N} \mapsto A$ such that the preimages of the relations of $\boldsymbol{A}$ under β are arithmetical (we call n an *index* for $\beta(n)$). If the structures $\boldsymbol{A}_i$ are arithmetical via β_i $(i = 0, 1)$, then we say a map $f : \boldsymbol{A}_0 \to \boldsymbol{A}_1$ is arithmetical if the corresponding relation on indices is.

In the following we discuss $\mathcal{A}_m$. Each arithmetical set C induces an arithmetical presentation of $[\mathbf{0}, \boldsymbol{c}]$, $\boldsymbol{c} = \deg_m(C)$, called the *canonical presentation*, in the following way: Let i be an index for the many-one degree of $\psi_i^{-1}(C)$ (a finite set if ψ_i is partial, see end of the introduction). Clearly, "$\psi_i^{-1}(C) \leqslant_m \psi_j^{-1}(C)$" is a $\Sigma_2^0(C \oplus \emptyset')$ relation of i, j.

We say that an u.s.l. W with least element 0 is *locally arithmetical* if there is an onto map $\alpha : \mathbb{N} \to W$ such that for each n, $[0, \alpha(n)]$ is arithmetical via a map β_n and, if $\alpha(n) \leqslant \alpha(m)$, then the inclusion map $[0, \alpha(n)] \to [0, \alpha(m)]$ is arithmetical with respect to these presentations. Clearly $\mathcal{A}_m$ itself is locally arithmetical, where $\alpha(n)$ is the many-one degree of the n-th arithmetical set C in some effective listing and β_n is the canonical presentation of $[\mathbf{0}, \boldsymbol{c}]$.

Theorem 1.1 *Up to isomorphism, $\mathcal{A}_m$ is the only locally arithmetical d.u.s.l. satisfying the extension property, where I, J are topped arithmetical d.u.s.l. and g, h are arithmetical maps.*

Sketch of proof. By a back and forth argument, any two such d.u.s.l. must be isomorphic. The extension property for $\mathcal{A}_m$ is obtained by analyzing the effective content of Ershov's construction. For simplicity, let us consider the case that $I = \{\mathbf{0}\}$. Thus we have to show that each arithmetical topped d.u.s.l. J is arithmetically isomorphic to an initial segment of $\mathcal{A}_m$. Here it suffices to look at Lachlan's characterization of the topped initial segments of $\mathcal{D}_m$ (Lachlan 1970), which served as a base for Ershov's work. Effectivizing the proof of Proposition VI.1.12. in Odifreddi (1989), we obtain an arithmetical sequence of strong indices for finite sets (D_i), $D_i \subseteq D_{i+1}$, which are subsets of J containing 0 and closed under join such that $\bigcup_i D_i = J$. Notice that each D_i is a distributive lattice. Now the construction in Lachlan (1970) (with χ_i being the inclusion map $D_i \to D_{i+1}$) produces an arithmetical set U such that J is arithmetically isomorphic to $[\mathbf{0}, \deg_m(U)]$. □

The characterization of $\mathcal{R}_m$ has been given by Denisov (1978) and is summarized in Nies (2000). As consequences of the extension properties, we notice that in all cases the degree structure possesses the maximum possible number of automorphisms, that any two minimal degrees are automorphic and that each proper final segment $\{\boldsymbol{x} : \boldsymbol{x} \geqslant \boldsymbol{e}\}$ is isomorphic to the whole structure.

The EP also leads to a characterization of the *k-orbits* for $\mathcal{D}_m$ and $\mathcal{A}_m$, i.e. for the orbits under the action of the automorphism group on k-tuples. For a k-tuple $\bar{\boldsymbol{a}}$ let $s_{\bar{\boldsymbol{a}}} = \sup(\{\boldsymbol{a}_i : i < k\})$.

Proposition 1.2 *Suppose $\bar{a}, \bar{b}$ are k tuples in $\mathcal{D}_m$ [$\mathcal{A}_m$]. Then $\bar{\boldsymbol{a}}, \bar{\boldsymbol{b}}$ are in the same k-orbit iff $[\mathbf{0}, s_{\bar{\boldsymbol{a}}}] \cong [\mathbf{0}, s_{\bar{\boldsymbol{b}}}]$ via an [arithmetical] isomorphism p such that $p(\boldsymbol{a}_i) = \boldsymbol{b}_i$.*

Sketch of proof. The direction from left to right is immediate for $\mathcal{D}_m$ and follows from Lemma 3.4 for $\mathcal{A}_m$. For the converse direction, one utilizes the EP in a further back and forth argument. □

Below we need the following concept.

Definition 1.3 *If I is an ideal in an u.s.l. U, b is a* strong minimal cover (s.m.c.) *of I if $I = [0, b)$. In the special case that $I = [0, c]$, we say that b is a s.m.c. of c.*

As a consequence of the EP (using in the arithmetical case techniques from the proof of Lemma 3.3 below), we obtain:

Proposition 1.4 *Each countable ideal of $\mathcal{D}_m$ has a s.m.c. For each $\boldsymbol{c} \in \mathcal{A}_m$, each arithmetical ideal of $\widehat{\boldsymbol{c}} = [\mathbf{0}, \boldsymbol{c}]$ has a s.m.c. in $\mathcal{A}_m$.* □

2. Automorphism bases for $\mathcal{D}_m$ and $\mathcal{A}_m$

Various theorems will have a version for $\mathcal{D}_m$ and one for $\mathcal{A}_m$, in which case we will write the changes for $\mathcal{A}_m$ in brackets "[]".

Theorem 2.1 (Nies 2000) *Each set $\{\mathbf{x} : \mathbf{x} \geqslant \mathbf{a}\}$ is an automorphism base of $\mathcal{D}_m$. The same holds for $\mathcal{A}_m$ and (if $\mathbf{a} < \mathbf{1}$) for $\mathcal{R}_m$.* □

The rest of this section is devoted to proving that many interesting subsets, like the minimal degrees, are *not* automorphism bases of $\mathcal{D}_m$ $[\mathcal{A}_m]$. We need some algebra of d.u.s.l.

Definition 2.2 *Suppose U is a d.u.s.l., $\pi : I \to J$ is an isomorphism of ideals and M is an ideal of U. We say that π* fixes M *if $\pi(x) = x$ for all $x \in I \cap M$ and $\pi^{-1}(x) = x$ for all $x \in J \cap M$ (and thus $M \cap I = M \cap J$).*

Lemma 2.3 *Suppose π fixes M. Then there is a unique isomorphism ρ between $I \vee M$ and $J \vee M$ which extends π and fixes M. We denote this map ρ by $\pi \vee M$.*

Proof. The map is unique, because any such ρ must satisfy $\rho(x \vee m) = \pi(x) \vee m$ (where $x \in I, m \in M$). To see that this equality describes a well-defined map, suppose $x \vee m = y \vee n$, where $x,y \in I, m,n \in M$. Then $x = \widetilde{y} \vee \widetilde{n}$ for some $\widetilde{y} \leqslant y$ and $\widetilde{n} \leqslant n$. Hence $\pi(x) \leqslant \pi(\widetilde{y}) \vee \widetilde{n} \leqslant \pi(y) \vee n$. By symmetry, $\pi(x) \vee m = \pi(y) \vee n$.

To see that ρ is an isomorphism, define a map $\sigma : J \vee M \to I \vee M$ by $\sigma(z \vee m) = \pi^{-1}(z) \vee m$. Then σ is well-defined and ρ and σ are inverses. □

We proceed to the main results of this Section. In the following, an ideal S of $\mathcal{A}_m$ is *locally arithmetical* if $S \cap [\mathbf{0}, \mathbf{c}]$ is an arithmetical subset of $[\mathbf{0}, \mathbf{c}]$ (with the canonical presentation), for each $\mathbf{c} \in \mathcal{A}_m$.

Theorem 2.4 *Suppose M is a proper ideal of $\mathcal{D}_m[\mathcal{A}_m]$ which is invariant under automorphisms [and locally arithmetical]. Then M is* not *an automorphism base.*

We begin with the proof for $\mathcal{D}_m$.

Lemma 2.5 *Suppose that $\pi_0 : I_0 \to J_0$ is an isomorphism of ideals of size $< 2^{\omega}$ which fixes M. Then π_0 extends to an automorphism π of $\mathcal{D}_m$ which fixes M.*

Theorem 2.4 follows from the Lemma: pick $\mathbf{z} \notin M$ and, by the EP, let $\mathbf{p}, \mathbf{q}$ be distinct strong minimal covers of $\mathbf{z}$. Let $I_0 = [\mathbf{0}, \mathbf{p}], J_0 = [\mathbf{0}, \mathbf{q}]$ and let $\pi_0 : I_0 \to J_0$ be the isomorphism fixing $[\mathbf{0}, \mathbf{z}]$. Then π_0 fixes M. So the automorphism π obtained from Lemma 2.5 shows that M is not an automorphism base.

Before proving the lemma, we give some concrete examples in $\mathcal{D}_m$.

Example 2.6 *For a countable ordinal α, let*

$$M_\alpha = [\{\mathbf{w} : \exists \beta \leqslant \alpha\ [\mathbf{0}, \mathbf{w}] \cong \beta\}]_{id}.$$

For instance, M_1 is the ideal generated by the minimal degrees. Clearly M_α is automorphism invariant. Moreover, $\alpha < \delta < \omega_1$ implies that M_α is a proper subideal of M_δ, and indeed that there is an automorphism fixing M_α, but not M_δ, pointwise: By Lachlan (1970) pick $\boldsymbol{u} \in \mathcal{D}_m$ such that $[\boldsymbol{0},\boldsymbol{u}) \cong \alpha$. As above, let $\boldsymbol{p},\boldsymbol{q}$ be distinct strong minimal covers of $\boldsymbol{u}$. Then $\boldsymbol{p},\boldsymbol{q} \in M_\delta - M_\alpha$. Let π_0 be the isomorphism $[\boldsymbol{0},\boldsymbol{p}] \to [\boldsymbol{0},\boldsymbol{q}]$ and extend it to an automorphism fixing M_α.

Proof of the Lemma. Fix some wellordering $\lhd$ of $\mathcal{D}_m$. We obtain π by a back and forth-construction through the ordinals $< 2^\omega$. Suppose $\pi_\alpha : I_\alpha \to J_\alpha$ has been defined such that π_α fixes M. If α is even, let w be the $\lhd$-least degree $\notin I_\alpha$, and let $X = \widehat{w} \cap M$. By the inductive hypothesis and Lemma 2.3, obtain the isomorphism $\pi_\alpha \vee X : I_\alpha \vee X \to J_\alpha \vee X$. Now let $I_{\alpha+1} = I_\alpha \vee \widehat{w}$. Apply the EP for the diagram

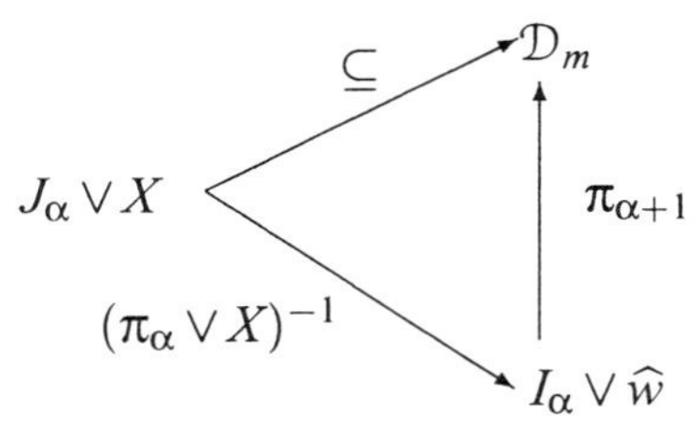

in order to obtain $\pi_{\alpha+1} : I_{\alpha+1} \to \mathcal{D}_m$. Let $J_{\alpha+1}$ be the range of $\pi_{\alpha+1}$. We verify that $\pi_{\alpha+1}$ still fixes M. Suppose that $z \in M \cap I_{\alpha+1}$ with the intention to show $\pi_{\alpha+1}(z) = z$. Then $z = w' \vee y$ for some $w' \leqslant w$ and $y \in I_\alpha \cap M$. Since $\pi_{\alpha+1}(y) = y$ by inductive hypothesis, it suffices to show that $\pi_{\alpha+1}(w') = w'$. But $w' \in X$, so, since the diagram commutes, $w' = \pi_{\alpha+1}((\pi_\alpha \vee X)^{-1}(w')) = \pi_{\alpha+1}(w')$.

To prove the second condition in Definition 2.3, it suffices now to show that $J_{\alpha+1} \cap M \subseteq I_{\alpha+1}$. Let $v = \pi_{\alpha+1}(w)$. Suppose $z \in J_{\alpha+1} \cap M$. Then $z = v' \vee y$ for some $v' \leqslant v, y \in J_\alpha$. By the inductive hypothesis $y \in I_\alpha$, so it suffices to show that $v' \in I_{\alpha+1}$. But M is automorphism invariant, so by Proposition 1.2 (with $k = 1$), $\pi_{\alpha+1}(\widehat{w} \cap M) = \widehat{v} \cap M$. So $v' = \pi_{\alpha+1}(w')$ for some $w' \in \widehat{w} \cap M = X$. This implies $w' = v'$, whence $v' \in X \subseteq I_{\alpha+1}$.

If α is odd we proceed as before but with I_α, J_α interchanged. At limit ordinals λ we take unions of partial isomorphisms. Note that $I_\lambda = \bigcup_{\alpha<\lambda} I_\alpha$ has cardinality $< 2^\omega$ since it is generated by $|\lambda|$ many principal ideals. □

To prove the Theorem for $\mathcal{A}_m$, we have to adapt Lemma 2.5 and its proof.

Lemma 2.7 *Suppose that $\pi_0 : I_0 \to J_0$ is an arithmetical isomorphism of initial intervals of $\mathcal{A}_m$ which fixes M. Then π_0 extends to an automorphism π of $\mathcal{D}_m$ which fixes M.*

The Theorem for $\mathcal{A}_m$ is now obtained as before. To prove the Lemma, go through a back an forth-construction of ω steps, using the EP for $\mathcal{A}_m$ and that M is locally arithmetical. □

Example 2.6 has an application to $\mathrm{Aut}(\mathcal{D}_m)$. Note that if S is an automorphism invariant subset of a structure $\boldsymbol{A}$, then the pointwise stabilizer of S is a normal subgroup in $\mathrm{Aut}(\boldsymbol{A})$.

Theorem 2.8 *The automorphism group of $\mathcal{D}_m$ possesses a strictly descending chain $(U_\alpha)_{\alpha<\omega_1}$ of normal subgroups.*

Proof. Let U_α be the stabilizer of M_α. This sequence is strictly descending by the remarks after Example 2.6. □

In a similar way, one obtains a descending chain of length ω_1^{CK} of normal subgroups in $\mathrm{Aut}(\mathcal{A}_m)$.

3. Atomic theories and prime models

Recall that a structure $\boldsymbol{A}$ is a *prime model* if $\boldsymbol{A}$ is an elementary substructure of each $\boldsymbol{B} \equiv \boldsymbol{A}$. For countable $\boldsymbol{A}$, being a prime model is equivalent to each realized k-type being principal, or again each k-orbit being $\emptyset$-definable. Thus, a proof that a countable $\boldsymbol{A}$ is prime helps us to understand the orbits. A complete theory T in a countable language has a prime model iff all it Lindenbaum algebras are atomic (such a T is itself called *atomic*).

3.1. Schemes

The following is discussed in more detail in Nies et al. (1998). A *scheme* for coding in an L-structure $\boldsymbol{A}$ is given by a list of L-formulas $\varphi_1, \dots, \varphi_n$ with a shared parameter list $\overline{p}$, together with a correctness condition $\alpha(\overline{p})$. If a scheme S_X is given, $X, X_0, X_1, \dots$ denote objects coded via S_X by a list of parameters satisfying the correctness condition.

Example 3.1 *A scheme S_N for coding models of some finitely axiomatized fragment PA^- of Peano arithmetic (in the language $L(+,\times)$) is given by the formulas $\varphi_{num}(x,\overline{p}), \varphi_+(x,y,z;\overline{p}), \varphi_\times(x,y,z;\overline{p})$ and a correctness condition $\alpha_0(\overline{p})$ which expresses, among others things, that φ_+ and $\varphi_\times$ define binary operations on the nonempty set $\{x : \varphi_{num}(x;\overline{p})\}$, and that $\{x : \varphi_{num}(x;\overline{p})\}$ with the corresponding operations satisfies the finitely many axioms of PA^-.*

For our applications it is sufficient to work with Robinson arithmetic Q. Then all coded models N have a standard part isomorphic to $\mathbb{N}$. If this standard part equals N we say that N is *standard*.

Example 3.2 *A scheme S_g for defining a function g is given by a formula $\varphi_1(x,y;\overline{p})$ defining the relation between arguments and values, and a correctness condition $\alpha(x,y;\overline{p})$ which says at least that a function is defined.*

3.2. Coded copies of $(\mathbb{N},+,\times)$

In the following "ar." abbreviates "arithmetical". First we need some facts on distributive upper semilattices. Recall that a sequence (u_i) in an u.s.l. U is *independent* if no u_i is below a finite supremum of other elements in the sequence. In the following, when we say the d.u.s.l. U is *embedded* into V, we mean hat U is isomorphic to an ideal of V.

Lemma 3.3 *Suppose U is a countable [ar.] d.u.s.l. and (a_i) is an [ar.] sequence in U. Then U can be embedded into a countable [ar.] d.u.s.l. V with greatest element which contains an [ar.] independent sequence (b_i) such that, for all i, b_i is a s.m.c. of a_i.*

Proof. Each countable [ar.] d.u.s.l. can be embedded, preserving $0,\vee$, into a countable [ar.] Boolean algebra $\mathcal{B}$. By the Stone representation theorem we can suppose that elements of $\mathcal{B}$ are [ar.] subsets of T where $T\subseteq 2^{<\omega}$ is an [ar.] tree, and the operations of $\mathcal{B}$ are the usual operations on sets. Coding the elements of T by odd numbers, we can assume that U is a subsemilattice of $(\mathcal{P}(\{2n+1 : n\in\mathbb{N}\}),\emptyset,\cup)$.

First we add a s.m.c. b_0 to only one element a_0. Let $b_0 = a_0\cup\{0\}$, and let V_0 be the collection of sets generated under finite unions by $U\cup\{b_0\}$. Clearly, each element of $V_0 - U$ is of the form $x\cup b_0$ for some $x\in U$. Thus U is an ideal of V_0 and b_0 is a s.m.c. of a_0. It remains to be shown that V_0 is distributive. Given $x\subseteq y\cup z$, we want to find $y'\subseteq y, z'\subseteq z$ such that $x=y'\cup z'$. If $y\cup z\in U$, this is possible by the distributivity of U. Otherwise, say $z\notin U$ so that $z=\tilde{z}\cup a_0\cup\{0\}$. Then $y\cup z=\tilde{y}\cup\tilde{z}\cup a_0\cup\{0\}$ where $\tilde{y}\in U, \tilde{y}\subseteq y$. If $x\in U$, then $x\subseteq\tilde{y}\cup\tilde{z}\cup a_0$, and we obtain x', y' by distributivity of U. Otherwise $x=\tilde{x}\cup a_0\cup\{0\}$, where $\tilde{x}\in U$. Then $\tilde{x}\subseteq\tilde{y}\cup\tilde{z}\cup a_0$ implies $\tilde{x}=y'\cup z''$ for some $y'\subseteq\tilde{y}$, $z''\subseteq\tilde{z}\cup a_0$. Let $z'=z''\cup a\cup\{0\}\in V$, then $x=y'\cup z'$.

Now continue this process, adding for $i>0$ a s.m.c. $b_i=a_i\cup\{2i\}$ in order to embed V_{i-1} into V_i, and let $V=\{\mathbb{N}\}\cup\bigcup_i V_i$. Clearly (b_i) is independent and V has a greatest element. Finally, in the arithmetical setting, V is arithmetical. □

The following Lemma on coding of copies of $(\mathbb{N},+,\times)$ again has versions for $\mathcal{D}_m$ and $\mathcal{A}_m$. The coding schemes used, which will be kept fixed from now on, do not change from one version to the other.

Lemma 3.4 *There are schemes S_M, S_f as in Example 3.1 and Example 3.2 with the following property. For any sequence $(\boldsymbol{a}_i)_{i\in\mathbb{N}}$ [any ar. sequence of degrees in an initial interval of $\mathcal{A}_m$], there is a standard M and a map f such that $\boldsymbol{a}_i=f(i^M)$. Moreover, the domain of M is an independent set. [Finally, S_M can be evaluated in an initial interval of $\mathcal{A}_m$.]*

Proof. We first consider $\mathcal{D}_m$. Let $U = [\{\boldsymbol{a}_i : i \in \mathbb{N}\}]_{id}$. We will embed U into a d.u.s.l. V encoding M, f as desired, and then apply the EP. The first step is to apply Lemma 3.3 to the sequence $\boldsymbol{a}_0, \boldsymbol{0}, \boldsymbol{a}_1, \boldsymbol{0}, \ldots$ in order to obtain an extension V' and a sequence of s.m.c. $(\boldsymbol{b}_i)$. The domain of M will be $\{\boldsymbol{b}_i : i \text{ even}\}$. By the proof of Thm 4.2 in Nies (1996), there is a scheme without parameters via which a copy M of $(\mathbb{N}, +, \times)$ can be encoded into a symmetric graph $(\mathbb{N}, E)$, where the even numbers serve as the domain of the coded model. Embed V' into a d.u.s.l. V by applying Lemma 3.3 to a sequence listing $\{\boldsymbol{b}_i \vee \boldsymbol{b}_j : Eij\}$ (notice that, by independence, all these suprema are distinct). Let $(\boldsymbol{c}_k)$ be the sequence of s.m.c. obtained in this way. Now apply the EP to U, V and identify the elements of V with many-one degrees. By Proposition 1.4, there are s.m.c. $\boldsymbol{b}, \boldsymbol{c}$ for the ideals $[\{\boldsymbol{b}_i : i \in \mathbb{N}\}]_{id}$ and $[\{\boldsymbol{c}_k : i \in \mathbb{N}\}]_{id}$, respectively. Then the sets $\{\boldsymbol{b}_i : i \in \mathbb{N}\}$ and $\{\boldsymbol{c}_k : k \in \mathbb{N}\}$ are definable in $[\boldsymbol{0}, \boldsymbol{b})$ and $[\boldsymbol{0}, \boldsymbol{c})$ as the maximal join-irreducible elements. Since we can now recover a copy of $(\mathbb{N}, E)$ in a first-order way, this gives a scheme S_M involving parameters $\boldsymbol{b}, \boldsymbol{c}$ to code M. The scheme S_f in the same parameters is determined by $f(\boldsymbol{x}) = \boldsymbol{a} \Leftrightarrow \boldsymbol{x} \in M$ *is a s.m.c. of* $\boldsymbol{a}$.

All the constructions involved are arithmetical, and V has a greatest element. This gives the Lemma for $\mathcal{A}_m$. □

As a consequence, we can quantify over countable subsets [arithmetical subsets of initial intervals]. Then, we may enrich the scheme S_M by a correctness condition implying that all coded M are isomorphic to $\mathbb{N}$: The standard part of any M is a set of the required type, so we can express in a first-order way that the standard part equals M. In the following we assume this additional correctness condition, and also add (quantifying over finite subsets of M) that the domain is an independent set. In the case of $\mathcal{D}_m$, arbitrary subsets Y of M can be represented by elements $\boldsymbol{d}$ of $\mathcal{D}_m$, since by this independence and the existence of a s.m.c. for $[Y]_{id}$, $Y = M \cap [\boldsymbol{0}, \boldsymbol{d}]$ for some $\boldsymbol{d}$. Thus, for each second-order formula $\psi(X)$ in the language of arithmetic, there is a formula $\beta(d, \widetilde{q})$ such that, if M is coded by parameters $\widetilde{\boldsymbol{q}}$, then $X \subseteq M$ satisfies ψ in M iff $\mathcal{D}_m \models \beta(\boldsymbol{d}, \widetilde{\boldsymbol{q}})$ for any $\boldsymbol{d}$ representing X.

3.3. Results for $\mathcal{D}_m$ and $\mathcal{A}_m$

The following Theorem connects three areas of Mathematical Logic.

Theorem 3.5 *If there is an analytical wellordering on $\mathcal{P}(\omega)$ (for instance if $V = L$), then* $\mathrm{Th}(\mathcal{D}_m)$ *is atomic.*

Proof. Suppose the set theoretical hypothesis holds. We must show that each nonempty definable k-ary relation R on $\mathcal{D}_m$ contains a $\emptyset$-definable k-orbit.

The functions ψ_i were defined at the end of the introduction. From here on, let us assume also that, for all i, ψ_{2i} is a total 1-1 map onto $\mathbb{N}^{[i]} = \{\langle n,i\rangle : n \in \mathbb{N}\}$. For $A \subseteq \mathbb{N}$, the following set of numbers describes $[\mathbf{0},\boldsymbol{a}]$:

$$\mathrm{Diag}(A) = \{\langle i,j\rangle : \psi_i^{-1}(A) \leqslant_m \psi_j^{-1}(A)\}. \tag{1}$$

In the following, we call a set $X \subseteq \mathbb{N}$ second-order definable if $\{X\}$ is projective. $X_0 \oplus \ldots \oplus X_{k-1}$ denotes the set $\{\langle n,j\rangle : n \in X_j \ \& \ j < k\}$.

Claim 3.6 *Suppose A_i $(i < k)$ are sets, $A = A_0 \oplus \ldots \oplus A_{k-1}$ and Diag(A) is second-order definable. Then the k-orbit of $(\boldsymbol{a}_0, \ldots, \boldsymbol{a}_{k-1})$ is $\emptyset$-definable.*

To prove the Claim, by Lemma 3.4, choose a standard M and f such that $f(i^M) = \deg_m(\psi_i^{-1}(A))$ and in particular, for $i < k$, $f(2i^M) = \boldsymbol{a}_i$.

By the remark after Lemma 3.4, replacing set quantifiers by degree quantifiers in a second-order definition of Diag(A), one obtains a formula $\beta(\boldsymbol{c},\widetilde{\boldsymbol{q}})$ such that, given $M_{\widetilde{\boldsymbol{q}}}$ (this notation indicates that M is coded by the tuple $\widetilde{\boldsymbol{q}}$), $\beta(\boldsymbol{c},\widetilde{\boldsymbol{q}})$ holds iff $\widehat{\boldsymbol{c}} \cap M_{\widetilde{\boldsymbol{q}}}$ is Diag(A) (viewed as a subset of M). Then $\overline{\boldsymbol{b}}$ is automorphic to $\overline{\boldsymbol{a}}$ iff

$\exists\widetilde{\boldsymbol{q}}\, \exists\widetilde{f} : M_{\widetilde{\boldsymbol{q}}} \to [\mathbf{0}, s_{\overline{\boldsymbol{b}}}]$ *onto* $\exists\boldsymbol{c}\ [\beta(\boldsymbol{c},\widetilde{\boldsymbol{q}})\ \&\ \forall i < k\ \widetilde{f}(2i^{M_{\widetilde{\boldsymbol{q}}}}) = \boldsymbol{b}_i\ \&$
$\{\langle \boldsymbol{i},\boldsymbol{j}\rangle \in M_{\widetilde{\boldsymbol{q}}} : \widetilde{f}(\boldsymbol{i}) \leqslant \widetilde{f}(\boldsymbol{j})\} = \widehat{\boldsymbol{c}} \cap M_{\widetilde{\boldsymbol{q}}}]$

(for the direction from right to left one uses Proposition 1.2). Since the statement can be formulated in the first-order language of u.s.l., this proves the Claim. Now, if S is a nonempty projective k-ary relation, then the class of sets $\{\mathrm{Diag}(A_0 \oplus \ldots \oplus A_{k-1}) : (A_0, \ldots, A_{k-1}) \in S\}$ is projective. By the set theoretic hypothesis, this class contains a 2nd-order definable element. If we let S be the (projective) relation on subsets of $\mathbb{N}$ corresponding to R, by the Claim we obtain a definable k-orbit in R. □

G. Hjorth and the author have proved that Th$(\mathcal{D}_m)$ is not atomic in a Cohen extension of the universe.

Theorem 3.7 *$\mathcal{A}_m$ is a prime model.*

Proof. We must show that each k-orbit is $\emptyset$-definable. Suppose $\overline{\boldsymbol{a}}$ is a k-tuple in $\mathcal{A}_m$, and that A_i is a set in $\boldsymbol{a}_i$. Let $A = A_0 \oplus \ldots \oplus A_{k-1}$. Then the set Diag$(A)$ given by (1) can be defined in $(\mathbb{N},+,\times)$ by a formula $\delta(n)$. We conclude that $\overline{\boldsymbol{b}}$ is automorphic to $\overline{\boldsymbol{a}}$ iff

$\exists\widetilde{M}\, \exists\widetilde{f} : \widetilde{M} \to [\mathbf{0}, s_{\overline{\boldsymbol{b}}}]$ onto $[\forall i < k\ \widetilde{f}(2i^{\widetilde{M}}) = \boldsymbol{b}_i\ \&$
$\{\langle \boldsymbol{i},\boldsymbol{j}\rangle \in \widetilde{M} : \widetilde{f}(\boldsymbol{i}) \leqslant \widetilde{f}(\boldsymbol{j})\} = \{\boldsymbol{n} \in \widetilde{M} : \widetilde{M} \models \delta(\boldsymbol{n})\}]$

(once again, we use Proposition 1.2). □

4. Proving that sets are automorphism bases

To prove a subset B of a structure $\boldsymbol{A}$ is an automorphism base, one might try to find a 1-1 B-definable map $H : \boldsymbol{A} \to B$. Then, if the automorphism F fixes B pointwise, it must be the identity. Our method uses this idea in a more general setting: a 1-1 map H maps elements of $\boldsymbol{A}$ (or, more generally, of an automorphism invariant subset C which is already known to be an automorphism base) to complex objects of one and the same "B-type" τ, which are constructed from elements of B. Such a map is first-order definable in a more general sense described below. We describe the verification that such a map H is 1-1 by finite games. Given distinct elements $\boldsymbol{x},\boldsymbol{y} \in C$, we try to show $H(\boldsymbol{x}) \neq H(\boldsymbol{y})$, while the opponent claims $H(\boldsymbol{x}) = H(\boldsymbol{y})$.

We first give an example with the structure $\mathcal{R}_T$, where $C = \mathcal{R}_T$ and $B = \mathrm{NC}$ is the set of noncappable degrees. For this we need the following splitting theorem, which can be readily proved using the finitary techniques in Soare (1987).

Proposition 4.1 *If $\boldsymbol{p}$ is low and $\boldsymbol{c} \not\leqslant \boldsymbol{p}$, then for each $\boldsymbol{x}$ there are low degrees $\boldsymbol{x}_0,\boldsymbol{x}_1$ such that $\boldsymbol{x} = \boldsymbol{x}_0 \vee \boldsymbol{x}_1$ and $\boldsymbol{c} \not\leqslant \boldsymbol{p} \vee \boldsymbol{x}_i$ for $i = 0,1$. Moreover, if $\boldsymbol{x} \in NC$, one can ensure that $\boldsymbol{x}_0,\boldsymbol{x}_1 \in NC$.* □

Proposition 4.2 *The map H given by $H(\boldsymbol{x}) = \{[\boldsymbol{p},\mathbf{1}] \cap NC, [\boldsymbol{q},\mathbf{1}] \cap NC : \boldsymbol{p} \vee \boldsymbol{q} = \boldsymbol{x}\}$ is 1-1. Hence NC is an automorphism base.*

Proof. Suppose $\boldsymbol{x} \neq \boldsymbol{y}$. We play the following game. Say $\boldsymbol{x} \not\leqslant \boldsymbol{y}$.

Round 1. We provide low degrees $\boldsymbol{p},\boldsymbol{q}$ such that $\boldsymbol{y} = \boldsymbol{p} \vee \boldsymbol{q}$. To support his claim, the opponent has to provide $\widetilde{\boldsymbol{p}},\widetilde{\boldsymbol{q}}$ such that $\boldsymbol{x} = \widetilde{\boldsymbol{p}} \vee \widetilde{\boldsymbol{q}}$, $[\widetilde{\boldsymbol{p}},\mathbf{1}] \cap \mathrm{NC} = [\boldsymbol{p},\mathbf{1}] \cap \mathrm{NC}$ and $[\widetilde{\boldsymbol{q}},\mathbf{1}] \cap \mathrm{NC} = [\boldsymbol{q},\mathbf{1}] \cap \mathrm{NC}$.

Round 2. At least one of $\widetilde{\boldsymbol{p}} \not\leqslant \boldsymbol{p}, \widetilde{\boldsymbol{q}} \not\leqslant \boldsymbol{q}$ holds, say the first. By Proposition 4.1, we choose $\boldsymbol{u} \in \mathrm{NC}$ such that $\boldsymbol{p} \leqslant \boldsymbol{u}$ but $\widetilde{\boldsymbol{p}} \not\leqslant \boldsymbol{u}$. (Let $\boldsymbol{x} = \mathbf{1}$ and $\boldsymbol{c} = \widetilde{\boldsymbol{p}}$. At least one $\boldsymbol{x}_i$ is noncappable, so let $\boldsymbol{u} = \boldsymbol{p} \vee \boldsymbol{x}_i$ for this i.) The opponent loses. □

Next we introduce the formalism of B-types and the corresponding maps.

Definition 4.3 *Suppose $\boldsymbol{A}$ is a structure and $B \subseteq \boldsymbol{A}$.*

(i) Each formula $\varphi(x,y)$ with parameters in B is a B-type.

(ii) If $\tau_1,\ldots,\tau_n$ are B-types and $\varphi(x,y_1,\ldots,y_n)$ is a formula with parameters in B, then $\tau = (\tau_1,\ldots,\tau_n;\varphi)$ is a B-type.

Definition 4.4 *Suppose $B \subseteq \boldsymbol{A}$ and τ is a B-type. We define a map $H_{\tau B}$ by induction on τ.*

(i) *If* τ *is* $\varphi(x,y)$*, let* $H_{\tau B}(\boldsymbol{x}) = \{\boldsymbol{y} \in B : \boldsymbol{A} \models \varphi(\boldsymbol{x},\boldsymbol{y})\}$.

(ii) *If* $\tau = (\tau_1, \ldots, \tau_n; \varphi)$*, then let* $H_{\tau B}(\boldsymbol{x}) = \{(H_{\tau_1 B}(\boldsymbol{y}_1), \ldots, H_{\tau_n B}(\boldsymbol{y}_n)) : \boldsymbol{A} \models \varphi(\boldsymbol{x},\boldsymbol{y}_1, \ldots, \boldsymbol{y}_n)\}$.

To see how this applies to the preceding example, for $i = 0,1$ let $\varphi_i(x_i,y)$ be $y \in \mathrm{NC}$ & $x_i \leqslant y$. Moreover, let $\psi(x,x_0,x_1)$ be $x = x_0 \vee x_1$. Then for the NC-type $\tau = (\varphi_0, \varphi_1; \psi)$, $H_{\tau\mathrm{NC}}$ is the map introduced in Proposition 4.2.

Suppose $F : A \to A$. We define $F(H_{\tau B}(\boldsymbol{x}))$ by applying F to the atomic components of $H_{\tau B}(\boldsymbol{x})$:

In Case (i) of Definition 4.3, let $F(H_{\tau B}(\boldsymbol{x})) = \{F(\boldsymbol{y}) : \boldsymbol{A} \models \varphi(\boldsymbol{x},\boldsymbol{y}) \;\&\; \boldsymbol{y} \in B\}$. In Case (ii), let

$$F(H_{\tau B}(\boldsymbol{x})) = \{(F(H_{\tau_1 B}(\boldsymbol{y}_1)), \ldots, F(H_{\tau_n B}(\boldsymbol{y}_n))) : \boldsymbol{A} \models \varphi(\boldsymbol{x},\boldsymbol{y}_1, \ldots, \boldsymbol{y}_n)\}.$$

Clearly, if F fixes B pointwise, then F fixes $H_{\tau B}(\boldsymbol{x})$. Moreover, by induction over the length of τ, one verifies that, whenever F is also an automorphism, then $H_{\tau B}(F(\boldsymbol{x})) = F(H_{\tau B}(\boldsymbol{x}))$.

The method is summarized in the following proposition.

Proposition 4.5 *Suppose* C *is an automorphism invariant automorphism base and* $H_{\tau B}\lceil C$ *is 1-1. Then* B *is an automorphism base.*

Proof. Suppose the automorphism F fixes B pointwise. Then, for each $\boldsymbol{x} \in C$, $H(F(\boldsymbol{x})) = F(H(\boldsymbol{x})) = H(\boldsymbol{x})$. Since $F(\boldsymbol{x}) \in C$, by the hypothesis on H this implies that $F(\boldsymbol{x}) = \boldsymbol{x}$. Thus F fixes C pointwise. □

Our first application of the method is to $\mathcal{R}_m$. Recall that $\widehat{\boldsymbol{x}}$ denotes $[\mathbf{0},\boldsymbol{x}]$.

Theorem 4.6 *Suppose* $D \subseteq \mathcal{R}_m$ *is* $\emptyset$*-definable and* $D \not\subseteq \{\mathbf{0},\mathbf{1}\}$*. Then* D *is an automorphism base of* $\mathcal{R}_m$.

As an example, for each finite distributive lattice L, the set $\{\boldsymbol{x} : [\mathbf{0},\boldsymbol{x}] \cong L\}$ is an automorphism base. In Nies (2000), this was proved for the special case of $L = \{0,1\}$. A similar result holds for each $L = (\omega^n, \leqslant)$, where $n < \omega$ (compare this to Example 2.6!).

Proof. Consider the first-order property

$$\gamma_D(y) \equiv \forall q[[0,y) \not\subseteq \widehat{q} \Rightarrow \exists d \in D\,(d < y \;\&\; d \not\leqslant q)]. \tag{2}$$

Many degrees enjoy this property:

Claim 4.7 $\mathcal{R}_m$ *satisfies* $\forall z \forall x[z \not\leqslant x \Rightarrow \exists e > x(z \not\leqslant e \;\&\; \gamma_D(e))]$.

Before proving the Claim, we apply it to provide an appropriate τD-map showing that D is an automorphism base: let

$$H_{\tau D}(\boldsymbol{x}) = \{[\mathbf{0},\boldsymbol{y}) \cap D : \gamma_D(\boldsymbol{y}) \,\&\, \boldsymbol{y} \geqslant \boldsymbol{x}\}.$$

To show that this map is 1-1 on $C = \mathcal{R}_m$, given $\boldsymbol{z},\boldsymbol{x}$ such that $\boldsymbol{z} \not\leqslant \boldsymbol{x}$, using the Claim we play $\boldsymbol{e} > \boldsymbol{x}$ such that $\boldsymbol{z} \not\leqslant \boldsymbol{e}$ and $\gamma_D(\boldsymbol{e})$. The opponent answers with $\boldsymbol{u} > \boldsymbol{z}$ such that $\gamma_D(\boldsymbol{u})$ and claims that $[\mathbf{0},\boldsymbol{u}) \cap D = [\mathbf{0},\boldsymbol{e}) \cap D$. But $[\mathbf{0},\boldsymbol{u}) \not\subseteq \widehat{\boldsymbol{e}}$, so this contradicts $\gamma_D(\boldsymbol{u})$.

Proof of the Claim. We can assume that D is a $\emptyset$-definable subset of $\mathcal{R}_m^- = \mathcal{R}_m\text{-}\{\mathbf{1}\}$ and $D \not\subseteq \{\mathbf{0}\}$. We make use of a technique which was introduced in Nies (2000) to prove that for some $\boldsymbol{e} < \mathbf{1}$, $[\mathbf{0},\boldsymbol{e})$ is an elementary submodel of $\mathcal{R}_m^-$. By recursion over $k \geqslant 0$, functions F_k on (indices of) c.e. many-one degrees were defined with the property that, for each $\boldsymbol{a} < \mathbf{1}$, $\boldsymbol{a} < F_k(\boldsymbol{a})$ and $[\mathbf{0}, F_k(\boldsymbol{a})) \prec_k \mathcal{R}_m^-$. (i.e., the inclusion map preserves Σ_k-formulas). In addition, $[\mathbf{0}, F_0(\boldsymbol{a})]$ is effectively isomorphic to $\mathcal{R}_m$.

Given $\boldsymbol{z},\boldsymbol{x}$ as in the Claim, we will let $\boldsymbol{e} = F_k(\boldsymbol{x})$, where $k \geqslant 1$ is a number such that D can be defined by a Σ_k-formula. Since we need $\boldsymbol{z} \not\leqslant \boldsymbol{e}$, we have to slightly extend the construction in Nies (2000) which leads to the functions F_r. For each r we want to achieve that $\boldsymbol{z} \not\leqslant \boldsymbol{a} \Rightarrow \boldsymbol{z} \not\leqslant F_r(\boldsymbol{a})$.

For $r = 0$ this is possible by Remark 2.4 (2) in Nies (2000). $F_{r+1}(\boldsymbol{a})$ is a degree $\boldsymbol{y}$ such that $[\mathbf{0},\boldsymbol{y}) = \bigcup_i [\mathbf{0}, F_r^{(i)}(\boldsymbol{a}))$. By Lemma 2.1 there, which goes back to Ershov and Lavrov, such a strong minimal cover can be chosen distinct from $\boldsymbol{z}$. Thus $\boldsymbol{z} \leqslant \boldsymbol{y}$ would imply $\boldsymbol{z} \leqslant F_r^{(i)}(\boldsymbol{a})$ for some i.

It remains to show $\gamma_D(\boldsymbol{e})$. Suppose $[\mathbf{0},\boldsymbol{e}) \not\subseteq \widehat{\boldsymbol{q}}$. Since $[\mathbf{0},\boldsymbol{e}) \prec_k \mathcal{R}_m^-$, there exists $\boldsymbol{d} \in (\mathbf{0},\boldsymbol{e}) \cap D$. If $\boldsymbol{d} \not\leqslant \boldsymbol{q}$, we are done. Otherwise we replace $\boldsymbol{d}$ by an automorphic image $\boldsymbol{d}'$ in $\widehat{\boldsymbol{e}}$ such that $\boldsymbol{d}' \not\leqslant \boldsymbol{q}$: Let $I = \widehat{\boldsymbol{e}} \cap \widehat{\boldsymbol{q}}$. By the hypothesis on $\boldsymbol{q}$, $I \neq [\mathbf{0},\boldsymbol{e})$. Therefore, for some $i > 0$, $\boldsymbol{b} = F_0(F_{k-1}^{(i)}(\boldsymbol{x})) \notin I$ and $\boldsymbol{b} < \boldsymbol{e}$. By the definition of F_0, $\widehat{\boldsymbol{b}}$ is effectively isomorphic to $\mathcal{R}_m$. We apply Lemma 2.1 in Nies (2000) to the Σ_3^0-ideal $I \cap \widehat{\boldsymbol{b}}$ of $\widehat{\boldsymbol{b}}$ and obtain an $\boldsymbol{a} < \boldsymbol{b}$ such that $I \cap \widehat{\boldsymbol{b}} \subseteq \widehat{\boldsymbol{a}}$. Now we apply the EP for $\mathcal{R}_m$, again with $\widehat{\boldsymbol{b}}$ in place of $\mathcal{R}_m$, to the following diagram:

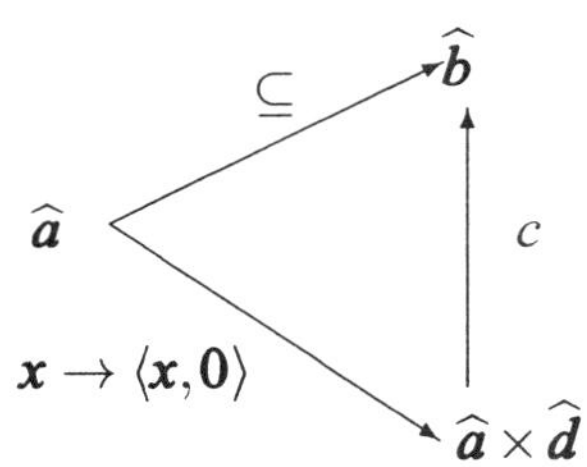

in order to obtain the map c. Let $\boldsymbol{d}' = c(\langle \mathbf{0},\boldsymbol{d}\rangle)$, then $\boldsymbol{d}' \leqslant \boldsymbol{b}$, $\boldsymbol{d}' \not\leqslant \boldsymbol{a}$ and therefore $\boldsymbol{d}' \not\leqslant \boldsymbol{q}$. By a back and forth-construction within $\widehat{\boldsymbol{e}}$, the map c can be extended to an automorphism of $\widehat{\boldsymbol{e}}$. Since $[\mathbf{0},\boldsymbol{e}) \prec_k \mathcal{R}_m^-$, this implies $\boldsymbol{d}' \in D$.

$\square$

Very little is known about the structure of $\mathrm{Aut}(\mathcal{R}_m)$. However, since $\mathcal{R}_m - \{\mathbf{1}\}$ is isomorphic to the Δ^0_2 m-degrees (Denisov 1978), one can infer the existence of an automorphism of order 2 which corresponds to complementation in the Δ^0_2-sets.

Our second application is to $\mathcal{R}_T$.

Theorem 4.8 (Ambos-Spies (ta)) *For each $\boldsymbol{c} \in \mathcal{R}_T - \{\mathbf{0}\}$, $[\mathbf{0}, \boldsymbol{c}]$ is an automorphism base.*

Proof. We will apply Proposition 4.5 with $C = \mathrm{NC}$ (a $\emptyset$-definable set which is an automorphism base by Proposition 4.2) and $B = \widehat{\boldsymbol{c}}$. We will introduce an appropriate $H_{\sigma\widehat{c}}$-map, where σ is a B-type. The map is based on Ambos-Spies' "downward splitting property": let

$$\mathrm{DSP} = \{\boldsymbol{a} \in \mathcal{R}_T : \forall \boldsymbol{b} \not\leqslant \boldsymbol{a}\, \forall \boldsymbol{d} \not\leqslant \boldsymbol{b} \vee \boldsymbol{a} [\widehat{\boldsymbol{d}} \cap \widehat{\boldsymbol{a}} \neq \widehat{\boldsymbol{d}} \cap \widehat{\boldsymbol{b}}]\}$$

This property corresponds to γ_D in the previous proof. Note that $\mathrm{DSP} \subseteq \mathrm{NC}$. The next Lemma shows that sufficiently many degrees are in DSP.

Lemma 4.9 (Ambos-Spies (ta)) *Suppose $\boldsymbol{p} \not\leqslant \boldsymbol{q}$, $\boldsymbol{p} \in NC$ and $\boldsymbol{q}$ is low. Then there is an $\boldsymbol{a} \in DSP$ such that $\boldsymbol{a} \leqslant \boldsymbol{p}$ and $\boldsymbol{a} \not\leqslant \boldsymbol{q}$.*

$H_{\sigma\widehat{c}}(\boldsymbol{x})$ is a complex object constructed from ideals $\widehat{\boldsymbol{c}} \cap \widehat{\boldsymbol{a}}$, where $\boldsymbol{a} \in \mathrm{DSP}$: let $H_{\sigma\widehat{c}}(\boldsymbol{x}) =$

$$\{\{\langle\{\widehat{\boldsymbol{c}} \cap \widehat{\boldsymbol{a}} : \boldsymbol{a} \in \widehat{\boldsymbol{p}}_0 \cap \mathrm{DSP}\}, \{\widehat{\boldsymbol{c}} \cap \widehat{\boldsymbol{a}} : \boldsymbol{a} \in \widehat{\boldsymbol{p}}_1 \cap \mathrm{DSP}\}\rangle : \boldsymbol{p}_0 \vee \boldsymbol{p}_1 = \boldsymbol{p} \,\&\, \boldsymbol{p}_i \in \mathrm{NC}\},$$
$$\{\langle\{\widehat{\boldsymbol{c}} \cap \widehat{\boldsymbol{a}} : \boldsymbol{a} \in \widehat{\boldsymbol{q}}_0 \cap \mathrm{DSP}\}, \{\widehat{\boldsymbol{c}} \cap \widehat{\boldsymbol{a}} : \boldsymbol{a} \in \widehat{\boldsymbol{q}}_1 \cap \mathrm{DSP}\}\rangle : \boldsymbol{q}_0 \vee \boldsymbol{q}_1 = \boldsymbol{q} \,\&\, \boldsymbol{q}_i \in \mathrm{NC}\} :$$
$$\boldsymbol{p} \vee \boldsymbol{q} = \boldsymbol{x} \,\&\, \boldsymbol{p}, \boldsymbol{q} \not\leqslant \boldsymbol{c} \,\&\, \boldsymbol{p}, \boldsymbol{q} \in \mathrm{NC}\}$$

We must show that $H_{\sigma\widehat{c}} \lceil \mathrm{NC}$ ia 1-1. Suppose $\boldsymbol{x}, \boldsymbol{y} \in \mathrm{NC}$ are given, and $\boldsymbol{x} \not\leqslant \boldsymbol{y}$. The game is as follows.

Round 1. By Proposition 4.1, we play a split $\boldsymbol{y} = \boldsymbol{p} \vee \boldsymbol{q}$, where $\boldsymbol{p}, \boldsymbol{q} \in \mathrm{NC} \cap \mathrm{Low}$ and $\boldsymbol{c} \not\leqslant \boldsymbol{p}, \boldsymbol{q}$. The opponent responds with a split $\boldsymbol{x} = \widetilde{\boldsymbol{p}} \vee \widetilde{\boldsymbol{q}}$ where $\widetilde{\boldsymbol{p}}, \widetilde{\boldsymbol{q}} \in \mathrm{NC}$ and $\boldsymbol{c} \not\leqslant \widetilde{\boldsymbol{p}}, \widetilde{\boldsymbol{q}}$.

Round 2. Say $\widetilde{\boldsymbol{p}} \not\leqslant \boldsymbol{p}$. Again by Proposition 4.1, we split $\widetilde{\boldsymbol{p}} = \boldsymbol{d}_0 \vee \boldsymbol{d}_1$, where $\boldsymbol{d}_i \in \mathrm{NC}$ $(i = 0, 1)$ and $\boldsymbol{c} \not\leqslant \boldsymbol{d}_i \vee \boldsymbol{p}$. The opponent responds with a split $\boldsymbol{p} = \boldsymbol{e}_0 \vee \boldsymbol{e}_1$, where $\boldsymbol{e}_i \in \mathrm{NC}$. Then $\boldsymbol{e}_i \in \mathrm{Low}$.

Round 3. Say $\boldsymbol{d}_0 \not\leqslant \boldsymbol{e}_0$. Using Lemma 4.9, we play $\boldsymbol{b}$ such that $\boldsymbol{b} \in \mathrm{DSP}$, $\boldsymbol{b} \leqslant \boldsymbol{d}_0$ and $\boldsymbol{b} \not\leqslant \boldsymbol{e}_0$. The opponent responds with $\boldsymbol{a}$ such that $\boldsymbol{a} \in \mathrm{DSP}$ and $\boldsymbol{a} \leqslant \boldsymbol{e}_0$, and claims that $\widehat{\boldsymbol{a}} \cap \widehat{\boldsymbol{c}} = \widehat{\boldsymbol{b}} \cap \widehat{\boldsymbol{c}}$.

But his claim is wrong: $\boldsymbol{b} \not\leqslant \boldsymbol{a}$ since $\boldsymbol{a} \leqslant \boldsymbol{e}_0$ while $\boldsymbol{b} \not\leqslant \boldsymbol{e}_0$. Moreover, $\boldsymbol{c} \not\leqslant \boldsymbol{a} \vee \boldsymbol{b}$ since $\boldsymbol{a} \vee \boldsymbol{b} \leqslant \boldsymbol{d}_0 \vee \boldsymbol{e}_0$ while $\boldsymbol{c} \not\leqslant \boldsymbol{d}_0 \vee \boldsymbol{p}$. Thus his claim contradicts $\boldsymbol{a} \in \mathrm{DSP}$.

Proof of Lemma 4.9. We sketch Ambos-Spies' construction of a c.e. set A such that $\boldsymbol{a} \in \mathrm{DSP}$. By the finitary character of the strategies, the construction can be easily expanded in order to prove the Lemma.

Given c.e. sets B, C, we enumerate sets $F \leqslant_T A, C$ and G_e ($e \in \mathbb{N}$) such that, under the assumption that $B \not\leqslant_T A$ and $C \not\leqslant_T B \oplus A$, the following requirements are met:

$$\begin{aligned} P_e: &\quad F \neq \{e\}^B \vee G_e \leqslant_T B, C \\ Q_{e,i}: &\quad F = \{e\}^B \Rightarrow G_e \neq \{i\}^A, \end{aligned}$$

where P_e has higher priority than any $Q_{e,i}$. The reductions $F \leqslant_T A, C$ are by direct permitting, $G_e \leqslant_T C$ (if applicable) by delayed permitting, while $G_e \leqslant_T B$ is is a proper T-reduction (but with no delay in correcting). We do not name these reductions and write e.g. $G_e \leqslant_T C(y)$ for the output on input y.

A *set-up* for $Q_{e,i}$ at stage s is a pair $x \in \omega^{[e]}, y \in \omega^{[\langle e,i \rangle]}$ such that $\{e\}^B(x) = \{i\}^A(y) = 0[s]$ and $u(A; i, y, s) < x < \gamma_B(y, s)$, where $\gamma_B(y, s)$ is the "standard marker" $max\{t \leqslant s : B_t\lceil y \neq B_{t+1}\lceil y\}$. The number x is targeted for F, while y is targeted for G_e. The strategy for $Q_{e,i}$ is the following: if the requirement is unsatisfied, a set-up x, y exists at stage s and $C_s\lceil y \neq C_{s+1}\lceil y$, then we say the strategy requires attention. At this stage s, attempt to meet P_e by putting x into F and A (note that this A-enumeration does not destroy $\{i\}^A(y)$). Moreover, initialize the strategies of lower priority than $Q_{e,i}$. This is a win for the higher priority requirement P_e (and hence all $Q_{e,i}$) unless, at a stage $t > s$, for the first time B changes below the use of $\{e\}^B(x)$. The use of the computation $G_e \leqslant_T B(y)$ at a stage r is defined to be $u(B, e, \gamma_B(y))[r]$ (where we assume that the use of $\{e\}^B$ is monotonic in the input). So this B-change makes $G_e \leqslant_T B(y)$ undefined. Also, at stage s we used the C-change to make $G_e \leqslant_T C(y)$ undefined, and redefine it only now at stage t. Thus we are allowed to put y into G_e and declare the requirement satisfied. If $F = \{e\}^B$, then the B-change must occur, so we always redefine $G_e \leqslant_T C(y)$ (if $Q_{e,i}$ is initialized before, we also redefine the reduction).

In the *construction* we start the highest priority strategy $Q_{e,i}$ which requires attention, and later do the G_e enumeration (which does not injure other strategies) when necessary. Also we update the functionals as described above.

The *verification* is as follows. Clearly the action of the $Q_{e,i}$ is finitary after it is no more initialized. Suppose that $F = \{e\}^B$. The reduction $G_e \leqslant C$ is total since it is always redefined and its use equals the input. The reduction $G_e \leqslant_T B$ is total since $\{e\}^B$ is and the standard markers reach a limit. Now suppose $Q_{e,i}$ is not met, i.e. $G_e = \{i\}^A$. If $B \not\leqslant_T A$, then infinitely many set-ups x, y for $Q_{e,i}$ must appear at stages s such that $\{e\}^B(x)$ and $\{i\}^A(y)$ are stable, and thus the set-up is permanent. Since $A \oplus B$ can recognize such permanent set-ups, $C \not\leqslant_T A \oplus B$ implies that infinitely often a $C\lceil y$ change must occur when a set-up x, y exists. Thus, after $Q_{e,i}$ is no more injured, we win the requirement. □

One can define composition of maps as above: for instance, if H is the map from Proposition 4.2, $H_{\rho\widehat{c}} = H \circ H_{\sigma\widehat{c}}(\boldsymbol{x})$ is $\{\{H_{\sigma\widehat{c}}(\boldsymbol{y}_1) : \boldsymbol{y}_1 \in [\boldsymbol{p}, \mathbf{1}] \cap \mathrm{NC}\},$

$\{H_{\sigma\widehat{c}}(y_2) : y_2 \in [\boldsymbol{q}, \mathbf{1}] \cap \mathrm{NC}\} : \boldsymbol{p} \vee \boldsymbol{q} = \boldsymbol{x}\}$. This is a 1-1 map from the whole structure into objects of type $\rho\widehat{\boldsymbol{c}}$. It follows that $\mathcal{R}_T$ is biinterpretable with $\mathbb{N}$ in parameters if, for some $\boldsymbol{c} \neq \mathbf{0}$, there is a parameter defined copy M of $(\mathbb{N}, +, \times)$ and a parameter definable 1-1 map $g : \widehat{\boldsymbol{c}} \mapsto M$.

References

Ambos-Spies, K.: (ta), *Automorphism bases in the recursively enumerable degrees*, to appear.

Degtev, A.: (1979), 'Several results on uppersemilattices and m-degrees', *Algebra and Logic* **18**, pp. 664–679.

Denisov, S.D.: (1978), 'The structure of the uppersemilattice of recursively enumerable m-degrees and related questions, I', *Algebra and Logic* **17**, pp. 418–443.

Ershov, Yu.L.: (1975), 'The upper semilattice of numerations of a finite set', *Algebra i Logika* **14**, pp. 258–284 (Russian).

Lachlan, A.H.: (1970), 'Initial segments of many-one degrees', *Canad. J. Math* **22**, pp. 75–85.

Lempp, S., Nies, A., Slaman, T.A.: (1998), 'The Π_3-theory of the enumerable Turing degrees is undecidable', *Trans. Amer. Math. Soc.* **350** (7), pp. 2719–2736.

Nies, A.: (ta), 'Definability in the computably enumerable degrees: Questions and result', to appear in *Contemporary Mathematics*.

Nies, A.: (2000), 'Model theory of the computably enumerable many-one degrees', to appear in *The Logic Journal of the IGPL*, http://www.dcs.kcl.ac.uk/journals/IGPL.

Nies, A.: (1996), 'Undecidable fragments of elementary theories', *Algebra Universalis* **35**, pp. 8–33.

Nies, A., Shore, R., Slaman, T.A.: (1998), 'Interpretability and definability in the recursively enumerable Turing-degrees', *Proc. Lond. Math. Soc.* **3** (77), pp. 241–291.

Odifreddi, P.: (1989), *Classical Recursion Theory (Volume I)*. North–Holland Publishing Co., Amsterdam.

Soare, R.I.: (1987), *Recursively Enumerable Sets and Degrees*. Perspectives in Mathematical Logic, Omega Series. Springer–Verlag, Heidelberg.

COMPUTABLE STRUCTURES: PRESENTATIONS MATTER*

Richard A. Shore
Department of Mathematics
Cornell University
Ithaca NY 14853

Abstract The computability properties of a relation R not included in the language of a computable structure $\mathcal{A}$ can vary from one computable presentation to another. We describe some classic results giving conditions on $\mathcal{A}$ or R that restrict the possible variations in the computable dimension of $\mathcal{A}$ (i.e. the number of isomorphic copies of $\mathcal{A}$ up to computable isomorphism) and the computational complexity of R. For example, what conditions guarantee that $\mathcal{A}$ is computably categorical (i.e. of dimension 1) or that R is intrinsically computable (i.e. computable in every presentation). In the absence of such conditions, we discuss the possible computable dimensions of $\mathcal{A}$ and variations (in terms of Turing degree) of R in different presentations (the degree spectrum of R). In particular, various classic theorems and more recent ones of the author, B. Khoussainov, D. Hirschfeldt and others about the possible degree spectra of computable relations on computable structures and the connections with computable dimension and categoricity will be discussed both in general model theoretic settings and in restricted classes of structures such as graphs, linear and partial orderings, lattices, Boolean algebras, Abelian and nilpotent groups, rings, integral domains, and real or algebraically closed fields.

1. Introduction

The general subject area of this paper is effective model theory. There are (at least) two versions of what is the basic subject matter of model theory. Along with Chang and Keisler (1990) one can take the "logical" point of view that it deals with the connections between formal language and their interpretations, or models. Alternatively, with Hodges (1993) we can say that it is the study of the construction and classification of structures within specified classes of structures. The first view starts with theories and so the effective version deals

*Partially supported by NSF Grant DMS-9802843.

P. Gärdenfors, J. Woleński and K. Kijania-Placek (eds.), In the Scope of Logic, Methodology and Philosophy of Science, Vol. I, 81-95.

with decidable theories, ones for which we have a computable procedure for deciding the validity of each sentence or its truth in a model under consideration. The second is an algebraic view that starts with structures and so the effective version deals with computable structures in the sense that we can compute the basic relations and operations on the structure. Each approach provides a rich area for investigation. In this paper we consider only certain topics from the second, structural or algebraic point of view. One that considers both is Khoussainov and Shore (1999). More general introductions can be found in *The Handbook of Recursive Algebra* (Ershov et al. 1998), especially the articles by Harizanov (1998) and Ershov and Goncharov (1998). This *Handbook* also contains other useful survey papers on aspects of effective model theory and algebra and an extensive bibliography. The one most closely related to the theme of this paper is Goncharov (1998). Another interesting survey is Millar (1999) in *The Handbook of Computability Theory* (Griffor 1999). A new book on the subject is Ash and Knight (2000).

As for approaches to effectiveness, one can profitably study the issues of computability in model theory and algebra in terms of a wide range of notions from polynomial-time to Borel. In this paper, we restrict ourselves to the fundamental notion of computability (and relative computability) as determined by Turing machines (with oracles). We begin with the basic definition.

Definition 1.1 *A structure $\mathcal{A}$ is* computable *if its domain A is computable and the functions and relations of $\mathcal{A}$ are uniformly computable (or, equivalently, the atomic diagram of $\mathcal{A}$, $D(\mathcal{A},a)_{a\in A}$, is computable). $\mathcal{A}$ is* computably presentable *if $\mathcal{A}$ is isomorphic to a computable structure $\mathcal{B}$ via a map $f: A \to B$ which we call a* computable presentation *of $\mathcal{A}$.*

- From now on, all structures will be computable unless otherwise specified.

The question we want to address in this paper is whether computability properties of structures depend on the choice of presentation and if so, how? From the classical mathematical viewpoint, two presentations $\mathcal{A}$ and $\mathcal{B}$ of a structure are, after all, the same structure, i.e. they are isomorphic, so one might ask how can they differ. On the other hand, it is a commonplace in computer science that the choice of representation of data can be crucial in determining the ease of its use. Thus, from the mathematical point of view, the surprise will be that there are differences at all. From the computer science point of view, the surprise is the extent of the differences. The choice of presentation can not only make operations faster or slower but can change them from computable to noncomputable in drastic ways. A classic and striking example is the notion of dependence in vector spaces (linear dependence) or algebraically closed fields (algebraic independence). While we know how to effectively determine whether elements are dependent or not in each setting in the standard presentations of, for example, the infinite dimensional vector space (or algebaically closed field) over the rationals, $\mathbb{Q}$, this cannot be done in all computable presentations of these well known structures (Metakides and Nerode 1977, 1979). The source of the problem is that, although $\mathcal{A}$ and $\mathcal{B}$ may be isomorphic, the isomorphism between them may not be computable. Thus a relation (like dependence), even though it is preserved under isomorphisms, is not explicitly part of the language and so may be computable in one computable presentation and not another. The point here is that the underlying notion of equivalence between computable structures that is relevant to issues about preserving computability properties is computable rather than classical isomorphism.

Definition 1.2 *$\mathcal{A}$ is* computably isomorphic *to $\mathcal{B}$, $\mathcal{A} \cong_c \mathcal{B}$, if there is a computable $f : A \to B$ which is an isomorphism. We also say then that $\mathcal{A}$ and $\mathcal{B}$ are* of the same computable isomorphism type.

Definition 1.3 *The* (computable) dimension *of a structure $\mathcal{A}$ is the number of its computable isomorphism types. $\mathcal{A}$ is* computably categorical *if its computable dimension is* 1*, i.e. every $\mathcal{B}$ isomorphic to $\mathcal{A}$ is computably isomorphic to $\mathcal{A}$.*

Example 1.4 $\langle \mathbb{Q}, \leqslant \rangle$ is computably categorical: The usual back and forth argument used to show that any two dense linear orderings without endpoints are isomorphic is effective and produces computable isomorphisms between any two computable presentations of $\mathbb{Q}$.

Example 1.5 $\langle \mathbb{N}, s \rangle$ is computably categorical: Given $\mathcal{B} \cong \mathbb{N}$, one defines the required computable $f : N \to B$ by recursion. First, $f(0)$ is the least element of B and then if $f(n) = b$ we let $f(n+1) = s(b)$.

On the other hand, if one considers the natural numbers with the usual ordering relation but without the successor function, the structure is not computably categorical. Indeed, as the above argument shows, any presentation $\mathcal{A}$ in which the successor function is computable is computably isomorphic to the standard presentation. Thus, there are presentations in which the successor function is not computable

Proposition 1.6 *$\langle \mathbb{N}, \leqslant \rangle$ has a computable presentation $\mathcal{A}$ in which the successor function is not computable and so $\langle \mathbb{N}, \leqslant \rangle$ is not computably categorical.*

Proof. We construct $\mathcal{A} = \langle A, \preceq \rangle$ by induction. Without loss of generality, we assume that we have a common enumeration of all the computations of the computable functions ϕ_n in which just one computation $\phi_n(x)$, for some

n and x, converges at each stage s. We begin with all the even numbers in A_0 in their usual order. At stage s of the construction, we see if ϕ_n converges on input $2n$ at stage s and gives output $2n+2$. If so, we add $2s+1$ to A_s to get A_{s+1} and put $2s+1$ between $2n$ and $2n+2$ in the ordering $\preceq$. If not, $A_{s+1} = A_s$. Of course, $A = \cup A_s$. It is easy to see that $\mathcal{A}$ is computable. (To see if some odd number $2s+1$ is in A just go to stage s of the construction. Similarly, if $x, y \in A$, we can determine if $x \preceq y$ at the stage by which both have been put into A.) It is also immediate that $\mathcal{A} \cong \mathbb{N}$ as we start out with an ordering on A_0 isomorphic to $\mathbb{N}$ and add at most one element between any two consecutive elements of the original ordering on A_0. On the other hand, the usual diagonal argument now shows that no ϕ_n is the successor function on $\mathcal{A}$. (If $\phi_n(2n)$ converges and equals $2n+1$ then the immediate successor of $2n$ in $\mathcal{A}$ is $2s+1$ where s is the stage at which $\phi_n(2n)$ converges. In every other case, the immediate successor of $2n$ in $\mathcal{A}$ is $2n+2$ which is not the value of $\phi_n(2n)$.) □

Example 1.7 As in Example 1.5, the integers $\mathbb{Z}$ as a group with $+$ or as a ring with $+$ and $\times$ is computably categorical as it is finitely generated. The rationals, $\mathbb{Q}$, are also computably categorical as a field as one can effectively generate them from 0, 1 and the field operations $+$ and $\times$ using a procedure that identifies a fraction p/q as the element r such that $p = q \times r$. A bit of Galois theory can then be used to show that $\bar{\mathbb{Q}}$, the algebraic closure of $\mathbb{Q}$, is computably categorical. On the other hand, the algebraic closure of $\mathbb{Q}$ extended by infinitely many transcendentals has computable dimension ω. The argument need here is more complicated than for $\langle N, \leqslant \rangle$ but follows from the results of Metakides and Nerode (1979) mentioned above as well as more general theorems that we will discuss below.

2. Categoricity and intrinsic computability

Since we have intimated that the source of variation in computability properties among different presentations is the fact that the isomorphism between them might not be computable, it is natural to conjecture that computable categoricity should obviate this problem and any relation or operation computable in one presentation of a computably categorical structure should be computable in all presentations. This conjecture is false as stated but is true for all the relations or operations in which a mathematician is likely to actually be interested. We begin with the formal definition of persistence of computability over presentations and a counterexample to our conjecture. We then show in what sense it is essentially correct.

Definition 2.1 (Ash and Nerode 1981) *If* $R \subseteq A^n$*, then* R *is* intrinsically computable (c.e.) *if* $f[R]$ *is computable (c.e.) for every isomorphism* $f : \mathcal{A} \to \mathcal{B}$.

Example 2.2 $\langle N, \leqslant \rangle$: The successor function is not intrinsically computable by Proposition 1.6.

Example 2.3 $\langle N, s \rangle$: Every computable relation is intrinsically computable. Suppose $R \subseteq N$ is computable and $f : N \to B$ is another computable presentation of N. By the argument of Example 1.5, f is computable. Of course, $\bar{R}$, the complement of R in A, is also computable. As f is a bijection $f[R]$ and $f[\bar{R}]$ are complementary in $\mathcal{B}$. As they are both the images of computable sets under a computable map they are computably enumerable and so actually computable. (The argument for n-ary relations R is essentially the same and we will frequently consider only unary relations when there is no real difference.)

On the other hand, computable categoricity does not guarantee that all computable relations are intrinsically computable.

Example 2.4 Consider the structure $\mathbb{Q}$ with the usual ordering and then additional dense subsets R and R' with R computable and R' not. It is easy to construct an isomorphism $f : \langle \mathbb{Q}, \leqslant, R \rangle \to \langle \mathbb{Q}, \leqslant, R' \rangle$ and so R is not intrinsically computable even though $\mathbb{Q}$ is computably categorical.

There are two views of what has gone wrong in this counterexample. One is that, although there is some computable map between the two presentations of $\mathbb{Q}$ given in this example (indeed the identity map is such an isomorphism as R is not in the language), the isomorphism f is not computable. The way to solve this problem is to strengthen the notion of computable categoricity.

Definition 2.5 *$\mathcal{A}$ is* computably stable *if every isomorphism $f : \mathcal{A} \to \mathcal{B}$ is computable.*

Thus the Examples above show that $\langle N, s \rangle$ is computably stable but $\langle \mathbb{Q}, \leqslant \rangle$, while computably categorical, is not computably stable. The stronger notion, however, clearly suffices to guarantee that the computability of any relation cannot change from one presentation to another. Perhaps surprisingly, it is easy to see that it is also necessary.

Proposition 2.6 (Ash and Nerode 1981) *$\mathcal{A}$ is computably stable if and only if every computable relation on $\mathcal{A}$ is intrinsically computable.*

Proof. The argument given in Example 2.3 proves sufficiency. For necessity, suppose every computable relation on $\mathcal{A}$ is intrinsically computable and consider the successor relation R on A as a computable set of natural numbers with the usual ordering. Suppose $\mathcal{B}$ is any other presentation of $\mathcal{A}$ given by $f : A \to B$. As the image of R is computable we can use it to computably calculate f as in Example 1.5. □

The other view of what is wrong with Example 2.4 is that the relation R considered is not invariant under automorphisms and so not mathematically relevant to the structure. This suggests the following proposition saying that computable categoricity is enough to guarantee intrinsic computability for all the relations of interest.

Proposition 2.7 *If $\mathcal{A}$ is computably categorical then every invariant (under automorphisms) computable relation on $\mathcal{A}$ is intrinsically computable.*

Proof. Suppose $\mathcal{A}$ is computably categorical, $R \subseteq A$ is computable, and g is an isomorphism from $\mathcal{A}$ to $\mathcal{B}$. We wish to show that $g[R]$ is computable. As $\mathcal{A}$ is computably categorical, there is a computable isomorphism $f : \mathcal{A} \to \mathcal{B}$. As in Example 2.3, $f[R]$ is computable. As R is invariant under automorphisms, in particular under $g^{-1}f$, $f[R] = g[R]$ and so $g[R]$ is also computable. □

Before considering specific conditions on structures that imply computable categoricity or stability and so guarantee that computational properties do not vary among presentations, we mention a number of general results on when various specified types of mathematical structures are computably categorical.

Theorem 2.8 (Goncharov 1973, LaRoche 1977, Remmel 1981, Goncharov and Dzgoev 1980) *A Boolean algebra is computably categorical if it has finitely many atoms. If not, it has dimension ω.*

Theorem 2.9 (Remmel 1981a, Goncharov and Dzgoev 1908) *A linear order is computably categorical if it has only finitely many adjacent pairs of elements. If not, it has dimension ω.*

Theorem 2.10 (Nurtazin 1974, Metakides and Nerode 1979) *A real or algebraically closed field of finite transcendence degree over $\mathbb{Q}$ is computably categorical. Ones of infinite degree have dimension ω.*

Theorem 2.11 (Goncharov 1980) *Every Abelian group has dimension* 1 *or* ω.

Theorem 2.12 (Goncharov, Lempp and Solomon 2002) *Every Archimedian ordered group has dimension* 1 *or* ω.

These results provide a type of dichotomy theorem for each of the classes of structures described. Every one is either computably categorical and so very well behaved in terms of computability properties or has the maximum possible number of computable isomorphism types. Moreover, in each case, there is a structural analysis that tells us into which category each model falls. As is indicated by these examples, when there is such a dichotomy the dividing line seems to be that some sort of finiteness property characterizes the computably categorical structures. We saw this explicitly in the analysis of $\langle N,s\rangle$ in Example 1.5 where the structure's being finitely generated in the usual sense was the key to computable stability. Of course, any finitely generated computable structure is computably stable by the same argument. As it turns out, with the proper version of what it means to be finitely generated and an additional assumption on the decidability of what for algebras amounts to the existence of solutions of systems of equations and inequations, this condition is also necessary.

Definition 2.13 *A structure $\mathcal{A}$ is n-*decidable *(for $n \in \mathbb{N}$) if the set of prenex sentences of $Th(\mathcal{A},a)_{a\in A}$ with $n-1$ alternations of quantifiers is computable. So, for example, $\mathcal{A}$ is* 1-decidable *if the set of prenex sentences of $Th(\mathcal{A},a)_{a\in A}$ with either only existential or only universal quantifiers is decidable.*

Theorem 2.14 (Ash and Nerode 1981, Goncharov 1975) *If $\mathcal{A}$ is 1-decidable, then $\mathcal{A}$ is computably stable if and only if there are constants $\overline{c} \in A$ and a computable sequence $\phi_i(\overline{c},x)$ of existential formulas such that for each i there is a unique $a \in A$ satisfying ϕ_i and each $a \in A$ satisfies some ϕ_i.*

Here the sequence of formulas ϕ_i provide the generating process from the finitely many parameters $\bar{c}$ as we can search for witnesses to the existential formulas to generate the unique element a satisfying each ϕ_i. This is the situation for $\mathbb{Q}$ in Example 1.7 where the formulas express the fact that each element is either a sum of 1's or the solution r of an equation of the form $p+r=0$ or $p=r\times q$ where p is a sum of 1's and q is such a sum or a solution to an equation of the first type. On the other hand, $\bar{\mathbb{Q}}$ is not finitely generated in the usual sense and the argument that it is computably categorical relies on specifying elements as roots of polynomials over $\mathbb{Q}$ and so determining them only up to isomorphism. Given the previously mentioned relationship between computable categoricity and the intrinsic computability of invariant relations, it is not unreasonable that we have a result for computable categoricity that is analogous to Theorem 2.14.

Theorem 2.15 (Goncharov 1975) *If $\mathcal{A}$ is 2-decidable then it is computably categorical if and only if there are $\overline{c} \in A$ and $\phi_i(\overline{c},x)$ as in Theorem 2.14 except that the ϕ_i determine elements not uniquely but only up to automorphism, i.e. if $\phi_i(\overline{c},a)$ and $\phi_i(\overline{c},b)$ then there is an automorphism of $\mathcal{A}$ taking a to b.*

The decidability assumptions in Theorem 2.14 and 2.15 are necessary (Goncharov 1977). Moreover, the type of dichotomy results given in Theorems 2.8-2.12 are also provided by decidability type conditions on the structures themselves.

Theorem 2.16 (Goncharov 1982, Nurtazin 1974, Goncharov 1977) *Every Δ_2-categorical structure (i.e. there is always a Δ_2 isomorphism between any two presentations) and every* 1*-decidable structure has* dim 1 *or* ω.

However, the dichotomy between computable categoricity and dimension ω does not hold for all types of structures. The first indication of such phenomena was in the following theorem:

Theorem 2.17 (Goncharov 1980a) *For each n, $1 \leqslant n \leqslant \omega$ there is a structure of dimension n.*

Indeed, there are several examples of common mathematical theories whose models can have all possible dimensions and, as we shall see in Theorem 4.1, exhibit the widest possible range of variation in computability properties.

In addition to trying to characterize when a structure is computably categorical or stable, it is natural to ask when individual relations are intrinsically computable. As with structures, there is a nice answer if we assume a certain amount of decidability.

Theorem 2.18 (Ash and Nerode 1981) *If $\langle \mathcal{A}, R \rangle$ is 1-decidable then R is intrinsically c.e. if and only if, for some finite list of parameters $\vec{a}$ from A and computable set of existential formulas ϕ_i, $\mathcal{A} \models R(\vec{x}) \Leftrightarrow \bigvee \phi_i(\vec{x}, \vec{a})$. Of course, R is intrinsically computable if and only if it and its complement are both intrinsically c.e.*

The possible routes to the failure of computable categoricity or stability or intrinsic computability are intimately connected with the possible complexity of a given computable relation R on $\mathcal{A}$ in various presentations of $\mathcal{A}$. We explore this topic in the next section.

3. Degree spectra

The question we want to address here is how complicated can $f[R]$ be for a computable (c.e.) relation R on $\mathcal{A}$ if R is not intrinsically computable. The idea of the range of variability is captured in the following definition from Harizanov (1987).

Definition 3.1 *If R is an n-ary relation on $\mathcal{A}$, the* degree spectrum *of R, $DgSp(R)$, is $\{\deg_T(f[R]) \mid f : \mathcal{A} \to \mathcal{B}$ is an isomorphism$\}$. ($\deg_T(X)$ is the Turing degree of X.)*

Of course, R is intrinsically computable if and only if $DgSp(R) = \{\mathbf{0}\}$. Many natural examples of relations that are not intrinsically computable have degree spectrum the set $\mathcal{C}$ of all c.e. degrees or $\mathcal{D}$, the set of all degrees. Again there are conditions that guarantee that R has such a spectrum (Harizanov 1987, 1991, Ash, Cholak and Knight 1997) but here we concentrate on the possibilities of producing more specific sets of degrees as degree spectra of computable relations. The general question is which sets of degrees can be realized as the degree spectrum of a computable relation. We mention a few examples.

Theorem 3.2 (Harizanov 1993) *There is an $\mathcal{A}$ and a relation R on $\mathcal{A}$ such $\mathcal{A}$ has exactly two computable presentations and $DgSp(R) = \{\mathbf{0}, \mathbf{c}\}$ with $\mathbf{c}$ noncomputable and Δ^0_2.*

Theorem 3.3 (Khoussainov and Shore 1998) *For any computable partially ordered set $\mathcal{P}$ (with least element) there exists a structure $\mathcal{A}$ of dimension the cardinality of $\mathcal{P}$ and a (computable) unary relation U on $\mathcal{A}$ such that $DgSp(U) \cong \mathcal{P}$ and every element of $DgSp(U)$ is c.e.*

Theorem 3.4 (Hirschfeldt 1999, 2000, 2001) *For every uniformly c.e. set $\mathcal{W}$ of c.e. degrees, there is an $\langle \mathcal{A}, R \rangle$ with $DgSp(R) = \mathcal{W}$.*

A more difficult problem (in the finite dimensional case) is to control both the degree spectrum of R and the computable dimension of $\mathcal{A}$ simultaneously. The following theorem subsumes all of the above results on finite degree spectra as well as Theorem 2.17 on computable dimension.

Theorem 3.5 (Hirschfeldt 1999, 2000, 2002, Khoussainov and Shore 2000) *For every finite set $\mathcal{W}$ of c.e. degrees, there is an $\langle \mathcal{A}, R \rangle$ such that $DgSp(R) = \mathcal{W}$ and* $\dim(\mathcal{A}) = |\mathcal{W}|$.

Many natural infinite classes of degrees other than $\mathcal{D}$ and $\mathcal{C}$ can also be realized as the degree spectra of computable relations. Here are some examples.

Theorem 3.6 (Hirschfeldt 1999, 2000) *Each of the following classes can be realized as the degree spectrum of a computable relation: the Σ^0_n degrees, the Δ^0_n degrees, $\mathcal{D}(\leqslant \mathbf{a})$ and $\mathcal{C}(\leqslant \mathbf{a})$ for any c.e. $\mathbf{a}$.*

The only known limitations on the degree spectrum of a computable relation R are imposed by the fact that for $R \subseteq A^n$, $\{f[R] |$ there is a computable $\mathcal{B}$ and an isomorphism $f : \mathcal{A} \to \mathcal{B}\}$ is Σ^1_1 in R. Thus, there are countable partial orders that cannot be realized in the c.e. degrees as the degree spectrum of any relation R on any structure $\mathcal{A}$. Similarly, such a partial order with least element cannot be realized anywhere in the Turing degrees as the degree spectrum of a computable relation. Indeed, not every finite set of degrees can be realized as a degree spectrum: Any degree spectrum containing a hyperarithmetic degree and a nonhyperarithmetic degree is uncountable and any unbounded spectrum contains a cone of degrees.

It turns out that controlling the degree spectrum is also connected to another long standing problem involving computable categoricity. In classical model theory, a countably categorical structure remains countably categorical when expanded by finitely many constants. Results of Goncharov (1975) and Millar (1986) showed that computably categorical structures remain computably categorical when expanded by a constant if they are sufficiently decidable. (Goncharov assumed 2-decidability and Millar reduced this to 1-decidability.) The question of whether the decidability assumption is necessary was finally answered in Cholak, Goncharov, Khoussainov and Shore (1999).

Theorem 3.7 (Cholak, Goncharov, Khoussainov and Shore 1999) *For each $k \in \omega$, $k \geqslant 2$, there is a computably categorical $\mathcal{B}$ whose expansion by any constant has computable dimension k.*

This result was then later derived as an easy corollary of the work in Khoussainov and Shore (1998) on degree spectra. Both analyses left open the question of whether expansion by a constant could change a computably categorical structure to one with infinite dimension. Hirschfeldt's (1999) proof of Theorem 3.5 supplied the techniques to answer this question as well.

Theorem 3.8 (Hirschfeldt, Khoussainov and Shore 2000) *There exists a computably categorical structure $\mathcal{A}$ whose expansion by any constant has dimension ω.*

4. Interpretations and algebraic examples

The original examples of structures with finite dimension and relations with finite degree spectra were always ad hoc constructions of families of computably enumerable sets or graphs. In the case of finite dimensionality, there was a sequence of results showing that this phenomena occurred in various classes of mathematical structures: partial orders and (implicitly) graphs in Goncharov (1980a) from which the case for lattices follows easily; groups and then two step nilpotent groups in Goncharov (1981) and Goncharov, Molokov and Romanovskii (1989); and integral domains in Kudinov (1997). Typically, these proofs proceeded by coding families of c.e. sets into structures in the desired classes. A more general coding method has now been devised for producing such results as well as many other phenomena involving degree spectra, categoricity and other computability issues in many classes of mathematical structures.

Theorem 4.1 (Hirschfeldt, Khoussainov, Shore and Slinko 2002) *For each of the following theories and each set $\mathcal{W}$ of c.e. degrees of size n, there is a model $\mathcal{A}$ and a relation R on $\mathcal{A}$ (which can be taken to be a substructure of $\mathcal{A}$) such that the dimension of $\mathcal{A}$ is n and the degree spectrum of R is $\mathcal{W}$.*

There is also a model $\mathcal{B}$ which is computably categorical but some expansion by a constant has dimension n and one which has dimension ω when expanded by a constant: graphs, lattices, partial orders, commutative semigroups, rings (with zero divisors), integral domains and nilpotent groups.

The analysis here begins with classical interpretability methods for theories as in Hodges (1993) and takes as its starting structures graphs (which are just symmetric irreflexive binary relations). Suppose we want to interpret arbitrary graphs $\mathcal{G}$ in a class $\mathfrak{A}$ of structures $\mathcal{A}$. The standard procedure calls for formulas $\phi_D(x)$ and $\phi_R(x,y)$ in the language of $\mathcal{A}$ that specify a domain $D^{\mathcal{A}}$ and an edge (SIB) relation $R^{\mathcal{A}}$ on $D^{\mathcal{A}}$ so that for each graph $\mathcal{G}$ there is an $\mathcal{A} \in \mathfrak{A}$ such that $\mathcal{G} \cong \langle D^{\mathcal{A}}, R^{\mathcal{A}} \rangle$. For our purposes we need especially effective translations. D and R must be intrinsically computable with a computable "natural transformation" from $D^{\mathcal{A}}$ to $\mathcal{G}$. In this way we can preserve computability properties across the interpretation. To deal with issues of isomorphisms, D and R must be invariant. To control dimension, we also want the interpreted structure $\mathcal{A}_{\mathcal{G}} = (D^{\mathcal{A}}, R^{\mathcal{A}})$ to determine $\mathcal{A}$ in an effective way and any isomorphism of $\mathcal{A}_{\mathcal{G}}$ to be extendible to one of $\mathcal{A}$. We guarantee this last condition by requiring that there be, for each $\mathcal{G}$, a computable family $\phi_i(\bar{a}, \bar{b}_i, x)$ of existential formulas with $\bar{a} \in A$ and $\bar{b}_i \in D^{\mathcal{A}}$ such that every $x \in \mathcal{A}_{\mathcal{G}}$ satisfies some $\phi_i(\bar{a}, \bar{b}_i, x)$ and no two distinct elements of $D^{\mathcal{A}}$ satisfy any one $\phi_i(\bar{a}, \bar{b}_i, x)$.

These conditions on interpretations suffice for the simplest of the classes mentioned in the theorem. For rings and integral domains we must generalize to interpreting equality as a new (invariant, intrinsically computable) equivalence relation $Q^{\mathcal{A}}$ and extend all the new conditions to deal with sets of representatives. The result for commutative semigroups follows from the interpretation into integral domains by considering a subset of the integral domain that forms a commutative semigroup under the multiplication operation of the ring. The interpretation into nilpotent groups is a two step one. The first uses the translation of graphs into integral domains (of finite characteristic $p > 2$) and the second uses Malcev's (1965) representation of a ring R as the center of the group of matrices of the form

$$\begin{pmatrix} 1 & a & c \\ 0 & 1 & b \\ 0 & 0 & 1 \end{pmatrix} \text{ with } a, b, c \in R$$

with the ring operations definable from the matrix multiplication in an appropriately effective way.

The especially effective interpretations produced to prove Theorem 4.1 can be used to transfer other phenomena from arbitrary computable structures to ones of the types listed in the theorem. We give two such applications.

Theorem 4.2 (Slaman 1998, Wehner 1998)*: There is a structure $\mathcal{A}$ without a computable presentation that has presentations of every nonzero degree.*

Corollary 4.3 *There are such lattices, partial orders, commutative semigroups, rings, nilpotent groups, integral domains.*

Theorem 4.4 (Ash 1986)*: For each even computable ordinal α there is a well-ordering $\mathcal{W}$ which is $\Delta_{\alpha+1}$-categorical but not Δ_α-categorical, i.e. if $\mathcal{V}$ is a computable presentation of $\mathcal{W}$ then there is an isomorphism between $\mathcal{W}$ and $\mathcal{V}$ which is $\Delta_{\alpha+1}$ but there is some presentation $\mathcal{V}$ for which there is no such isomorphism which is Δ_α.*

Corollary 4.5 *There are such lattices, partial orders, commutative semigroups, rings, nilpotent groups, integral domains.*

We close this paper by mentioning some diverse topics within mathematical logic that are related to the methods and goals of the analysis of computability properties of relations on computable structures discussed here. First, there are clear analogies between the methods and results on intrinsic computability and degree spectra and those of reverse mathematics. This subject systematically analyzes which existence assumption axioms are needed to prove standard theorems of classical mathematics. (The best source for Reverse Mathematics is Simpson (1999) although the connections with computable model theory and algebra are not always made explicit.) The fact that some relation on a structure such as "commuting with every element of a group" is not intrinsically computable typically means that the existence of the set of such elements, the center of the group, is not provable in RCA_0 the standard base theory of reverse mathematics. The fact that the spectrum of the relation includes $\mathbf{0}'$, typically means that the proof of existence requires ACA_0 (arithmetic comprehension). The other standard systems of reverse mathematics have similar computability theoretic analogs and usually proofs from one area carry over to analogous ones in the other.

Second, the phenomenon of dichotomies between structure theory and non-structure theorems has become commonplace in both model theory and descriptive set theory. Of course, each area has its own version of what one requires to have a structure theorem. Usually, the requirements from the computability standpoint are the most stringent. In model theory, the relevant issues include ones about transferring model-theoretic phenomena from structures of one class such as graphs where certain properties are easy to arrange to others such as groups where they are less obvious (as in Mekler (1981)). In descriptive set theory, the analysis frequently centers around the issue of completeness of various properties at different levels of a hierarchy with respect to Borel reducibilities (as, for example, in Friedman and Stanley (1989), Camerlo and

Gao (2001), Hjorth and Kechris (1996)). Thus the types of interpretations described in this section for translating results on degree spectra will supply ones that can be used in both model theory and descriptive set theory. On the other hand, even when not entirely effective the work in these other areas can at times be used in our setting. One example is the interpretations of graphs in fields in Friedman and Stanley (1989). Although not quite satisfying the conditions we need in Theorem 4.1, their interpretation can be used, for example, to derive the analog of Theorem 4.4 for fields as long as α is sufficiently large. We look forward to the development of further connections among all these topics.

References

Ash, C.J.: (1986), 'Recursive labelling systems and stability of recursive structures in hyperartihmetical degrees', *Trans. Amer. Math. Soc.* **298**, pp. 497–514.

Ash, C.J., Cholak, P., Knight, J.F.: (1997), 'Permitting, forcing and copying of a given recursive relation', *Ann. Pure Appl. Logic* **86**, pp. 219–236.

Ash, C.J., Knight, J.F.: (2000), *Computable Structures and the Hyperarithmetical Hierarchy*, Elsevier, Amsterdam.

Ash, C.J., Nerode, A.: (1981), 'Intrinsically recursive relations', in J.N. Crossley (ed.) *Aspects of Effective Algebra* (Clayton, 1979), Upside Down A Book Co., Yarra Glen, Australia, pp. 26–41.

Camerlo, R., Gao, S.: (2001) 'The completeness of the isomorphism relation of countable Boolean algebras', *Trans. Amer. Math. Soc.* **353**, pp. 491–518.

Chang, C.C., Keisler, H.J.: (1990), *Model Theory*, third edition, Studies in Logic and the Foundations of Mathematics **73**, North-Holland, Amsterdam-New York.

Cholak, P., Goncharov S.S., Khoussainov, B., Shore R.A.: (1999), 'Computably Categorical Structures and Expansions by Constants', *Journal of Symb. Logic* **64**, pp. 13–37.

Ershov, Yu.L., Goncharov, S.S., Nerode, A., Remmel, J.B. (eds.), Marek, V.W. (assoc. ed.): (1998), *Handbook of Recursive Mathematics*, Elsevier, New York.

Ershov, Yu.L., Goncharov, S.S.: (1998), 'Elementary theories and their constructive models', in Ershov et al. (1998), pp. 115–166.

Friedman, H., Stanley, L.: (1989), 'A Borel reducibility theory for classes of countable structures', *Journal of Symb. Logic* **54**, pp. 894–914.

Goncharov, S.S.: (1973), 'Constructivizability of superatomic Boolean algebras' (Russian), *Algebra i Logika* **12**, pp. 31–40; trans. in *Algebra and Logic* **12**, pp. 17–22.

Goncharov, S.S.: (1975), 'Autostability, and computable families of constructivizations' (Russian), *Algebra i Logika* **14**, pp. 647–680; trans. in *Algebra and Logic* **14**, pp. 392–409.

Goncharov, S.S.: (1977), 'The quantity of nonautoequivalent constructivizations', *Algebra and Logic* **16**, pp. 169–185

Goncharov, S.S.: (1980), 'Autostability of models and abelian groups' (Russian), *Algebra i Logika* **19**, pp. 23–44, 132; trans. in *Algebra and Logic* **19**, pp. 13–27.

Goncharov, S.S.: (1980a), 'The problem of the number of nonautoequivalent constructivizations' (Russian), *Algebra i Logika* **19**, pp. 621–639, 745; trans. in *Algebra and Logic* **19**, pp. 401–414.

Goncharov, S.S.: (1981), 'Groups with a finite number of constructivizations', *Soviet Math. Dokl.* **23**, pp. 58–61.

Goncharov, S.S.: (1982), 'Limit equivalent constructivizations' (Russian), in *Mathematical Logic and the Theory of Algorithms*, Trudy Inst. Mat. **2**, "Nauka" Sibirsk. Otdel., Novosibirsk, pp. 4–12.

Goncharov, S.S.: (1998), 'Autostable models and their algorithmic dimension', in Ershov et al. (1998), pp. 261–288.

Goncharov, S.S., Dzgoev, V.D.: (1980), 'Autostability of models' (Russian), *Algebra i Logika* **19**, pp. 45–58; trans. in *Algebra and Logic* **19**, pp. 28–37.

Goncharov, S.S., Khoussainov, B.: (1997), 'On the spectrum of the degrees of decidable relations' (Russian), *Dokl. Akad. Nauk* **352**, pp. 301–303; trans. in *Dokl. Math.* **55**, pp. 55–57.

Goncharov, S.S., Lempp, S., Solomon, D.R.: (2002) 'The computable dimension of ordered Abelian groups', to appear.

Goncharov, S.S., Molokov, A.V., Romanovskii, N.S.: (1989), 'Nilpotent groups of finite algorithmic dimension', *Sib. Math. J.* **30**, pp. 63–68.

Griffor, E. (ed.): (1999), *Handbook of Computability Theory*, North-Holland, Amsterdam-New York..

Harizanov, V.S.: (1987), *Degree spectrum of a Recursive Relation on a Recursive Structure*, Ph.D. Thesis, University of Wisconsin, Madison WI.

Harizanov, V.S.: (1991), 'Some effects of Ash-Nerode and other decidability conditions on degree spectra', *Ann. Pure Appl. Logic* **55**, pp. 51–65.

Harizanov, V.S.: (1993), 'The possible Turing degree of the nonzero member in a two element degree spectrum', *Ann. Pure Appl. Logic* **60**, pp. 1–30.

Harizanov, V.S.: (1998), 'Pure computable model theory', in Ershov et al. (1998), pp. 3–114.

Hirschfeldt, D.: (1999), *Degree Spectra of Relations on Computable Structures*, Ph.D. Thesis, Cornell University.

Hirschfeldt, D.: (2000), 'Degree Spectra of Relations on Computable Structures', *Bull. Symb. Logic* **6**, pp. 192–212.

Hirschfeldt, D.: (2001), 'Degree spectra of intrinsically c.e. relations', *Journal of Symbolic Logic* **66**, pp. 441–469.

Hirschfeldt, D.: (2002), 'Degree spectra of relations of finite computable dimension', *Ann. Pure Applied Logic*, to appear.

Hirschfeldt, D., Khoussainov, B., Shore R.A.: (2003), 'From computable categoricity to computable dimension ω by expansion by a constant', to appear.

Hirschfeldt, D., Khoussainov, B., Shore R.A., Slinko, A.: (2002), 'Degree spectra and computable categoricity in algebraic structures', *Ann. Pure Applied Logic*, to appear.

Hjorth, G., Kechris, A.S.: (1996), 'Borel equivalence relations and classifications of countable models', *Ann. of Pure and Applied Logic* **82**, pp. 221–272.

Hodges, W.: (1993), *Model Theory*, Encyclopedia of Mathematics and its Applications **42**, Cambridge University Press, Cambridge, England.

Khoussainov, B., Shore, R.A.: (1998), 'Computable Isomorphisms, Degree Spectra of Relations, and Scott Families', *Ann. of Pure and Applied Logic* **93**, pp. 153–193.

Khoussainov, B., Shore R.A.: (1999), 'Effective model theory: the number of models and their complexity', in S.B. Cooper, J.K. Truss (eds.), *Models and Computability, Invited papers from Logic Colloquium '97*, LMSLNS **259**, Cambridge University Press, Cambridge, England, pp. 193–240.

Khoussainov, B., Shore, R.A.: (2000), 'On the Solution of the Goncharov-Ash Problem and the Spectrum Problem in the Theory of Computable Models', *Dokl. Akad. Nauk SSSR* **371**, pp. 30–31; trans. *Doklady Math.* **61**, pp. 178–179.

Kudinov, O.: (1997), 'An integral domain with finite algorithmic dimension', personal communication.

LaRoche, P.: (1977), 'Recursively represented Boolean algebras', *Notices Amer. Math. Soc.* **24**, A-552.

Malcev, A.I.: (1965), 'On a correspondence between rings and groups', *Amer. Math. Soc. Translations (series 2)* **45**, pp. 221–232.

Mekler, A.H.: (1981), 'Stability of nilpotent groups of class 2 and prime exponent', *Journal of Symb. Logic* **46**, pp. 781–788.

Metakides, G., Nerode, A.: (1977), 'Recursively enumerable vector spaces', *Ann. Math. Logic* **11**, pp. 147–171.

Metakides, G., Nerode, A.: (1979), 'Effective content of field theory', *Ann. Math. Logic* **17**, pp. 289–320.

Millar, T.: (1986), 'Recursive categoricity and persistence', *J. Symbolic Logic* **51**, pp. 430–434.

Millar, T.: (1999), 'Abstract recursive model theory', in Griffor (1999), pp. 507–532.

Nurtazin, A.T.: (1974), 'Strong and weak constructivizations and computable families' (Russian), *Algebra i Logika* **13**, pp. 311–323, 364; trans. in *Algebra and Logic* **13**, pp. 177–184.

Remmel, J.B.: (1981), 'Recursive isomorphism types of recursive Boolean algebras', *J. Symbolic Logic* **46**, pp. 572–594.

Remmel, J.B.: (1981a), 'Recursively categorical linear orderings', *Proc. Amer. Math. Soc.* **83**, pp. 387–391.

Simpson, S.G.: (1999), *Subsystems of Second Order Arithmetic*, Springer-Verlag, Berlin.

Slaman, T.A.: (1998), 'Relative to any nonrecursive set', *Proc. Amer. Math. Soc.* **126**, pp. 2117–2122.

Wehner, S.: (1998), 'Enumerations, countable structures and Turing degrees', *Proc. Amer. Math. Soc.* **126**, pp. 2131–2139.

IV

AXIOMATIC SET THEORY

EFFECTIVE CARDINALITY

Greg Hjorth
Department of Mathematics
UCLA, LA CA90095-1555
greg@math.ucla.edu

In this brief note I want to talk a little about a fairly recent development in modern set theory. When I say recent, I mean that it dates back to some time around the late 1980's, at which point I understand it had Alexander Kechris and Alain Louveau as its main advocates. In an earlier form there are hints of these ideas in work of Harvey Friedman, who had tried to seriously consider what the world must look like to a mathematician who declares that the only things in their ontology are the objects which can be viewed as *Borel*. And Vladimir Kanovei in Kanovei (1991) has pointed to even earlier precursors of this idea in the works of the classical analysts. But even granting such instances I think it was not until Harrington, Kechris and Louveau (1990) and the papers which followed that set theorists began a systematic study and that the central definitions were isolated.

Usually when one talks about a new idea in a mathematical discipline such as set theory, one has in mind a technique which enables old problems to be solved for the first time, or at the very least provides new proofs for previously known theorems. While there perhaps have been a few isolated cases of this taking place, the new idea I want to discuss represents instead a new way of looking at existing objects. It represents a different way of thinking about "how many" objects there are in a collection when that collection is infinite.

Before going any further we should at least consider the traditional notion of cardinality which operates among set theorists. The formulation of this notion at the end of the 19th century represents George Cantor's great triumph, and it has been justifiably dominant ever since.

1. Cardinality in ZFC

The usual notion of cardinality rests heavily on our ability to *well order* any given set. One can then compare the cardinality of sets by, in effect, comparing

P. Gärdenfors, J. Woleński and K. Kijania-Placek (eds.), In the Scope of Logic, Methodology and Philosophy of Science, Vol. I, 99-115.

their shortest possible well ordering. Rather than become involve in defining the notion of a well order, I will instead take a different but equivalent course through the definitions.

Definition 1.1 A set α is said to be an *ordinal* if:

1. it is *transitive* – that is to say, if $x \in \alpha$ and $y \in x$, then $y \in \alpha$;
2. is linearly ordered by the membership relation – that is to say, if γ, δ are in α, then either

   $\gamma \in \delta$, or

   $\gamma = \delta$, or

   $\delta \in \gamma$.

It turns out that any two ordinals α and β are *comparable*: that is to say, either $\alpha = \beta$, or

$$\alpha \in \beta,$$

or

$$\beta \in \alpha.$$

Moreover an initial segment of the ordinals are provided by the usual (finite) counting numbers.

Examples of ordinals 1.2. $\emptyset$, the set having no members, which set theorists customarily identify with 0.

$1 = \{0\}$, the set whose only member is 0.

$2 = \{0, 1\}$, gives the usual set theoretical definition of 2. Then we keep going with $3 = \{0, 1, 2\}$, $4 = \{0, 1, 2, 3\}$, and so on.

We reach the first infinite ordinal with the set of natural numbers:

$$\omega = \{0, 1, 2, \ldots\}.$$

This again leads to a whole new ladder of ordinals:

$$\omega + 1 = \{0, 1, 2, \ldots, \omega\},$$

$$\omega + 2 = \{0, 1, 2, \ldots, \omega, \omega + 1\},$$

$$\omega + 3 = \{0, 1, 2, \ldots, \omega, \omega + 1, \omega + 2\},$$

and onwards:

$$\omega + \omega = \{0, 1, 2, \ldots, \omega, \omega + 1, \omega + 2, \ldots\},$$

$$\omega+\omega+1=\{0,1,2,\ldots,\omega,\omega+1,\omega+2,\ldots,\omega+\omega\},$$

$$\omega\times\omega=\{0,1,2,\ldots,\omega,\omega+1,\ldots\omega+\omega,\ldots\omega+\omega-\omega,\ldots\},$$

ad infinitum.

Definition 1.3 A function $f:X\to Y$ is said to be a *bijection* if

1. f is one-to-one: for all $x_1,x_2\in X$ with $x_1\neq x_2$ we have

$$f(x_1)\neq f(x_2);$$

2. f is onto: for all $y_0\in Y$ there exists some $x_0\in X$ with

$$f(x_0)=y_0.$$

So for example, if we rummaged through my apartment and collected all the left shoes together in a set X and all the right shoes into a set Y, then hopefully we could provide a bijection by associating to each left shoe

$$x_0$$

the corresponding right shoe

$$f(x_0)$$

which one needs to form a matching pair.

Definition 1.4 An ordinal is said to be a *cardinal* if it can not be placed in a bijection with any earlier ordinal.

All the finite ordinals of the $n=\{0,1,2,\ldots,n-1\}$ are cardinals.

$$\omega=\{0,1,2,3,\ldots\}$$

is the first infinite cardinal.

On the other hand $\omega+1=\{0,1,2,\ldots,\omega\}$ is *not* a cardinal since we may place it in a bijection with ω:

$$f(\omega)=0,$$

$$f(0)=1,$$

$$f(1)=2,$$

and more generally

$$f(n) = n + 1.$$

Essentially as a consequence of the axiom of choice one has:

Fact 1.5 Every set X may be placed in a bijection with some ordinal.

Definition 1.6 We let the *cardinality* of a set X be the least ordinal with which it may be placed in a bijection. We denote that least ordinal by $|X|$.

We say that a set has *cardinality n* if it may be placed in a bijection with the finite set $n = \{0, 1, 2, \ldots, n-1\}$. We say it has cardinality $\aleph_0$ if it may be placed in a bijection with ω, that it has cardinality $\aleph_1$ if it may be placed in a bijection with the least cardinal after ω, $\aleph_2$ if it may be placed in a bijection with the least cardinal after $\aleph_1$, and so on.

Many examples of infinite sets have size $\aleph_0$:

Example 1.7

1. The set of possible poems in English has size $\aleph_0$. (Assume here we raise no objection to poems going arbitrarily long.)

2. The set of integers, positive and negative, has cardinality $\aleph_0$. For instance I can take as a bijection $f(0) = 0$,

$$f(-n) = 2n + 1,$$

$$f(n) = 2n + 2.$$

3. The set of all rationals ($\frac{n}{m}$ where n, m are integers, suitably reduced) has cardinality $\aleph_0$.

However not all important infinite sets have size $\aleph_0$. At the end of the last century George Cantor showed that

$$|\mathbb{R}| \geqslant \aleph_1.$$

In other words, the least ordinal which may be placed in a bijection with the set of all real numbers is strictly bigger than ω.

Cantor went on to conjecture the *continuum hypothesis*, to the effect that the set of reals has size exactly $\aleph_1$. The issue of the truth of this conjecture became the first of the famous Hilbert problems.

Note here that the cardinality of $\mathbb{R}$ is the same as that $\mathcal{P}(\omega)$, the collection of sets of the natural numbers. (For the purposes of the rest of this article I want to largely identify the real number line $\mathbb{R}$ with $\mathcal{P}(\omega)$, the set of all subsets of the natural numbers, ω. These two objects are not literally the same, but one can find definable bijections between the two, and from the point of view of set theory they have essentially the same status.)

It turns out that ZFC, the usual axiomatization of mathematical reasoning is unable to provide an answer.

Theorem 1.8 (Gödel 1938) ZFC cannot prove $|\mathbb{R}| > \aleph_1$.

In other words, the continuum hypothesis is consistent with ZFC. A quarter of a century later it was shown to be independent.

Theorem 1.9 (Cohen 1963) ZFC cannot prove $|\mathbb{R}| = \aleph_1$.

Part of the difficulty with this problem is that although the axiom of choice promises us that there will exist some bijection between some ordinal and $\mathbb{R}$, it gives no information at all about the nature of this bijection or this ordinal. It is a pure existence axiom.

In the second half of the century various set theorists for various reasons have proposed $|\mathbb{R}| = \aleph_2$, and it may be that eventually some combination of plausibility arguments and heuristic considerations will lead to wide spread acceptance of this as *true*.

None the less it does seem that the status of the continuum hypothesis will not be settled in the way that mathematical questions are usually settled, by either a proof or counterexample being presented under ZFC (or perhaps ZFC extended by some large cardinal assumption).

2. The notion of effective cardinality

A rather different notion is provided by restricting our attention to only those sets and functions which lay some claim on being *reasonably definable* from the most concrete of mathematical objects, such as the natural numbers or the subsets of the natural numbers. This is the *new idea* I want exposit.

Clearly it will take some effort to make precise what we mean by *reasonably definable* from $\mathcal{P}(\omega)$. Here there are two quite different definitions in wide currency. The first of these is much more generous than the second, but the structural properties of both approaches are similar.

2.1. Cardinality in $L(\mathbb{R})$

We let $L(\mathbb{R})$ denote the smallest class of sets containing all the reals, all the ordinals, and closed under certain very basic set theoretical operations. (Such as the pairing function: $(x,y) \mapsto \{x,\{x,y\}\}$. Product. Boolean operations. And such like.) So in some loose sense this might be considered the universe of everything we can define in a ground up fashion from the reals.

It turns out that $L(\mathbb{R})$ models all the axioms of ZFC, except possibly choice. Under the additional and widely supported axiom of $\mathrm{AD}^{L(\mathbb{R})}$ the model $L(\mathbb{R})$ has a canonical structure, and there are no known independence results along the lines of the continuum hypothesis.

Definition 2.1.1 For $A, B \in L(\mathbb{R})$ we say that the $L(\mathbb{R})$-cardinality of A is no greater than that of B, written

$$|A|_{L(\mathbb{R})} \leqslant |B|_{L(\mathbb{R})},$$

if there is a one-to-one function $\rho \in L(\mathbb{R})$

$$\rho : A \hookrightarrow B.$$

Similarly $|A|_{L(\mathbb{R})} < |B|_{L(\mathbb{R})}$ if

$$|A|_{L(\mathbb{R})} \leqslant |B|_{L(\mathbb{R})}$$

holds but $|B|_{L(\mathbb{R})} \leqslant |A|_{L(\mathbb{R})}$ fails.

One of the original impetuses for set theory was the study of the real number line, and indeed it turns out that many of the objects of greatest interest can be realized as quotients of $\mathbb{R}$ by an appropriately chosen equivalence relation. Moreover it turns out that every set in $L(\mathbb{R})$ can be placed in a bijection with a set of the form

$$\bigcup_{\alpha \in \kappa} A_\alpha,$$

where the $(A_\alpha)_{\alpha\in\kappa}$ is a sequence in $L(\mathbb{R})$ for which there is (again in $L(\mathbb{R})$) a corresponding sequence $(E_\alpha)_{\alpha\in\kappa}$ of equivalence relations on $\mathbb{R}$ and bijections

$$\pi_\alpha : X_\alpha \cong \mathbb{R}/E_\alpha.$$

Since the ordinals are relatively well understood, we thus have that the study of cardinality in $L(\mathbb{R})$ is largely the study of the quotients of $\mathbb{R}$ by equivalence relations. (For the proofs of this and other basic facts about cardinality in $L(\mathbb{R})$ the reader could look in Hjorth (1995) and chapter nine of Hjorth (2000).)

Here I am using $\mathbb{R}/E$ to indicate the collection of E-equivalence classes: $\{[x]_E : x \in \mathbb{R}\}$, where for each $x \in \mathbb{R}$ we write

$$[x]_E = \{y \in \mathbb{R} : yEx\}$$

for the equivalence class of x.

Examples 2.1.2 In all of the following, we have an equivalence relation on a reasonably nice space X. In most cases the $X = \mathbb{R}$; at the very least we require X to be a space which exists in $L(\mathbb{R})$ and which $L(\mathbb{R})$ can place in a bijection with the reals.

1. id($\mathbb{R}$), the identity relation on $\mathbb{R}$, setting each x to be only equivalent to itself;

2. E_v, the Vitali equivalence relation, given by

$$xE_vy$$

   if $x-y$ is a rational number;

3. E_0 on 2^ω, the space of infinite binary sequences, given by

$$\langle x_0, x_1, x_2 \ldots \rangle = \vec{x}E_0\vec{y} = \langle y_0, y_1, \ldots \rangle$$

   if from some point on they agree:

$$\exists N \in \omega \, \forall k > N(x_k = y_k);$$

4. we may naturally identify $\mathcal{P}(\omega)$ with 2^ω by associating with each $X \subset \omega$ its characteristic function:

$$f_X : \omega \to \{0, 1\},$$

   with $f_X(n) = 1$ if $n \in X$, $= 0$ otherwise. Then it is not hard to see that E_0 on 2^ω corresponds to the equivalence relation Fin, where

$$X \, \mathrm{Fin} \, Y$$

   if there are only finitely many natural numbers in exactly one of these sets; that is to say, their *symmetric difference* is finite.

Here we have

$$|\mathbb{R}|_{L(\mathbb{R})} < |\mathbb{R}/E_v|_{L(\mathbb{R})}.$$

Moreover one has

$$|\mathbb{R}/E_v|_{L(\mathbb{R})} \leqslant |\mathcal{P}(\omega)/\mathrm{Fin}|_{L(\mathbb{R})}$$

and

$$|\mathcal{P}(\omega)/\mathrm{Fin}|_{L(\mathbb{R})} \leqslant |\mathbb{R}/E_v|_{L(\mathbb{R})}.$$

In ZFC one obtains that the ordinals are arranged in a linear ordering, and thus the subclass of the ordinals corresponding to cardinals is again linearly ordered. For any two cardinals κ and λ, either they are equal, or one is bigger than the other.

The situation is completely different for $L(\mathbb{R})$. For instance one has that $|\mathbb{R}|_{L(\mathbb{R})}$ is incomparable with $\aleph_1$ in $L(\mathbb{R})$. More generally, there is no ordinal with the same cardinality as $\mathbb{R}$ inside $L(\mathbb{R})$.

It turns out that for reasonably simple equivalence relations, one has that the cardinality of $\mathbb{R}/E$ is less than $\mathbb{R}/F$ if and only if there is a kind of reduction of E to F.

Definition 2.1.3 A subset of $\mathbb{R}^2$, two dimensional euclidean space, is said to be *Borel* if it is generated from the collection of open squares of the form

$$\{(x_0,x_1) : r_0 < x_0 < s_0, r_1 < x_1 < s_1\},$$

by the operations of complementation, countable intersections (that is to say, if $(A_n)_{n\in\omega}$ are Borel, then so too is $\bigcap_{n\in\omega} A_n$), and countable unions. An equivalence relation E on $\mathbb{R}$ is said to be a *Borel equivalence relation* if the set $\{(x_0,x_1) : x_0 E x_1\}$ is Borel as a subset of $\mathbb{R}^2$.

All Borel sets appear in $L(\mathbb{R})$.

Definition 2.1.4 For E and F equivalence relations on $\mathbb{R}$ we write

$$E \leqslant_{L(\mathbb{R})} F$$

if there is a function $\pi : \mathbb{R} \to \mathbb{R}$ in $L(\mathbb{R})$ such that for all $x_1, x_2 \in L(\mathbb{R})$ we have

$$x_1 E x_2 \Leftrightarrow \pi(x_1) F \pi(x_2).$$

Fact 2.1.5 (Folklore) For E and F Borel equivalence relations one has

$$E \leqslant_{L(\mathbb{R})} F$$

if and only if

$$|\mathbb{R}/E|_{L(\mathbb{R})} \leqslant |\mathbb{R}/F|_{L(\mathbb{R})}.$$

Using a variation of the usual example of a set in $L(\mathbb{R})$ that allows no uniformization – namely, $\{(x,y) : x \notin \mathrm{OD}(y)\}$ – one can show that this is untrue for general $E, F \in L(\mathbb{R})$. While a set theorist would expect that some such counterexample should exist, it actually seems surprisingly difficult to construct. The somewhat technical argument I sketch in the appendix is the only one I know, and it rests heavily on a recent theorem due to Adams and Kechris.

2.2. Borel cardinality

Definition 2.2.1 A function $f:\mathbb{R}\to\mathbb{R}$ is said to be *Borel* if its graph

$$\{(x_0,x_1) : f(x_0)=x_1\}$$

is Borel as a subset of $\mathbb{R}^2$.

Any continuous function is Borel, and more generally the functions which typically arise in calculus, most undergraduate mathematics, and many branches of analysis, are Borel.

Given the broad role the notion of *Borel* plays in many branches of mathematics, it seems natural to restrict our attention from all the sets in $L(\mathbb{R})$ to just those that are Borel. We are then faced with the task of defining a notion of *Borel cardinality*. In view of various facts such as 2.1.5 most people in this field have come to accept the following explication.

Definition 2.2.2 For E and F Borel equivalence relations we say that $\mathbb{R}/E$ has *Borel cardinality* no greater than $\mathbb{R}/F$ if there is a Borel function $f:\mathbb{R}\to\mathbb{R}$ such that for all $x_1,x_2\in\mathbb{R}$

$$x_1Ex_2 \Leftrightarrow f(x_1)Ff(x_2).$$

In other words, there is a one-to-one function

$$\hat{f}:\mathbb{R}/E\hookrightarrow\mathbb{R}/F$$

which lifts up to a Borel function.

We write $E\leqslant_B F$ to indicate that there is such an f with for all $x_1,x_2\in\mathbb{R}$

$$x_1Ex_2 \Leftrightarrow f(x_1)Ff(x_2).$$

Naturally we may write $E<_B F$ if $E\leqslant_B F$ holds but $E<_B F$ fails.

Here one has

$$\mathrm{id}(\mathbb{R}) <_B E_v.$$

That is to say the quotient object arising from the Vitali equivalence relation has strictly bigger Borel cardinality than that of the reals. More generally all the results mentioned at 2.1.2, along with the overwhelming majority of specific cases that have been subject to serious study, hold for the Borel notion of cardinality exactly as they do for $L(\mathbb{R})$.

3. Why study this notion?

3.1. Distrust of the more abstract set theoretical objects

One might for instance take the position that the mathematical objects for which we have the strongest evidence are firstly the natural numbers, then the subsets of the natural numbers, which for our purposes can be identified with $\mathbb{R}$, and finally those objects which are easily produced from the reals, such as the Borel sets and quotients by Borel equivalence relations.

This is not quite a rejection of abstract objects, since it is not clear that the number 5 is any less abstract than the first "supercompact cardinal". Nevertheless there might be some interest in developing a parsimonious theory of cardinality, which refuses to borrow on any but the most concrete and well established of mathematical concepts.

3.2. Dissatisfaction with independence

One reaction to the independence of the continuum hypothesis might be a kind of despair. If we are unable to ever know the answer to this kind of question, then perhaps we should dispose of it altogether.

While it would be unreasonable to conclude that a problem is meaningless or somehow bogus simply because we have no method to determine the answer, and while it is not even accepted among set theorists that future principles of reasoning will never determine the ZFC cardinality of $\mathbb{R}$, it is also an attractive feature of the theory Borel cardinalities and of the theory of $L(\mathbb{R})$ cardinality that these are largely immunized against independence.

3.3. Recurrence of fundamental issues

It turns out that some of the questions arising in the study of Borel cardinalities have reappeared in many different contexts. I will give two examples, one from a mainstream mathematical field and one from the branch of mathematical logic known as *recursion* (or *computability*) theory.

3.3.1 Infinite dimensional group representations. In 1961 Glimm proved Mackey's conjecture, an outstanding open problem in the theory of infinite dimensional group representations, by showing that in some sense the representations of a locally compact group have an appropriately nice structure if and only if the irreducible representations considered up to unitary equivalence have Borel cardinality below $\mathbb{R}$. His oblique proof was eventually replaced by a much simpler argument due to Effros, which in turn reduced to the following:

Theorem (Effros) Let E be a Borel equivalence relation where we additionally assume:

1. E is the orbit equivalence relation induced by the continuous action of a Polish group; and
2. E is F_σ (defined by the countable union of closed sets in the product space).

Then either:

(I) $E \leqslant_B \mathrm{id}(\mathbb{R})$; or

(II) $E_v \leqslant_B E$.

In other words, E is either no more complicated than the identity relation on the reals or at least as complicated as the Vitali equivalence relation given by rational translation on $\mathbb{R}$.

In 1990, (Harrington, Kechris and Louveau 1990), Harrington, Kechris, and Louveau proved this without the additional assumptions 1. and 2. above for general Borel equivalence relations. Thus one has the reduction of a landmark theorem to an entirely general fact regarding Borel cardinality.

3.3.2 The structure of the Turing degrees. An important part of recursion theory is the study of subsets of ω considered up to *Turing equivalence*, where one might loosely think of two subsets as having the same Turing degree if they have same information, from the point of view of some machine that is allowed only finitary operations. This has further led in particular to the still open Martin's conjecture, which asserts that the functions in $L(\mathbb{R})$ from the Turing degrees to the Turing degrees have relatively few possibilities.

Letting $\cong_T$ denote the equivalence relation on Turing equivalence on subsets of ω (which for our purposes can be identified with elements of $\mathbb{R}$) Slaman and Steel proved a number of results, which in hindsight can be seen as impinging on the study of the Borel and $L(\mathbb{R})$ cardinality of $\omega/\cong_T$, as part of efforts to clarify Martin's conjecture.

Appealing to Effros' theorem, and certain easy facts about the low level Borel equivalence relations, one can obtain from their results that

$$E_v <_B \cong_T .$$

$$|\mathbb{R}/E_v|_{L(\mathbb{R})} < |\mathcal{P}(\omega)/\cong_T|_{L(\mathbb{R})}.$$

If either of these strict inequalities had failed to hold it would have at once lead to an immediate failure of Martin's conjecture.

Determining the exact $L(\mathbb{R})$ cardinality of

$$\mathcal{P}(\omega)/\cong_T$$

remains totally open and closely connected with the structure of the Turing degrees. Kechris has conjectured that it should be the same as the of $2^{F_2}/F_2$, where F_2 is the free group on two generators acting on the space

$$2^{F_2} = \{\varphi : F_2 \to \{0,1\}\}$$

by shift, and this competing conjecture would in turn refute the Martin conjecture.

Further discussion of this line of research can be found in Dougherty and Kechris (1999).

3.4. New ways of considering old knowledge

We already obtain some sense of this with the results of 3.3.2. One might want to view

$$E_v <_B \cong_T$$

as an indication that the Turing degrees are *complicated.* For instance it is already sufficient to show that there is no Borel linear ordering on the Turing degrees, and thus that incomparable degrees exist.

Let me mention one other result along these lines.

Definition 3.4.1 One defines the Borel hierarchy of subsets of $\mathbb{R}$ like so: The $\underset{\sim}{\Sigma}^0_1$ sets in $\mathbb{R}$ are the open sets – those that are the unions of intervals of the form

$$\{x : r_1 < x < r_2\}$$

where $r_1, r_2 \in \mathbb{R}$. Then the $\underset{\sim}{\Pi}^0_1$ sets are their complements.

More generally, for $\alpha < \aleph_1$, we let $\underset{\sim}{\Sigma}^0_\alpha$ be the countable unions of sets which are all $\underset{\sim}{\Pi}^0_\beta$ some $\beta < \alpha$ and then define $\underset{\sim}{\Pi}^0_\alpha$ to be those sets whose complements in $\mathbb{R}$ are $\underset{\sim}{\Sigma}^0_\alpha$.

Classically it was shown that for all $\alpha < \beta < \aleph_1$ we have

$$\underset{\sim}{\Sigma}^0_\alpha \neq \underset{\sim}{\Sigma}^0_\beta.$$

In Hjorth (1998) this was strengthened to a result about effective cardinality:

3.4.2 Theorem (Hjorth) For all $\alpha < \beta < \aleph_1$ we have

$$|\underset{\sim}{\Sigma}^0_\alpha|_{L(\mathbb{R})} < |\underset{\sim}{\Sigma}^0_\beta|_{L(\mathbb{R})}.$$

3.5. The intrinsic interest of the structure of the effective cardinals

At the low levels, the Borel equivalence relations and the cardinals of $L(\mathbb{R})$ display a dramatic and interesting structure, encapsulated in the following "dichotomy theorems". In the first and third of these one has that the cardinality of $\mathbb{R}$ is the canonical obstruction to being well orderable, where we can respectively consider well orders in the Borel or $L(\mathbb{R})$ context. The second and fourth are more subtle; the second at least can be viewed as providing an analysis of when we may think of a Borel cardinality as corresponding to a subset of the reals.

It is also interesting here to observe the parallels between our different contexts. 3.5.3 and 3.5.4 can be thought of as $L(\mathbb{R})$ echoes of the original theorems 3.5.1 and 3.5.2 for Borel.

Theorem 3.5.1 (Silver) Let E be a Borel equivalence relation. Then exactly one of the following two things must happen:

(I) $E \leqslant_B \mathrm{id}(\omega)$.

(II) $\mathrm{id}(\mathbb{R}) \leqslant_B E$.

In other words, there are either only countably many equivalence classes or there are as many equivalence classes as real numbers.

Theorem 3.5.2 (Harrington, Kechris, Louveau) Let E be a Borel equivalence relation. Then exactly one of the following two things must happen:

(I) $E \leqslant_B \mathrm{id}(\mathbb{R})$.

(II) $E_v \leqslant_B E$.

In other words, E is either as complicated as the Vitali equivalence relation or the collection of E-equivalence classes may be identified with a subset of $\mathbb{R}$.

Theorem 3.5.3 (Woodin) Let $A \in L(\mathbb{R})$. Then exactly one of the following two things must happen:

(I) $|A|_{L(\mathbb{R})} \leqslant |\alpha|_{L(\mathbb{R})}$, some ordinal α.

(II) $|\mathbb{R}|_{L(\mathbb{R})} \leqslant |A|_{L(\mathbb{R})}$.

So in other words, every set in $L(\mathbb{R})$ is either well orderable or has a subset which admits a bijection with $\mathbb{R}$.

Theorem 3.5.4 (Hjorth) Let $A \in L(\mathbb{R})$. Then exactly one of the following two things must happen:

(I) $|A|_{L(\mathbb{R})} \leqslant |\mathcal{P}(\alpha)|_{L(\mathbb{R})}$.

(II) $|\mathcal{P}(\omega)/\mathrm{Fin}|_{L(\mathbb{R})} = |\mathbb{R}/E_\nu|_{L(\mathbb{R})} \leqslant |A|_{L(\mathbb{R})}$.

In other words, any set in $L(\mathbb{R})$ is either injectable into the subsets of some ordinal or it is as complicated as the subsets of ω considered up to finite difference.

4. Further results

Globally the Borel equivalence relations are a hopeless mess. For instance

Theorem 4.1 (Louveau, Veličković) There is an assignment

$$S \mapsto E_S$$

of Borel equivalence relations to subsets of ω such that for all $S, T \subset \omega$ we have that $T \setminus S$ (the elements of T not in S) is finite if and only if

$$E_T \leqslant_B E_S.$$

The overall structure seems to be incomprehensible.

One response would be to leave all structural questions to the side. Accept the effective cardinals on their face value, and instead try to categorize various classes of mathematical objects in terms of the more prominent cardinalities.

This has been discussed extensively in Hjorth (2000), and papers such at Hjorth and Kechris (1996), Thomas and Veličković (200x), Hjorth and Kechris (200x), Camerlo and Gao (200x), and Dougherty and Kechris (1999) can indeed be thought of as trying to calculate the effective cardinality of certain standard mathematical objects.

An alternative approach is to continue the structural investigation but with a much more limited focus; we do not try to enumerate every possible effective cardinal, but instead we can try to understand certain crucial parts and dividing lines. For instance we can try to understand which cardinals exist below some specific effective cardinal or what kinds of obstructions prevent us, for instance, reducing a cardinal to *isomorphism on countable structures,* or some *countable Borel equivalence relation,* that is to say a Borel equivalence relation all of whose classes are countable.

Question 4.2 (Hjorth, Kechris) Let $\mathrm{Mod}(\sigma)$ be the space of countable models of a (possibly infinitary) sentence σ. Suppose the isomorphism relation $\cong_{\mathrm{Mod}(\sigma)}$ is Borel.

Must exactly one of the following hold?

(I) There is a countable Borel equivalence relation E with

$$\cong_{\mathrm{Mod}(\sigma)} \leqslant_B E.$$

(II) For $(E_v)^\omega$ the infinite product of the Vitali equivalence relation viewed as an equivalence relation on $\mathbb{R}^\omega$ we have

$$(E_v)^\omega \leqslant_B \cong_{\mathrm{Mod}(\sigma)} .$$

Many special cases of this have been verified. At most one of 1. and 2. can hold and there is no Borel cardinality strictly between E_v and $(E_v)^\omega$.

Presumably if this is true then there should be an $L(\mathbb{R})$ version for the hereditarily well orderable sets, but this has not been seriously examined.

Appendix

Here we give the counterexample from Section 2.1. The necessary terminology and background, along with a more comprehensive list of references, can be found in Adams and Kechris (1999), Hjorth and Kechris (1995), Hjorth (2000), and Harrington, Kechris and Louveau (1990).

Example Following Adams and Kechris (1999) let $(E_x)_{x\in\mathbb{R}}$ be a family of equivalence relations on $\mathbb{R}$ with associated Borel probability measures $(\mu_x)_{x\in\mathbb{R}}$ on $\mathbb{R}$ such that:

1. These sequences exist in $L(\mathbb{R})$.
2. Each E_x has all its equivalence classes countable.
3. The relation $\{(x,y,z) : yE_xz\}$ is Borel.
4. For x_1,x_2 distinct we have $E_{x_1}|_M$ is not Borel reducible to E_{x_2} for any μ_{x_1}-measure one set $M \subset \mathbb{R}$.
5. Each E_x is μ_x-ergodic, in the sense that if $B \subset \mathbb{R}$ is E_x-invariant and μ_x-measurable, then it is either null or conull.
6. Each E_x equivalence class is null.

We then can define two different ways to thread this family into a single equivalence relation. First we may define E as an equivalence relation on $\mathbb{R}\times\mathbb{R}$ by

$$(x_1,y_1)E(x_2,y_2)$$

if $x_1 = x_2$ and $y_1E_{x_1}y_2$. Then we can define F on $\mathbb{R}\times\mathbb{R}\times\mathbb{R}$ by

$$(x_1,y_1,z_1)F(x_2,y_2,z_2)$$

if either

(i) z_1 is ordinal definable from (x_1,y_1) and z_2 is ordinal definable from (x_2,y_2); or

(ii) $z_1 \notin \mathrm{OD}(x_1,y_1)$, $z_2 \notin \mathrm{OD}(x_2,y_2)$, and $x_1 = x_2$, and $y_1 E_{x_1} y_2$.

Clearly

$$|\mathbb{R}\times\mathbb{R}/E|_{L(\mathbb{R})} \leqslant |\mathbb{R}\times\mathbb{R}\times\mathbb{R}/F|_{L(\mathbb{R})}.$$

On the other hand if

$$\theta : \mathbb{R}\times\mathbb{R} \to \mathbb{R}\times\mathbb{R}\times\mathbb{R}$$

performed a reduction of the equivalence relations in $L(\mathbb{R})$, then it would be $\mathrm{OD}(x)$, some $x \in \mathbb{R}$. We then have by 2. and 3. above that

$$\forall^{\mu_x} y \in \mathbb{R}(\theta(x,y) = (x,\rho_0(y),\rho_1(y))),$$

some $\rho_0(y),\rho_1(y) \in \mathbb{R}$. Let N be this μ_x-measure 1 set.

Claim For any $y_0 \in N$, the set

$$\{y : \rho_0(y) = y_0\}$$

is included in $\mathrm{OD}(x,y_0)$.

Proof of claim Since E_x is an equivalence relation with all its equivalence classes countable, it follows that $\{y : \rho_0(y) = y_0\}$ is a countable set. Since this set is countable, ordinal definable from (x,y_0), and we are working in $L(\mathbb{R})$ under the assumption of AD, it follows that it is included in $\mathrm{OD}(x,y_0)$.

(Claim□)

But now we can just choose z_1,z_2 in this conull set N with

$$z_1 \neg E_x z_2,$$

and hence

$$(x,\rho_0(z_1),\rho_1(z_1))\neg F(x,\rho_0(z_2),\rho_1(z_2)),$$

and thus for some $i \in \{1,2\}$ we have

$$\rho_1(z_i) \notin \mathrm{OD}(x,\rho_0(z_i)),$$

with a contradiction to the claim.

References

Adams, S., Kechris, A.S.: (1999), 'Linear algebraic groups and countable Borel equivalence relations,' preprint.

Camerlo, R., Gao, S.: (200x), 'The classification of countable Boolean algebras,' to appear.

Cantor, G.: (1874), 'Über eine Eigenschaft des Ingebriffs aller reellen algebraischen Zahlen', *J. f. Math.* **77**, pp. 258–262.

Cohen, P.: (1963), 'The independence of the continuum hypothesis,' *Proceedings of the National Academy of Sciences of the United States of America*, vol. 50, pp. 1143–1148.

Dougherty, R. and Kechris, A.S.: (1999), 'How many Turing degrees are there?,' to appear in *Proceedings of the AMS Summer Research Conference on Computability Theory and Applications*, American Mathematical Society, Boulder, Colorado.

Effros, E.: (1975), 'Transformation groups and C^*-algebras,' *Annals of Mathematics* **81**, pp. 38-55.

Glimm, J.: (1961), 'Type I C^*-algebras', *Annals of Mathematics* **73**, pp. 572–612.

Gödel, K.: (1938), 'The consistency of the axiom of choice and the generalized continuum hypothesis,' *Proceedings of the National Academy of Sciences of the United States of America*, vol. 24, pp. 556-557.

Harrington, L.A., Kechris, A.S., Louveau, A.: (1990), 'A Glimm-Effros dichotomy for Borel equivalence relations,' *Journal of the American Mathematical Society* **3**, pp. 903-928.

Hjorth, G.: (1995), 'A dichotomy for the definable universe', *Journal of Symbolic Logic* **60**, pp. 1199–1207.

Hjorth, G.: (1998), 'An absoluteness principle for Borel sets,' *Journal of Symbolic Logic* **63**, pp. 663–693.

Hjorth, G.: (2000), *Classification and orbit equivalence relations,* American Mathematical Society, Rhode Island.

Hjorth, G., Kechris, A.S.: (1995), 'Analytic equivalence relations and Ulm-type classifications,' *Journal of Symbolic Logic* **60**, pp. 1273-1300.

Hjorth, G., Kechris, A.S.: (1996), 'Borel equivalence relations and classifications of countable models,' *Annals of Pure and Applied Logic* **82**, pp. 221–272.

Hjorth, G., Kechris, A.S.: (1997), 'New dichotomies for Borel equivalence relations,' *Bulletin of Symbolic Logic* **3**, pp. 329–346.

Hjorth, G., Kechris, A.S.: (200x), 'The complexity of the classification of Riemann surfaces and complex manifolds,' to appear in the *Illinois Journal of Mathematics*.

Kanovei, V.: (1991), 'The cardinality of the set of equivalence classes,' *Mathematik Zametki* **49**, pp. 55–62.

Louveau, A., Veličković, B.: (1994), 'A note on Borel equivalence relations,' *Proceedings of the American Mathematical Society* **120**, pp. 255–259.

Silver, J.H.: (1980), 'Counting the number of equivalence classes of Borel and coanalytic equivalence relations,' *Annals of Mathematical Logic* **18**, pp. 1–28.

Slaman, T., Steel, J.: (1988), 'Definable functions on degrees,' *Cabal Seminar 81-85*, Springer Lecture Notes, vol. 1333, pp. 37-55.

Thomas, S., Veličković, B.: (200x), 'On the complexity of the isomorphism relation for finitely generated groups,' to appear in the *Journal of Algebra*.

V

LOGIC AND COMPUTATION

ON HOARE LOGIC, KLEENE ALGEBRA, AND TYPES

Dexter Kozen
Department of Computer Science
Cornell University
Ithaca, New York 14853-7501, USA
kozen@cs.cornell.edu

Abstract We show that propositional Hoare logic is subsumed by the type calculus of typed Kleene algebra augmented with subtypes and typecasting. Assertions are interpreted as typecast operators. Thus Hoare-style reasoning with partial correctness assertions reduces to typechecking in this system.

1. Introduction

In previous work (Kozen 2000), we have shown that Kleene algebra with tests (KAT) subsumes propositional Hoare logic (PHL). Thus the specialized syntax and deductive apparatus of Hoare logic are inessential and can be replaced by simple equational reasoning. We have also shown that KAT provides a complete deductive system for Hoare-style inference rules involving partial correctness assertions and that PHL is *PSPACE*-complete.

In other recent work (Kozen 1998), we have introduced a simple and natural type system for Kleene algebra (KA) in which objects have types $\mathsf{s} \to \mathsf{t}$. The use of types was motivated by the desire to handle nonsquare matrices, although there are other useful interpretations.

In this paper we extend the type system of KA to KAT by adding rules for subtypes and typecasting. Tests are interpreted as typecast operators. We then observe that a Hoare partial correctness assertion $\{b\}\,p\,\{c\}$ can be regarded as a type judgement $p : \mathsf{b} \to \mathsf{c}$. We show that under this encoding, all the inference rules of PHL can be derived in the type calculus. Thus Hoare-style reasoning with partial correctness assertions is essentially a matter of typechecking in this system. Moreover, the typing rules of KAT can be soundly encoded in pure (typeless) KAT.

The interplay of types and assertions, programs and proofs has been observed in many contexts and at many levels, from constructive mathematics

P. Gärdenfors, J. Woleński and K. Kijania-Placek (eds.), In the Scope of Logic, Methodology and Philosophy of Science, Vol. I, 119-133.

and programming language semantics to program analysis and compiler certification. Perhaps the most far-reaching example is the Curry-Howard isomorphism, or propositions-as-types principle, and its ramifications (Bates and Constable 1985, Constable 1988, Constable 1989, Girard 1971, Girard, Lafont and Taylor 1989, Martin-Löf 1982, Reynolds 1974). The present work reveals yet another aspect of this phenomenon.

2. Definitions

2.1. Hoare logic

Hoare logic is a system for reasoning inductively about well-structured programs. Comprehensive surveys can be found in Apt (1981), Cousot (1990).

A common choice of programming language in Hoare logic is the language of **while** programs. The first-order version of this language contains a simple assignment $x := e$, conditional test **if** b **then** p **else** q, sequential composition $p\,;q$, and a looping construct **while** b **do** p.

The basic assertion of Hoare logic is the *partial correctness assertion* (PCA) $\{b\}\,p\,\{c\}$, where b and c are formulas and p is a program. Intuitively, this statement asserts that whenever b holds before the execution of the program p, then if and when p halts, c is guaranteed to hold of the output state. It does not assert that p must halt.

Semantically, programs p in Hoare logic are usually interpreted as binary input/output relations $p^{\mathcal{M}}$ on a domain of computation $\mathcal{M}$, and assertions b are interpreted as subsets $b^{\mathcal{M}}$ of $\mathcal{M}$ (Cook 1978, Pratt 1978). The definition of the relation $p^{\mathcal{M}}$ is inductive on the structure of p; for example, $(p\,;q)^{\mathcal{M}} = p^{\mathcal{M}} \circ q^{\mathcal{M}}$, the ordinary relational composition of the relations corresponding to p and q. The meaning of the PCA $\{b\}\,p\,\{c\}$ is the same as the meaning of the DL formula $b \to [p]\,c$, where $\to$ is ordinary propositional implication and the modal formula $[p]\,c$ is interpreted in $\mathcal{M}$ as the set of states s such that for all $(s,t) \in p^{\mathcal{M}}$, the output state t satisfies c.

Hoare logic provides a system of specialized rules for deriving valid PCAs, one for each programming construct. The verification process is inductive on the structure of programs. The traditional Hoare inference rules are:

Assignment rule:

$$\{b\,[x/e]\,\}\,x := e\,\{b\} \tag{1}$$

Composition rule:

$$\frac{\{b\}\,p\,\{c\}, \quad \{c\}\,q\,\{d\}}{\{b\}\,p\,;q\,\{d\}} \tag{2}$$

Conditional rule:

$$\frac{\{b \wedge c\}\, p\, \{d\}, \quad \{\neg b \wedge c\}\, q\, \{d\}}{\{c\}\ \textbf{if}\ b\ \textbf{then}\ p\ \textbf{else}\ q\ \{d\}} \tag{3}$$

While rule:

$$\frac{\{b \wedge c\}\, p\, \{c\}}{\{c\}\ \textbf{while}\ b\ \textbf{do}\ p\ \{\neg b \wedge c\}} \tag{4}$$

Weakening rule:

$$\frac{b' \to b, \quad \{b\}\, p\, \{c\}, \quad c \to c'}{\{b'\}\, p\, \{c'\}}. \tag{5}$$

The propositional fragment of Hoare logic (PHL) consists of atomic proposition and program symbols, the usual propositional connectives, **while** program constructs, and PCAs built from these. Atomic programs are interpreted as arbitrary binary relations on a set $\mathcal{M}$ and atomic propositions are interpreted as arbitrary subsets of $\mathcal{M}$. The deduction system consists of the composition, conditional, while, and weakening rules (2)–(5) and propositional logic. The assignment rule (1) is omitted, since there is no first-order relational structure over which to interpret program variables; in practice, its role is played by PCAs over atomic programs that are postulated as assumptions.

2.2. Kleene algebra

Kleene algebra (KA) is the algebra of regular expressions (Conway 1971, Kleene 1956). The axiomatization used here is from Kozen (1994). A *Kleene algebra* is an algebraic structure $(K, +, \cdot, ^*, 0, 1)$ that is an idempotent semiring under $+, \cdot, 0, 1$ and that satisfies

$$1 + pp^* = p^* \tag{6}$$
$$1 + p^*p = p^* \tag{7}$$
$$q + pr \leqslant r \to p^*q \leqslant r \tag{8}$$
$$q + rp \leqslant r \to qp^* \leqslant r \tag{9}$$

where $\leqslant$ refers to the natural partial order on K:

$$p \leqslant q \stackrel{\text{def}}{\iff} p + q = q.$$

The operation $+$ gives the supremum with respect to the natural order $\leqslant$. Instead of (8) and (9), we might take the equivalent axioms

$$pr \leqslant r \to p^*r \leqslant r \tag{10}$$
$$rp \leqslant r \to rp^* \leqslant r. \tag{11}$$

These axioms say essentially that * behaves like the Kleene asterate operator of formal language theory or the reflexive transitive closure operator of relational algebra.

Terms in the language of Kleene algebra are built from variables $x, y, \ldots$, binary operators $+$ and $\cdot$, unary operator *, and constants 0 and 1. Terms are often called *regular expressions* and are denoted $p, q, \ldots$. Atomic formulas are equations between terms. The expressions $p \leqslant q$ and $q \geqslant p$ are abbreviations for $p + q = q$.

Kleene algebra is a versatile system with many useful interpretations in semantics, verification, and algorithm design and analysis. Standard models include the family of regular sets over a finite alphabet; the family of binary relations on a set; and the family of $n \times n$ matrices over another Kleene algebra. A more unusual interpretation is the min,+ algebra used in shortest path algorithms.

The following are some typical identities that hold in all Kleene algebras:

$$(p^*q)^*p^* = (p+q)^* \tag{12}$$

$$p(qp)^* = (pq)^*p \tag{13}$$

$$(pq)^* = 1 + p(qp)^*q \tag{14}$$

$$p^* = (pp)^*(1+p). \tag{15}$$

All the operators are monotone with respect to $\leqslant$. In other words, if $p \leqslant q$, then $pr \leqslant qr$, $rp \leqslant rq$, $p + r \leqslant q + r$, and $p^* \leqslant q^*$ for any r.

The completeness result of Kozen (1994) says that all true identities between regular expressions interpreted as regular sets of strings are derivable from the axioms of Kleene algebra. In other words, the algebra of regular sets of strings over the finite alphabet Σ is the free Kleene algebra on generators Σ. The axioms are also complete over relational models.

See Kozen (1994) for a more thorough introduction.

2.3. Kleene algebra with tests

Kleene algebras with tests (KAT) were introduced in Kozen (1996), Kozen (1997) and their theory was further developed in Cohen, Kozen and Smith (1996), Kozen and Smith (1996). A Kleene algebra with tests is just a Kleene algebra with an embedded Boolean subalgebra. That is, it is a two-sorted structure

$$\mathcal{K} = (K, B, +, \cdot, {}^*, {}^-, 0, 1)$$

such that

- $(K, +, \cdot, {}^*, 0, 1)$ is a Kleene algebra,

- $(B, +, \cdot, \bar{}, 0, 1)$ is a Boolean algebra, and
- $B \subseteq K$.

The Boolean complementation operator $\bar{}$ is defined only on B. Elements of B are called *tests*. The letters p, q, r, s denote arbitrary elements of K and a, b, c denote tests. If Σ is an alphabet representing atomic Kleene elements and B is an alphabet representing atomic Boolean elements, then $T_{\Sigma,\mathsf{B}}$ and T_{B} denote the set of all terms and the set of all Boolean terms, respectively.

This deceptively simple definition actually carries a lot of information in a concise package. The operators $+, \cdot, 0, 1$ each play two roles: applied to arbitrary elements of K, they refer to nondeterministic choice, composition, fail, and skip, respectively; and applied to tests, they take on the additional meaning of Boolean disjunction, conjunction, falsity, and truth, respectively. These two usages do not conflict – for example, sequential testing of b and c is the same as testing their conjunction – and their coexistence admits considerable economy of expression.

The encoding of the **while** program constructs is as in PDL (Fischer and Ladner 1979):

$$p\,;q \stackrel{\text{def}}{=} pq \tag{16}$$

$$\textbf{if } b \textbf{ then } p \textbf{ else } q \stackrel{\text{def}}{=} bp + \bar{b}q \tag{17}$$

$$\textbf{while } b \textbf{ do } p \stackrel{\text{def}}{=} (bp)^*\bar{b}. \tag{18}$$

For applications in program verification, the standard interpretation would be a Kleene algebra of binary relations on a set and the Boolean algebra of subsets of the identity relation. One could also consider trace models, in which the Kleene elements are sets of traces (sequences of states) and the Boolean elements are sets of states (traces of length 0). As with KA, one can form the algebra $\mathrm{Mat}(\mathcal{K}, n)$ of $n \times n$ matrices over a KAT $\mathcal{K} = (K, B)$; the Boolean elements of this structure are the diagonal matrices over B. There is also a language-theoretic model that plays the same role in KAT that the regular sets of strings over a finite alphabet play in KA, namely the family of regular sets of *guarded strings* over a finite alphabet Σ with guards from a set B. This is the free KAT on generators Σ, B; that is, the equational theory of this structure is exactly the set of all equational consequences of the KAT axioms. Moreover, KAT is complete for the equational theory of relational models (Kozen and Smith 1996).

In Kozen (2000), it was shown that KAT subsumes PHL in the following sense. A partial correctness assertion $\{b\}\,p\,\{c\}$ is encoded as an equation

$bp\bar{c} = 0$, or equivalently, $bp = bpc$. If a rule

$$\frac{\{b_1\}\, p_1\, \{c_1\},\ \ldots,\ \{b_n\}\, p_n\, \{c_n\}}{\{b\}\, p\, \{c\}}$$

is derivable in PHL, then its translation, the universal Horn formula

$$b_1 p_1 \bar{c}_1 = 0 \wedge \cdots \wedge b_n p_n \bar{c}_n = 0 \quad \rightarrow \quad bp\bar{c} = 0,$$

is a theorem of KAT. More generally, one can show that all relationally valid Horn formulas of the form

$$r_1 = 0 \wedge \cdots \wedge r_n = 0 \quad \rightarrow \quad p = q$$

are theorems of KAT (Kozen 2000).

2.4. Typed Kleene algebra

Typed Kleene algebra was introduced in Kozen (1998). It is motivated primarily by the desire to interpret regular expressions as matrices of various shapes, possibly nonsquare. For example, in the completeness proof of Kozen (1991), Kozen (1994), it must be argued that a few essential theorems of Kleene algebra, such as

$$ax \leqslant xb \quad \rightarrow \quad a^* x \leqslant xb^*, \tag{19}$$

still hold when the symbols are interpreted as matrices of various sizes and shapes, provided there is no type mismatch. The equational implication (19) holds in $\mathrm{Mat}(\mathcal{K}, n)$, the $n \times n$ matrices over a Kleene algebra $\mathcal{K}$, simply by virtue of the fact that $\mathrm{Mat}(\mathcal{K}, n)$ is a Kleene algebra. However, for the purposes of Kozen (1991), Kozen (1994), we need to know that it holds even when a is interpreted as an $m \times m$ matrix, x is interpreted as an $m \times n$ matrix, and b is interpreted as an $n \times n$ matrix for any m and n.

To handle nonsquare matrices and other similar typed applications, we introduced a typing discipline in which regular expressions p have types of the form $\mathsf{s} \rightarrow \mathsf{t}$, where s and t are elements of an abstract set Ω. Every expression has a *most general typing* (mgt) under which the expression is well-typed and which is refined by every other typing for which this is true. For example, the most general typing of the expression ab^*c is $a : \mathsf{u} \rightarrow \mathsf{v}$, $b : \mathsf{v} \rightarrow \mathsf{v}$, $c : \mathsf{v} \rightarrow \mathsf{w}$, where u, v, and w are distinct. Most general typings exist and are unique up to a bijection. These ideas give rise to a theory called *typed Kleene algebra*.

In our principal interpretation, $\Omega = \mathbb{N}$ and the type judgement $p : \mathsf{s} \rightarrow \mathsf{t}$ indicates that p is a matrix with row and column dimensions s and t, respectively. There are other useful interpretations as well: sets of traces in a labeled transition system, binary relations with specified domains and ranges, regular sets of guarded strings (Kozen 1997, Kozen and Smith 1996), semiadditive categories (Manes 1992).

Let Ω be a set and $\omega : \{x, y, \ldots\} \to \Omega^2$. Elements of Ω are denoted $\mathsf{s}, \mathsf{t}, \mathsf{u}, \mathsf{v}, \ldots$ and are called *pretypes*. Elements of Ω^2 are called *types* and are denoted $\mathsf{s} \to \mathsf{t}$. (In Kozen (1998) we also included a type 2 for Boolean values.)

The map ω is called a *type environment*. If $\omega(x) = \mathsf{s} \to \mathsf{t}$, we write $x : \mathsf{s} \to \mathsf{t}$ and say that x *has type* $\mathsf{s} \to \mathsf{t}$ *under* ω.

We can use the following calculus to derive types for certain expressions from ω. A *type judgement* is an expression

$$p : \mathsf{s} \to \mathsf{t}$$

where p is a regular expression and $\mathsf{s} \to \mathsf{t}$ is a type. Given a type environment ω, types for compound terms and formulas are inferred inductively according to the following rules:

$$\frac{p : \mathsf{s} \to \mathsf{t} \quad q : \mathsf{s} \to \mathsf{t}}{p + q : \mathsf{s} \to \mathsf{t}} \tag{20}$$

$$\frac{p : \mathsf{s} \to \mathsf{t} \quad q : \mathsf{t} \to \mathsf{u}}{pq : \mathsf{s} \to \mathsf{u}} \tag{21}$$

$$\frac{p : \mathsf{s} \to \mathsf{s}}{p^* : \mathsf{s} \to \mathsf{s}} \tag{22}$$

$$0 : \mathsf{s} \to \mathsf{t} \tag{23}$$

$$1 : \mathsf{s} \to \mathsf{s} \tag{24}$$

Note that 0 has all types and 1 all square types (types of the form $\mathsf{s} \to \mathsf{s}$ for some $\mathsf{s} \in \Omega$).

Every type environment ω extends uniquely to a minimal set of type judgements closed under these rules. This unique extension is also denoted ω and is called a *typing*. An expression p is *well-typed* under the typing ω if ω contains a type judgement $p : \mathsf{s} \to \mathsf{t}$. A set of expressions is said to be *well-typed* under ω if every expression in the set is well-typed under ω.

Not all expressions are well-typed under all typings. For example, if $x : \mathsf{s} \to \mathsf{t}$ and $\mathsf{s} \neq \mathsf{t}$, the expression x^* is not well-typed. Moreover, the type of an expression under a typing ω is not unique; for example, if $x : \mathsf{s} \to \mathsf{t}$, then $x0 : \mathsf{s} \to \mathsf{u}$ for all u. However, the type of a variable is unique.

A class of models called *typed Kleene algebras* was defined in Kozen (1998); this semantics is reviewed below in Section 4.

In Kozen (1998) it was shown that a wide class of theorems of untyped Kleene algebra are also theorems of typed Kleene algebra under their most general typings.

3. Encoding Hoare logic

In Kozen (2000) we showed that partial correctness assertions can be regarded as equations in the language of KAT and that the usual rules of propositional Hoare logic are theorems of KAT under this encoding.

In this section we extend the type calculus of KA as described in Section 2.4 to account for tests. We augment the system with subtypes, intersection types, and rules for subtyping and typecasting. Tests are interpreted as typecast operators. We then show how PHL is subsumed by this type calculus. To complete the triangle, we show how to encode the rules of the type calculus as valid universal Horn formulas in the language of pure (typeless) KAT. These two encodings compose to give the encoding of Kozen (2000).

Let $\mathcal{K} = (K, B)$ be an arbitrary KAT. We impose a type structure on $\mathcal{K}$ as follows. Take the Boolean algebra B as the set of pretypes; we write $b \in B$ in a different font b when using it as a pretype. As in Section 2.4, a type judgement is an expression of the form $p : \mathsf{b} \to \mathsf{c}$.

We regard the natural order $\leqslant$ on B as a subtype order and the conjunction operation on B as a type intersection operator on pretypes. We postulate the following subtyping rule:

$$\frac{\mathsf{b}' \leqslant \mathsf{b}, \quad p : \mathsf{b} \to \mathsf{c}, \quad \mathsf{c} \leqslant \mathsf{c}'}{p : \mathsf{b}' \to \mathsf{c}'}. \tag{25}$$

(We will reconcile this view with the flat type structure of typed KA in Section 4 below.)

We regard a test $b \in B$ as a *typecast* or *coercion* operator that takes an object of type c and casts it down to an object of type bc. This is reflected in the following typing rule:

$$b : \mathsf{c} \to \mathsf{bc} \tag{26}$$

Note that in the presence of the subtype rule (25), this subsumes the rules $1 : \mathsf{s} \to \mathsf{s}$ and $0 : \mathsf{s} \to \mathsf{t}$ of typed KA.

The rule (26) represents an idealized form of the behavior of typecast operators and runtime type checks in modern programming languages. For example, consider the following Java fragment:

```
class High {}
class Low extends High {}
...
void fun(High y) {
   Low x = null;
   try {
      x = (Low)y;
   } catch (ClassCastException e) {}
}
...
fun(new Low());
```

The typecast operator `(Low)` is applied to an object whose runtime type `Low` is a proper subtype of its type `High` as determined by the static type environment. If the cast is unsuccessful, then a `ClassCastException` is thrown. But successful or not, the type of the expression `(Low)y` is `Low`, and after the cast it is type-correct to assign the object to a variable of that type.

The type calculus of KAT consists of the rules (20)–(24) of typed KA, the subtype rule (25), and the typecast rule (26).

To encode PHL in the type calculus of KAT, we encode the PCA $\{b\}\,p\,\{c\}$ by the type judgement

$$p : \mathsf{b} \to \mathsf{c}. \tag{27}$$

Using (16)–(18) and (27), we obtain the following translations of the Hoare rules (2)–(5):

Composition rule:

$$\frac{p : \mathsf{b} \to \mathsf{c}, \quad q : \mathsf{c} \to \mathsf{d}}{pq : \mathsf{b} \to \mathsf{d}} \tag{28}$$

Conditional rule:

$$\frac{p : \mathsf{bc} \to \mathsf{d}, \quad q : \bar{\mathsf{b}}\mathsf{c} \to \mathsf{d}}{bp + \bar{b}q : \mathsf{c} \to \mathsf{d}} \tag{29}$$

While rule:

$$\frac{p : \mathsf{bc} \to \mathsf{c}}{(bp)^*\bar{b} : \mathsf{c} \to \bar{\mathsf{b}}\mathsf{c}} \tag{30}$$

Weakening rule:

$$\frac{\mathsf{b}' \leqslant \mathsf{b}, \quad p : \mathsf{b} \to \mathsf{c}, \quad \mathsf{c} \leqslant \mathsf{c}'}{p : \mathsf{b}' \to \mathsf{c}'} \tag{31}$$

We now show that these rules can be derived in the type calculus of KAT.

Theorem 3.1 *The rules* (28)–(31) *are derived rules of the type calculus* (20)–(26).

Proof. The rule (28) is just the composition rule (21) and the rule (31) is just the subtype rule (25), so in these two cases there is nothing to prove.

For (29), we have

$$\cfrac{\cfrac{b : \mathsf{c} \to \mathsf{bc}^{\,(a)} \qquad p : \mathsf{bc} \to \mathsf{d}^{\,(b)}}{bp : \mathsf{c} \to \mathsf{d}}\;(e) \qquad \cfrac{\bar{b} : \mathsf{c} \to \bar{\mathsf{b}}\mathsf{c}^{\,(c)} \qquad q : \bar{\mathsf{b}}\mathsf{c} \to \mathsf{d}^{\,(d)}}{\bar{b}q : \mathsf{c} \to \mathsf{d}}\;(f)}{bp + \bar{b}q : \mathsf{c} \to \mathsf{d}}\;(g)$$

Here (a) and (c) are instances of the typecast rule (26), (b) and (d) are the premises, (e) and (f) are applications of the composition rule (21), and (g) is an application of the sum rule (20).

For (30), we have

$$\cfrac{\cfrac{\cfrac{b : \mathsf{c} \to \mathsf{bc}^{\,(a)} \qquad p : \mathsf{bc} \to \mathsf{c}^{\,(b)}}{bp : \mathsf{c} \to \mathsf{c}}\;(c)}{(bp)^* : \mathsf{c} \to \mathsf{c}}\;(d) \qquad \bar{b} : \mathsf{c} \to \bar{\mathsf{b}}\mathsf{c}^{\,(e)}}{(bp)^*\bar{b} : \mathsf{c} \to \bar{\mathsf{b}}\mathsf{c}}\;(f)$$

Here (a) and (e) are instances of the typecast rule (26), (b) is the premise, (c) and (f) are applications of the composition rule (21), and (d) is an application of the iteration rule (22). □

Now we encode the rules of the type calculus as universal Horn formulas in the language of pure (typeless) KAT; thus the type calculus is redundant. Interpret the type judgement $p : \mathsf{b} \to \mathsf{c}$ as one of the two equivalent equations

$$bp\bar{c} = 0 \quad \text{or} \quad bp = bpc. \tag{32}$$

The typing rule

$$\frac{p_1 : \mathsf{b}_1 \to \mathsf{c}_1, \; \ldots, \; p_n : \mathsf{b}_n \to \mathsf{c}_n}{p : \mathsf{b} \to \mathsf{c}}$$

becomes the universal Horn formula

$$b_1 p_1 \bar{c}_1 = 0 \wedge \cdots \wedge b_n p_n \bar{c}_n = 0 \quad \to \quad bp\bar{c} = 0. \tag{33}$$

Theorem 3.2 *The typing rules* (20)–(26)*, encoded as Horn formulas according to* (32) *and* (33)*, are all theorems of* KAT*. In other words, the type calculus of* KAT *is sound under the interpretation* (32).

Proof. Translating the rules (20)–(26) according to (32) and (33), we obtain

$$bp\bar{c} = 0 \wedge bq\bar{c} = 0 \quad \to \quad b(p+q)\bar{c} = 0 \tag{34}$$
$$bp = bpc \wedge cq = cqd \quad \to \quad bpq = bpqd \tag{35}$$
$$bp = bpb \quad \to \quad bp^* = bp^*b \tag{36}$$
$$b0\bar{c} = 0 \tag{37}$$
$$b1\bar{b} = 0 \tag{38}$$
$$b' \leqslant b \wedge bp\bar{c} = 0 \wedge c \leqslant c' \quad \to \quad b'p\bar{c'} = 0 \tag{39}$$
$$cb\overline{bc} = 0, \tag{40}$$

respectively. These are all easy exercises in KAT. The most difficult is (36), which we argue explicitly. The inequality $bp^*b \leqslant bp^*$ holds by monotonicity of multiplication, thus it suffices to show

$$bp \leqslant bpb \quad \to \quad bp^* \leqslant bp^*b.$$

By (9), it suffices to show

$$bp \leqslant bpb \quad \to \quad b + bp^*bp \leqslant bp^*b.$$

But if $bp \leqslant bpb$, then $bp \leqslant pb$ by monotonicity, therefore

$$b + bp^*bp \;\leqslant\; bb + bp^*pb \;=\; b(1+p^*p)b \;=\; bp^*b. \qquad \square$$

4. Embedding typed KA in KAT

Although based on the type discipline of typed KA (Kozen 1998) as described in Section 2.4, the type discipline of KAT as described in Section 3 looks quite different. The latter assumes a Boolean algebra structure on pretypes, whereas the former is flat. In this section we show that the two type disciplines are compatible by showing that every typed KA has a natural embedding in a KAT such that the type structure is preserved. The embedding is an extension of a natural embedding of a typed KA in a typeless KA described in Kozen (1998).

First we review the semantics of typed KA from Kozen (1998). Briefly, a *typed Kleene algebra* is structure in which

- each element has a unique type of the form $\mathsf{s} \to \mathsf{t}$;
- there is a collection of polymorphic typed operators $+$, $\cdot$, *, 0, 1 and binary relation $=$ whose application is governed by the typing rules;

- all well-typed instances of the Kleene algebra axioms hold.

Formally, a *typed Kleene algebra* is a structure

$$\mathcal{K} \quad = \quad (K, \Omega, \omega, +, \cdot, {}^*, 0, 1, =)$$

where K and Ω are sets and $\omega : K \to \Omega^2$; we write $\omega(p) = \mathsf{s} \to \mathsf{t}$. Elements of K are denoted $p, q, r, \ldots$

For $\mathsf{s}, \mathsf{t} \in \Omega$, define

$$K_{\mathsf{st}} \quad = \quad \{p \in K \mid \omega(p) = \mathsf{s} \to \mathsf{t}\}.$$

The operators $+$, $\cdot$, *, 0, 1 and relation $=$ have the following polymorphic types:

$$\begin{aligned}
+ \quad &: \quad \Lambda \mathsf{s}, \mathsf{t} \in \Omega.(\mathsf{s} \to \mathsf{t}) \times (\mathsf{s} \to \mathsf{t}) \to (\mathsf{s} \to \mathsf{t}) \\
\cdot \quad &: \quad \Lambda \mathsf{s}, \mathsf{t}, \mathsf{u} \in \Omega.(\mathsf{s} \to \mathsf{t}) \times (\mathsf{t} \to \mathsf{u}) \to (\mathsf{s} \to \mathsf{u}) \\
* \quad &: \quad \Lambda \mathsf{s} \in \Omega.(\mathsf{s} \to \mathsf{s}) \to (\mathsf{s} \to \mathsf{s}) \\
0 \quad &: \quad \Lambda \mathsf{s}, \mathsf{t} \in \Omega.(\mathsf{s} \to \mathsf{t}) \\
1 \quad &: \quad \Lambda \mathsf{s} \in \Omega.(\mathsf{s} \to \mathsf{s}) \\
= \quad &: \quad \Lambda \mathsf{s}, \mathsf{t} \in \Omega.(\mathsf{s} \to \mathsf{t}) \times (\mathsf{s} \to \mathsf{t}) \to 2.
\end{aligned}$$

This means for example that $+$ consists of a family of functions $+_{\mathsf{st}} : K_{\mathsf{st}}^2 \to K_{\mathsf{st}}$, one for each choice of $\mathsf{s}, \mathsf{t} \in \Omega$. The operator $+_{\mathsf{st}}$ can only be applied to arguments of type $\mathsf{s} \to \mathsf{t}$ and produces a sum of type $\mathsf{s} \to \mathsf{t}$. The polymorphic constant 0 represents a family of elements 0_{st}, one for each choice of $\mathsf{s}, \mathsf{t} \in \Omega$. The polymorphic constant 1 represents a family of square elements 1_{ss}, $\mathsf{s} \in \Omega$.

To be a typed Kleene algebra, $\mathcal{K}$ must also satisfy all well-typed instances of the Kleene algebra axioms. For example, the multiplicative associativity property $p(qr) = (pq)r$ must hold whenever the expression $p(qr) = (pq)r$ is well typed; that is, whenever $p : \mathsf{s} \to \mathsf{t}$, $q : \mathsf{t} \to \mathsf{u}$, and $r : \mathsf{u} \to \mathsf{v}$ for some $\mathsf{s}, \mathsf{t}, \mathsf{u}, \mathsf{v} \in \Omega$.

We now show how to construct a KAT $\mathcal{M}$ from a given typed KA $\mathcal{K}$ with pretypes Ω. Let B be the smallest Boolean subalgebra on 2^Ω containing all singleton sets. Thus B consists of the finite and cofinite subsets of Ω. (If Ω is finite, then B is just 2^Ω.) We denote elements of B by $b, c, \ldots$ Consider the $\Omega \times \Omega$ matrices P of finite support with $P_{\mathsf{st}} \in K_{\mathsf{st}}$. As argued in Kozen (1998), this is a typeless 1-free KA, and $\mathcal{K}$ embeds homomorphically into it under the map $p \mapsto P$ such that $P_{\mathsf{st}} = p$, where $\omega(p) = \mathsf{s} \to \mathsf{t}$, and $P_{\mathsf{uv}} = 0_{\mathsf{uv}}$ elsewhere. (The word "embedding" is used here in the sense of *typed embedding* (Kozen 1998); although all 0_{st} are mapped to the zero matrix, no pair of distinct elements of the same type are collapsed.) Now include the Boolean algebra B in the form of diagonal matrices; the matrix corresponding to $b \in B$ has as its

s^{th} diagonal element either 1_{ss} or 0_{ss} according as $s \in b$ or $s \notin b$, respectively. Closing under the KA operations, the resulting matrices are no longer necessarily of finite support, but all rows and columns are still of finite support, so that multiplication is defined. Moreover, the diagonal elements are almost all 1 or almost all 0, thus each matrix decomposes into a block diagonal matrix of two blocks, one a finite square matrix and the other a square zero or identity matrix, thus * is defined. This gives a KAT $\mathcal{M}$ whose Boolean elements are the diagonal matrices over 0,1 corresponding to elements of B.

In $\mathcal{M}$, the interpretation (32) says that $P : \mathsf{b} \to \mathsf{c}$ iff

$$s \in b \text{ and } P_{st} \neq 0 \quad \Rightarrow \quad t \in c; \tag{41}$$

in other words, the $b \times \bar{c}$ submatrix of P is the zero matrix.

It is instructive to understand the significance of the typing rules (20)–(26) in light of (41). For example, the rule (22) says that if $P : \mathsf{b} \to \mathsf{b}$, then $P^* : \mathsf{b} \to \mathsf{b}$. The matrix P can be decomposed (after permuting the rows and columns) into quadrants

$$P = \begin{bmatrix} A & B \\ C & D \end{bmatrix}$$

where A, B, C, and D are the $b \times b$, $b \times \bar{b}$, $\bar{b} \times b$, and $\bar{b} \times \bar{b}$ submatrices of P, respectively. The type judgement $P : \mathsf{b} \to \mathsf{b}$ says that $B = 0$. But by the definition of * for matrices,

$$\begin{bmatrix} A & B \\ C & D \end{bmatrix}^* = \begin{bmatrix} (A+BD^*C)^* & (A+BD^*C)^*BD^* \\ (D+CA^*B)^*CA^* & (D+CA^*B)^* \end{bmatrix},$$

and $(A+BD^*C)^*BD^* = 0$, therefore P^* satisfies the same property.

Similarly, $I : \mathsf{b} \to \mathsf{b}$ for any b, where I is the identity matrix, since if $s \in b$ and $I_{st} \neq 0$, then $s = t$, therefore $t \in b$; thus (41) holds.

Finally, the typecast rule (26) says that for any Boolean element b, if B is the diagonal matrix corresponding to b, then $B : \mathsf{c} \to \mathsf{bc}$. By (41), if $s \in c$ and $B_{st} \neq 0$, then we should have $t \in bc$. But $B_{st} \neq 0$ iff $s = t$ and $s \in b$, therefore $t \in bc$.

The following theorem describes the relationship between the typing disciplines of $\mathcal{K}$ and $\mathcal{M}$.

Theorem 4.1 *Let $\mathcal{K}$ be an arbitrary typed* KA *with pretypes Ω, and let $\mathcal{M}$ be the* KAT *constructed from it as described above. Let $p \in \mathcal{K}$ such that $p : \mathsf{s} \to \mathsf{t}$, and let P be its image in $\mathcal{M}$. Then either*

(i) $p = 0_{st}$, *in which case $P : \mathsf{b} \to \mathsf{c}$ for all* $\mathsf{b}, \mathsf{c} \in 2^{\Omega}$; *or*

(ii) $p \neq 0_{st}$, *in which case $P : \{\mathsf{s}\} \to \{\mathsf{t}\}$ but not $P : \{\mathsf{s}\} \to \{\mathsf{u}\}$ for any* $\mathsf{u} \neq \mathsf{t}$ *(thus by* (25), *not $P : \{\mathsf{s}\} \to \emptyset$), and $P : \{\mathsf{u}\} \to \emptyset$ for all* $\mathsf{u} \neq \mathsf{s}$.

Proof. (i) If $p = 0_{\mathsf{st}}$, then P is the zero matrix, thus by (41), $P : \mathsf{b} \to \mathsf{c}$ for all $\mathsf{b}, \mathsf{c} \in 2^{\Omega}$.

(ii) If $p \neq 0_{\mathsf{st}}$ and $p : \mathsf{s} \to \mathsf{t}$, then $P_{\mathsf{st}} = p \neq 0$ and $P_{\mathsf{uv}} = 0$ elsewhere. This says that the sth row of P is nonzero in column t and zero elsewhere, therefore by (41) $P : \{\mathsf{s}\} \to \{\mathsf{t}\}$ but not $P : \{\mathsf{s}\} \to \{\mathsf{t}\}$ for any $\mathsf{u} \neq \mathsf{t}$. All other rows are zero, therefore by (41), $P : \{\mathsf{u}\} \to \emptyset$ for $\mathsf{u} \neq \mathsf{s}$. □

Acknowledgements

I am indebted to Robert Constable, Neal Glew and Greg Morrisett for valuable ideas and comments. This work was supported by the National Science Foundation under grant CCR-9708915.

References

Apt, K.R.: (1981), 'Ten years of Hoare's logic: a survey – part 1', *ACM Trans. Prog. Lang. Syst.* **3**, pp. 431–483.

Bates, J.L., Constable, R.L.: (1985), 'Proofs as programs', *ACM Trans. Program. Lang. Syst.* **7**(1), pp. 53–71.

Cohen, E., Kozen, D., Smith, F.: (1996), 'The complexity of Kleene algebra with tests', *Technical Report 96-1598*, Computer Science Department, Cornell University.

Constable, R.L.: (1988), 'Themes in the development of programming logics circa 1963–1987', *Ann. Rev. Comput. Sci.* **3**, pp. 147–165.

Constable, R.L.: (1998), 'Types in mathematics, logic, and programming', in S.R. Buss (ed.), *Handbood of Proof Theory*, Elsevier, Amsterdam, chapter X, pp. 683–786.

Conway, J.H.: (1971), *Regular Algebra and Finite Machines*, Chapman and Hall, London, U.K.

Cook, S.A.: (1978), 'Soundness and completeness of an axiom system for program verification', *SIAM J. Comput.* **7**(1), pp. 70–90.

Cousot, P.: (1990), 'Methods and logics for proving programs', in J. van Leeuwen (ed.), *Handbood of Theoretical Computer Science*, Vol. B, Elsevier, Amsterdam, pp. 841–993.

Fischer, M.J., Ladner, R.E.: (1979), 'Propositional dynamic logic of regular programs', *J. Comput. Syst. Sci.* **18**(2), pp. 194–211.

Girard, J.-Y.: (1971), 'Une extension de l'interprétation de Gödel à l'analyse, et son application à l'élimination des coupures dans l'analyse et la théorie des types', *Proc. 2nd Scand. Logic Symp.*, Springer-Verlag, Berlin, pp. 63–69.

Girard, J.-Y., Lafont, Y., Taylor, P.: (1989), *Proofs and Types*, Cambridge University Press, Cambridge.

Kleene, S.C.: (1956), 'Representation of events in nerve nets and finite automata', in C.E. Shannon, J. McCarthy (eds.), *Automata Studies*, Princeton University Press, Princeton, N.J., pp. 3–41.

Kozen, D.: (1991), 'A completeness theorem for Kleene algebras and the algebra of regular events', *Proc. 6th Symp. Logic in Comput. Sci.*, IEEE, Amsterdam, pp. 214–225.

Kozen, D.: (1994), 'A completeness theorem for Kleene algebras and the algebra of regular events', *Infor. and Comput.* **110**(2), pp. 366–390.

Kozen, D.: (1996), 'Kleene algebra with tests and commutativity conditions', in T. Margaria, B. Steffen (eds.), *Proc. Second Int. Workshop Tools and Algorithms for the Construction and Analysis of Systems (TACAS'96)*, Vol. 1055 of *Lecture Notes in Computer Science*, Springer-Verlag, Passau, Germany, pp. 14–33.

Kozen, D.: (1997), 'Kleene algebra with tests', *Transactions on Programming Languages and Systems* **19(3)**, pp. 427–443.

Kozen, D.: (1998), 'Typed Kleene algebra', *Technical Report 98-1669*, Computer Science Department, Cornell University.

Kozen, D.: (2000), 'On Hoare logic and Kleene algebra with tests', *Trans. Computational Logic* **1(1)**, pp. 60–76.

Kozen, D., Smith, F.: (1996), 'Kleene algebra with tests: Completeness and decidability', in D. van Dalen, M. Bezem (eds.), *Proc. 10th Int. Workshop Computer Science Logic (CSL'96)*, Vol. 1258 of *Lecture Notes in Computer Science*, Springer-Verlag, Utrecht, The Netherlands, pp. 244–259.

Manes, E.: (1992), *Predicate transformer semantics*, Cambridge University Press, Cambridge.

Martin-Löf, P.: (1982), 'Constructive mathematics and computer programming', *6th Int. Congr. Logic, Methodology, and Philosophy of Science*, North-Holland, Amsterdam, pp. 153–175.

Pratt, V.R.: (1978), 'A practical decision method for propositional dynamic logic', *Proc. 10th Symp. Theory of Comput.*, ACM, pp. 326–337.

Reynolds, J.C.: (1974), 'Towards a theory of type structure', *Proc. Colloque sur la Programmation*, Vol. 19 of *Lect. Notes in Comput. Sci.*, Springer-Verlag, Berlin, pp. 408–423.

VI

LOGIC, LANGUAGE AND COGNITION

LOGIC AND COGNITION

Human Performance in Default Reasoning

Francis Jeffry Pelletier
Department of Philosophy, University of Alberta
Edmonton, Alberta, Canada T6G 2E5
jeffp@cs.ualberta.ca

Renée Elio
Department of Computing Science, University of Alberta
Edmonton, Alberta, Canada T6G 2H1
ree@cs.ualberta.ca

1. Background

There has long been a history of studies investigating how people ("ordinary people") perform on tasks that involve deductive reasoning. The upshot of these studies is that people characteristically perform some deductive tasks well but others badly. For instance, studies show that people will typically perform MP ("modus ponens": from 'If *A* then *B*' and '*A*', infer '*B*') and biconditional MP (from: '*A* if and only if *B*' and '*A*', infer '*B*') correctly when invited to make the inference and additionally can discover of their own accord when such inferences are appropriate. On the other hand, the same studies show that people typically perform MT ("modus tollens": from 'If *A* then *B*' and 'not-*B*', infer 'not-*A*') and biconditional MT badly. They not only do not recognize when it is appropriate to draw such inferences, but also they will balk at doing them even when they are told that they can make it. Related to these shortcomings seems to be the inability of people to understand that contrapositives are equivalent (that 'If *A* then *B*' is equivalent to 'If not-*B* then not-*A*'). [Studies of people's deductive inference-drawing abilities have a long history, involving many studies in the early 20th century concerning Aristotelian syllogisms. But the current spate of studies draws much of its impetus from Wason (Wason 1968; see also Wason & Johnson-Laird 1972).] The general conclusion seems to be that there are specific areas where "ordinary people" do not perform very logically. This conclusion will not come as a surprise to teach-

P. Gärdenfors, J. Woleński and K. Kijania-Placek (eds.), In the Scope of Logic, Methodology and Philosophy of Science, Vol. I, 137-154.

ers of elementary logic, who have long thought that the majority of "ordinary people" are inherently illogical and need deep and forceful schooling in order to overcome this flaw.

There has been a similar history – although somewhat shorter – of studies investigating how people perform on tasks that involve probabilistic reasoning. The upshot of these studies is that people characteristically make a number of errors because they do not pay attention to certain relevant information and because they give undue importance to certain kinds of irrelevant information. The study of the types of errors people make on this probabilistic reasoning has come to be called "heuristics and biases" because the underlying assumption is that people have inbuilt or inherent biases as part of their mental equipment that causes them to ignore or give undue status to some information. (Usually it is claimed that these biases stem from people having certain rule-of-thumb heuristics which allow them to evaluate information quickly and efficiently in "normal" circumstances but which will lead people astray in other circumstances. Much of this work stems from Kahneman & Tversky: see Kahneman et al. 1982 and Evans 1987).

Using a similar methodology, researchers have found that people's decision-making reasoning deviates from the prescriptions of decision theory in a number of regards. Many of the conclusions reached here mirror the "heuristics and biases" approach used in the probabilistic reasoning studies: people simply have tendencies to ignore certain information because of the (evolutionary) necessity to make these types of decisions quickly. This gives rise to "biases" in judgments concerning what they "really" want to do. (See Baron 1994 and Yates 1990).

So, there is a long tradition of finding "ordinary people" wanting in their abilities to perform logical inferences, whether in deductive logic, probabilistic reasoning, or decision making. Yet perhaps we can see something of a difference in the three cases. In the second and third cases there is no universal agreement as to what the "correct" underlying theory is. In the case of decision making there are alternatives to decision-theoretic formalizations. And in the case of probabilistic reasoning there are at least the two alternatives of subjective probability and frequency theories. Some researchers have argued that people are *not* in fact so bad at probabilistic reasoning when the information is presented in the right way and we recognize what the "really right" conclusions are. Gigerenzer and his colleagues (see Gigerenzer 1991, 1994, 1998 for example) think that all the flaws discovered by researchers in this tradition are due to their adherence to a false theory of probability – subjective probabilities rather than the correct frequency theory. And Cohen (1981) proposed a new system of probability theory that was intended to capture the sort of probabilistic reason that people actually follow. (His "Baconian probability theory").

In the probability theory and decision theory cases there is disagreement over what the normatively correct underlying theory is. Hence, when people's performance is measured there can be questions about whether they are performing "correctly" or not. If Theory_1 is the normatively right theory for the field, then they are performing badly; but if Theory_2 is the normatively right theory in the field, then they are performing correctly. Related to this is yet another idea: One might say that in certain fields there simply *is no other standard* than what people in fact do. In determining how North Americans distinguish green from blue, for example, there is no real concept of right and wrong in the task. Some people draw the boundary in one place, some in another place; we can tell whether there are identifiable subgroups who all draw it at one location while other subgroups draw it in another place; and we can tell whether someone is different from the majority in where s/he draws the line. But there is no real notion of "draws the blue/green boundary incorrectly"; all that exists is how people in fact do it. This attitude is often called "psychologistic" because it locates the object of the study (or the normativity of the theory) in the psychology of people and denies that there is any "external standard of correctness" for the field.

One can imagine that decision theory and even probability theory are psychologistic in this way: there is no notion of "correct inference" other than what people actually do. The correct probability theory (for example) could be some sort of generalization of how people actually reason probabilistically; and therefore when we say that someone is making a mistake in probabilistic reasoning, all we mean is that the person deviates from what people normally do in this regard. Cohen's (1981) view is essentially of this nature; he denies that people as a species can be shown to be irrational because 'rationality' is the sort of notion that is *defined* as what people's underlying capacities have them do in the relevant circumstances.

It is somewhat more difficult to imagine the psychologistic response being made in the realm of deductive reasoning. The reason for this is that we in fact have a perfectly clear "outside standard" of what counts as correct deductive reasoning, namely, *preserves truth from premises to conclusion*. If there were agreement in the probability or decision theories about what the "outside standard" is, then the psychologistic response would be seen as less plausible in those fields also. But once the possibility of a psychologistic attitude is brought up in the deductive arena, we will find some theorists arguing for a "relevance logic" or a "fuzzy logic" or [...], on the grounds that "this is the way people *actually* reason".

2. Introduction

Non-Monotonic Logics is the name given to any logic (i.e., any formal, symbolic system) with this feature:

$$S_1, S_2, \ldots, S_n \vdash C \text{ and yet it is false that } S_1, S_2, \ldots, S_n, S_{n+1} \vdash C$$

That is, it is possible to go from a valid argument to an invalid argument merely by adding more information to the premises (and not deleting any other premise). This formal property is related to properties of counterfactual conditionals (where adding more information to the antecedent of such a conditional can make the truth-value of the entire conditional change from true to false), and to *prima facie* reasoning in ethics and law (where conclusions of such reasoning tell us what is right or legal in the absence of contravening factors). And researchers have claimed that similar patterns of reasoning occur in many different fields. As a whole, this type of reasoning (whether or not it is formalized so as to exhibit the non-monotonic logic feature) is called "default reasoning".

Probability theory allows for the probabilistic value of "*A*, given *B*" to be greater than the the value of "*A*, given both *B* and *C*", and that to be greater than the value of "*A*, given *B* and *C* and *D*", etc., so there are many attempts to accommodate all default reasoning within probability theory. (Pearl 1988, Bacchus 1991). Because this way of looking at default reasoning presumes to assign a numerical value between 0 and 1 to sentences (given a background), this is called a *quantitative* theory of default reasoning. Besides probability theory, another quantitative theory of default reasoning is fuzzy logic.

Opposed to the quantitative theories are the *qualitative* theories. These theories do not assign some number between 0 and 1 to sentences, but rather give sentences one of two values: either True or False. (Or sometimes a third value, Uncertain/Undefined, is allowed; and sometimes a fourth value. [See Belnap 1978, where the four values are: True, False, Both True and False, Neither True nor False). But if there are too many possible values then such a qualitative theory slides into a quantitative theory.] Classical examples of the qualitative theories of default reasoning are Reiter's (1980) *default logic*, McCarthy's (1980) *circumscription*, and Moore's (1985) *autoepistemic logic* (among others).

Even within just the qualitative or quantitative camp, there are a variety of differences among the different theories, both in their overall outlook and methodology and in the class of inferences that they each sanction as being valid default arguments. It is to this issue that we wish to address our next remarks.

3. How should we decide on valid default inferences?

In humans, not all of our information about the world is stored in an explicit form in our minds; instead much of it is *implicit* in that it is a consequence of, or follows from, other information which *is* stored explicitly. The reason for this psychological truism is that it would be incredibly inefficient, incredibly wasteful of memory, etc., if everything that we knew were to be mentally entered in "longhand" for our inspection. Instead only some of our knowledge is explicitly stored, and much other information can be recovered or inferred from this explicit store. For instance, none of us has explicitly stored the fact that there is no roadrunner in our refrigerator; yet we know this nonetheless. And the way we know it is that it (somehow) follows from explicitly stored facts about our refrigerator and the locations of roadrunners. The goal of the constructors of "knowledge bases" is to mimic this manner of storage. It is commonly believed by these researchers that this is the only way that major advances will occur in computational information storage.

One of the ways we infer these implicit pieces of knowledge is by default reasoning. Most researchers in this area think this is the most pervasive form of reasoning in our everyday life. It occurs whenever we wish to draw new conclusions from old information, where the old information does not *deductively* guarantee the truth of the new conclusions. For example, from our general knowledge that birds typically can fly, we conclude that our neighbor's new pet bird can fly ... despite the acknowledged possibility that he might have managed to buy a roadrunner. Two quotations taken almost at random from researchers in the area show the attitude which is pretty universally adopted.

> Most of what we know about the world, when formalized, will yield an incomplete theory precisely because we cannot know everything – there are gaps in our knowledge. The effect of a default rule is to implicitly fill in some of those gaps by a form of plausible reasoning [...] Default reasoning may well be the rule, rather than the exception, in reasoning about the world since normally we must act in the presence of incomplete knowledge [...] Moreover, [...] most of what we know about the world has associated exceptions and caveats. (Reiter 1978)

> It is commonly acknowledged that an agent need not, indeed cannot, have absolute justification for all of his beliefs. An agent often assumes, for example, that a certain member of a particular kind has a certain property simply because it is typically true that entities of that kind have that property [...] Such default reasoning allows an agent to come to a decision and act in the face of incomplete information. It provides a way of cutting off the possibly endless amount of reasoning and observation that an agent might perform [...]. (Selman & Kautz 1989)

We see here that the artificial intelligence community (or at least the knowledge representation subdivision of it) believes that humans explicitly store only a portion of the information that they know, and that when called upon, they typically use some sort of default reasoning to infer the rest.

What might also be noted in these quotes, and any other quotes that one might wish to bring up from this literature, is that they all justify the search for a formal theory of default reasoning by *appeal to human performance.* "It is because humans do such-and-such task so well ... " or "People can get around in the world with no problems ... " are typical of the justification given for the whole enterprise. This is *much* different than the case of logic or arithmetic where there is some "external standard" that is independent of how humans reason logically or arithmetically. In the cases of logic and arithmetic we would probably *not wish* our artificial agents to mirror human performance, because we have the desire that these agents should compute the answers *correctly according to the external standard.* But in the case of default reasoning there is no standard which is external to how humans perform their default reasoning. Another way of putting the point, using the language developed above, is that default reasoning (unlike logic and arithmetic) is *psychologistic* in nature; it is by definition the study of how people perform on certain types of problems and in certain types of situations. *There simply is no other standard against which to judge how good some abstract system of default inference is, nor any other standard with which to compare artificial agents that embody such a system.* And so it follows that any judgment of the correctness of a proposed system of default reasoning will be as a test of the system against the way humans perform on that reasoning problem. (This outlook on default reasoning, augmented by many quotations from AI researchers who would appear to believe the same, can be found in Pelletier & Elio 1997).

4. The non-monotonic Benchmark Problems

Despite the fact that all the researchers in the field appeal to the goodness of human performance as their justification for employing default reasoning mechanisms in artificial agents, the truth is that none of them in fact have ever investigated how people actually employ default reasoning. A consequence of this is that the different proposed formalisms validate different sets of inferences, and there is no agreed-upon method to decide which inferences should be sanctioned as "really legitimate." Recognizing that there was no accepted background for finding the extent of legitimate default inferences, Lifschitz (1989) set out a number of inferences that were supposed to be valid in any proposed default reasoning system. Different groups of these problems were addressed to different aspects of the default reasoning process, so that perhaps not every reasoning system needed to accommodate all problems; but for any area that a system proposed a mechanism, it was to be able to deal at least with the inferences relevant to that area. We call these problems "the Benchmark

Problems", and the accepted answers that were proposed for these problems in Lifschitz (1989) the "AI answers" to the Benchmark Problems.

Although one might accept the legitimacy of establishing a set of benchmark problems in this way because these are the areas in which default reasoning is to be employed by artificial agents, one might nonetheless wonder about the legitimacy of determining the AI answers as a matter of agreement among the various researchers. After all, if it is true that the realm of default reasoning is psychologistic and that therefore the correct answers are determined by the way "ordinary people" will (on the whole) use the method, then the fact that some elite subgroup of people think the answers should be such-and-so is not a good justification. For one thing, their opinions might be colored by how their systems perform. More importantly, an individual's intuitions are not a reliable guide to how the population as a whole treats the phenomenon. In Pelletier & Elio (1997) we have investigated the various reasons researchers give for allowing their own intuitions to be their sole guide in this regard and for not engaging in large-scale investigations of how it is manifested in the population as a whole. We did not find any of these reasons very compelling, and recommended that researchers undertake such studies in order to determine the correct direction for their formal theories to follow.

The present paper sets out the preliminaries for such work, and gives some very tentative results. Despite the tentative nature of the results, we think they should give default reasoning researchers pause in their confidence that they have in fact fathomed the true nature of the reasoning process they are trying to model.

We will discuss only one area here from Lifschitz's list of Benchmark Problems, namely the ones he calls "Basic Default Inference", his Problems #1–#4. As stated in his article they are:

Figure 1. "Basic default inference" (from Lifschitz's Benchmark Problems)

1. Blocks *A* and *B* are heavy.
Heavy blocks are normally located on this table.
<u>*A* is not on this table.</u>
B is on this table.

2. Blocks *A* and *B* are heavy.
Heavy blocks are normally located on this table.
A is not on this table.
<u>*B* is red</u>
B is on this table.

3. Blocks *A* and *B* are heavy.
Heavy blocks are normally located on this table.
Heavy blocks are normally red.
A is not on this table.
<u>*B* is not red.</u>
B is on this table.
A is red.

4. Blocks *A* and *B* are heavy.
Heavy blocks are normally located on this table.
<u>*A* is possibly an exception to this rule.</u>
B is on this table.

Each of Problems 1–4, what Lifschitz called the basic default reasoning problems, concerns two objects governed by one or more default rules. Additional information is given to indicate that one of the objects (at least) does not follow one of the default rules. We refer to this as the *exception object* (for that default rule). The problem then asks for a conclusion about the remaining object. We refer to this as the object-in-question.

It is clear from Figure 1 that the four default reasoning problems are variations on the same theme. Problem 2 includes the extra information that Block *B* is red, but we are still to conclude that it is on the table. Problem 3 mentions two default rules: the exception object violates the default concerning location, and the object-in-question, Block *B*, violates the default rule about color, but we are still to conclude that Block *B* follows the default location rule. Problem 4 states only that Block *A might* violate the default rule, in contrast with Problem 1, which states with certainty that there is an exception to the rule.

For all these problems, the given default conclusion is that Block *B* obeys the default rule concerning location. According to the AI answers – that is, according to the collective wisdom of researchers into nonmonotonic theories – the existence of an exception object for a default rule, or additional information about that exception object, should have no bearing on a conclusion drawn about any other object when using that rule. Extra information about the object in question itself (e.g., Block *B*'s color) should also have no bearing on whether a default rule about location applies. And being an exception object for some *other* default rule should have no bearing on whether it does or does not follow the present default rule.

An implicit but important assumption here is that a logic for manipulating assertions of this form can be developed for non-monotonic reasoning without regard for the semantic content of the lexical items. Given these formal problems about blocks and tables, it seems easy to accept the idea that object color should have no "logical" bearing on whether a default about object location is applied. Yet it is equally easy to imagine real-world scenarios in which it makes (common) sense that an object's color might be predictive of or at least related to an object's default location (e.g., how an artist or designer might organize work items in a studio). We do not wish to confound people's *logical* abilities with their ability to "look up" information they have stored in memory. And so we would want to test them on scenarios that they have some "feel" for but which they have no stored information about.

5. An experiment into basic default reasoning

We investigated two factors concerning the exception object that intuitively seemed likely to influence plausible conclusions about the object in question: the specificity of information about how the exception object violates the default rule, and the apparent similarity of the exception object to the object in question. It is well known (Evans 1987) that posing information in negative form is more difficult for people to deal with in such tasks as drawing inferences than is the same information posed in positive form. Thus we wished to investigate whether the presence of an explicit negation (as the Benchmark Problems had) caused more difficulties in processing these problems than giving positive information concerning the object (which would entail the negative information). The negative form can be seen to "unspecific" with respect to information conveyed: saying that a block is *not* on the table does not say where it is, but saying that the block is on the floor not only implies that it is not on the table but also gives the specific information as to its location. Interestingly, the issue of how negation should be handled in the Prolog programming language is also an issue of some complexity, perhaps mirroring the difficulties that people have in using negative information.

Most people who have thought about default reasoning have considered the possibility that the willingness of someone to draw a given inference perhaps is dependent on the specific lexical items or situations being discussed, and it therefore has nothing to do with logic – that is, nothing to do with the logical form of the inferences to be drawn. In particular for Problems 1–4, one might speculate that the amount of perceived similarity between the exception object and the object-in-question would influence whether subjects thought the object in question should behave like the exception object or should follow the default rule. Thus one of the variables to be manipulated was how much similarity was stated to hold between the two objects.

Subjects. Eighty subjects enrolled in an introductory psychology course participated for partial credit towards an optional experiment-participation component of their course.

Design. There were two between-subject independent variables. The first was the *specificity* of the information about the exception object. In Benchmarks 1–4, the manner in which an object violates the default rule is unspecified (e.g., *Block A is not on the table*) We call this the non-specific form. The specific form of the problem identified a particular state for the exception (e.g., *Block A is in the cabinet*). The second between-subject variable was *who* the agent solving the problem was supposed to be: a human (actually, the subject) or a robot. Interviews with pilot-study subjects indicated that this made a difference in the kinds of answers generated. (That is, people might take a different view about the inferences *they* would make with these default logic questions were *they* in the scenario described in the experiment from those inferences they would want or expect *an intelligent robot* to produce). We had no a priori prediction or intuition about the human vs. robot dimension, but it seemed an interesting meta-cognitive issue to explore.

There was one within-subject variable: *object similarity*. Each subject answered both a low similarity and a high-similarity version of Benchmarks 1–4. The low similarity version had sentences corresponding to just those assertions in the original Benchmark. The high similarity version had additional statements describing commonalities shared by the exception object and object in question. Figure 3 illustrates two of the four combinations of specificity and similarity for Benchmark #2.

Figure 2. Alternative Forms of Benchmark 2

Low similarity / Non-specific Form / Human

You know	There is a Craftsman electric drill and there is also a Black and Decker electric drill.
	Electric drills are normally stored in the utility cabinet.
	The Black and Decker drill is a cordless model.
You also know	The Craftsman drill is not in the utility cabinet.

Question: What is reasonable to decide about where the Black and Decker drill is?

High similarity / Non-specific form/ Robot

Robot knows	Western Construction and ConCo Consulting have each submitted confidential bids for contract work.
	Confidential bids are normally kept in the Department Head's office.
	The bid by ConCo Consulting was prepared by an outside consultant.
Robot also knows	The bid by Western Construction is in the auditor's office.
	The Western Construction and the ConCo bids were considerably lower than the other bids that were received.
	Both these companies have good track records for consulting work.
	Their bids were received 2 hours after the deadline date, which was Friday at noon.

Question: What is reasonable to decide about where the ConCo bid is?

It is important to appreciate that the additional statements in the high-similarity condition do not enjoin us to conclude that the two objects behave identically with respect to the default rule. On the other hand, we designed these assertions under the guiding principle that they could, conceivably, constitute an explanation as to why an object might not obey a default rule. For example, in the "contract bids" cover story, the extra assertions describe a set of circumstances that could explain why a particular bid is not where it is supposed to be by default: it came in late, it was a very low bid, and it comes from a good company. Maybe these facts together constitute a reason why it is sitting in the auditor's office. They certainly *entail* the conclusion that any bid should not follow the default rule and a problem solver would have to do a good bit of work to weave these into an explanation.

Two different cover stories were developed for each Benchmark. We counterbalanced which cover story was used as the low similarity version and which was used as the high-similarity version across subjects.

Procedure. Subjects were randomly assigned to receive either human-specific, robot-specific, human-nonspecific, or robot-nonspecific problems. The eight problems (four Benchmarks under two similarity versions) were randomly ordered and presented in booklet form. To lessen the chance that subjects would detect the underlying similarity among the problems, we put one filler problem between each of the randomly-ordered benchmark problems. These filler problems were similar in format and also asked for common-sense reasoning conclusions. The instructions emphasized that there were no right or wrong answers to these problems, and that the goal of the experiment was to discover something about how people make (or how robots should make) plausible conclusions in situations for which there is only general information. Subjects generated their own answers and were told that "can't tell" was also an acceptable answer.

Results. We coded subjects' answers about the object-in-question according to one of four answer categories: (a) it followed the AI answer, (b) it followed the exception object, (c) it was some other answer, or (d) "can't tell." The AI answer for Benchmark 2 (drills) is that the Black and Decker drill is "in the utility cabinet"; the exception-answer is "on the workbench." An answer that would be coded as "other" might be "in the mail between the Dept Head and President" for the contract-bid example. We converted the subject data into the proportion of answers generated in each category for each problem; under this scheme, answer category becomes another factor.

Let us take a first quick look at the percentage of answers that followed the AI answer:

Table 1. Percent of answers agreeing with AI Answer on the Benchmark Problems

Problem	%AI Answer
1	78 %
2	75 %
3	56 %
4	77 %

It is clear that Problem #3 is significantly different from Problems #1, #2, and #4. (All results we describe as distinct are significant at the $p = .05$ level). This suggests that the more default rules an object is known to violate, the more "generally exceptionable" the object is, and the more likely the object is to violate other defaults.

Turning our attention to the issue of similarity we see:

Table 2. Percent agreeing with AI answer as a function of Similarity

Condition	%AI Answer
Hi Similarity:	64%
Lo Similarity	80%

Although this effect is qualified by an interaction with NEG, this suggests that the object-in-question is more likely to "follow" some other, exceptional object if additional information creates a scenario under which the objects are *seen* as similar – even if this information does *not* state, imply, or "make it really more likely" that the object-in-question should also be an exception.

As remarked, there is an interaction between the degree of perceived similarity of the exception object and the object-in-question, on the one hand, and whether the information given is presented in a negative form (" ... the object is *not* ... ") or in a positive form ("the object *is* ...") that implies the negative form.

Figure 3. %AI answer's interaction between Similarity and Pos/Neg

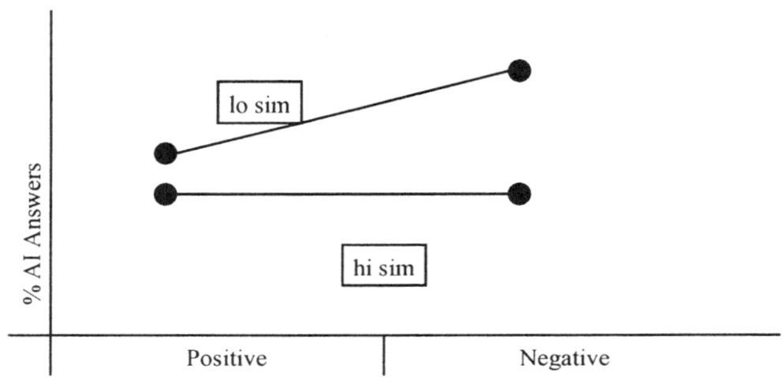

Not only is the Low Similarity condition generally more likely to yield the AI answer, but also Positive and Negative affect the Hi/Low Similarity differently. Presumably this has something to do with the fact that negation amounts to some sort of "lack of specificity" about the object (we know that it is 'not manifesting some condition' but we do not know which condition it *is* manifesting), and when this is added to the Low Similarity condition (we do not know anything about whether the object-in-question shares any properties with the exception object) we will more often ignore the exception object's possible influence on the object-in-question.

There was no main effect for either POS/NEG or for Robot/Human, although there was a significant interaction between the two. Subjects more often gave the AI answer for Humans when the information was POS than when it was NEG, but they more often gave the AI for Robots when the information was NEG than when it was POS. Figure 4 displays this result:

Figure 4. Interaction of Pos/Neg with Human/Robot on % of AI answers

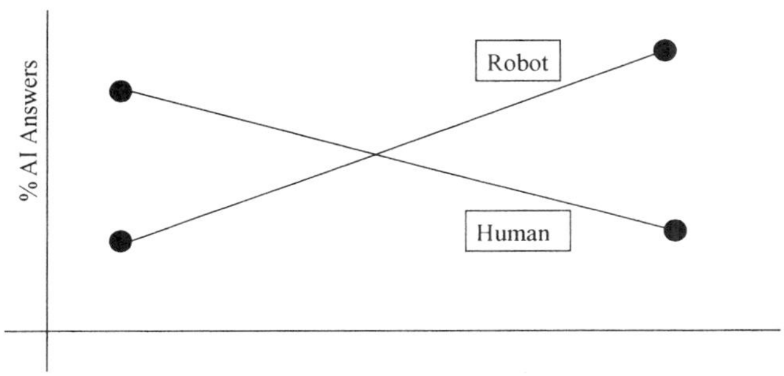

We will return to this very puzzling result shortly. But first we will take a somewhat different look at the data collected in this experiment.

Recall that we coded subjects' answers about the object-in-question according to one of four answer categories: (a) it followed the AI answer, (b) it followed the exception object, (c) it was some other answer, or (d) "can't tell."

Table 3. Proportion of Answers as a Function of Condition and Answer Category

	Answer Category			
	AI	Exception	Other	Can't Tell
Problem Form				
Positive				
high similarity	.450	.195	.190	.165
low similarity	.605	.090	.090	.215
Negative				
high similarity	.525	.060	.230	.185
low similarity	.760	.050	.045	.140

We can see from these data that Low Similarity tends to get the AI answer much more than High Similarity, and that within the Low Similarity condition the NEG subjects gave many more AI answers than the POS subjects. When we look at what happens with subjects who do *not* give the AI answer, the POS subjects tend to give more "can't tell" answers in the Lo Similarity condition than they do in the Hi Similarity condition. For Hi Similarity, the NEG subjects tend to say "other", whereas POS subjects are more likely to say that the object-in-question follows the exception object. We note that, with the possible exception of Pollock (1987), *no formal theory of defeasible reasoning predicts any effect about knowledge of an exception object upon the object-in-question, nor do any of them predict an effect for Similarity of an exception object with the object-in-question.* This seems a serious shortcoming of existing theories.

There is also a three-way interaction among answer category, specificity, and similarity in this study. First, when similarity is high, there was no significant impact of the specificity manipulation on the proportion of AI answers given (or for that matter, of the other answer categories). But when similarity is low, i.e., subjects are given the vanilla benchmark problems to solve, fewer AI answers are given under the specific form than under the nonspecific form. One interpretation of this result is that the impact of the extra information in the high-similarity case just outweighs any effect of specific or non-specific information about the exception. Second, we had predicted that the similarity and specificity manipulations would increase the proportion of "like the exceptional object" answers. Although the effect was in the right direction for the specific condition (.195 versus .090) and did approach significance, the increase in this particular answer was not reliable. Thirdly, the high proportion of "can't tell" answers under the high similarity/nonspecific case (.230) may be due to an artifact of the materials: perhaps the Hi Similarity subjects did not want to give the AI answer but there was no way in which they could easily say it was "like the exception" since no specific information about the exception was given. Instead, they drew their answers from the other information given in the high-similarity sentences added to the problem.

Returning now to the Robot/Human manipulation, let us take a look at the data (Table 4). Note (again) that Problem #3 is significantly different from the others. This problem is interesting because each of the objects violates one of the two default rules; our results on this Benchmark suggest that if people find an object atypical in one way, they may find it plausible to conclude it will be atypical in other ways. Note also that when subjects imagine themselves as agents in the problems (the Human condition), Problem #3 is again different from the others, and that most of this difference is due to the high occurrences of "other" answers. In contrast, note that the difference in Problems #2 and #4 between the Human and the Robot condition is due our subjects wanting to

Table 4. Proportion of Answers as a Function of Problem and Problem Solving

Problem Solver:	Human Benchmark Problem				Robot Benchmark Problem			
	1	2	3	4	1	2	3	4
Answer Category								
AI	.625	.600	.437	.612	.762	.587	.525	.737
exception	.175	.137	.175	.100	.075	.025	.025	.037
other	.062	.125	.237	.112	.087	.162	.175	.112
can't tell	.137	.137	.150	.175	.075	.225	.275	.112

have the Robot say "can't tell" much more often than they require it of themselves. We enshrine this last feature as:

The Asimov Effect: People believe robots should be cautious (saying they "Can't tell") in cases when they themselves would be willing to give a definite answer.

Among the results of this experiment that we found interesting was the role that Hi vs. Lo Similarity played. Our view is that subjects interpret any "extra" information as possibly giving a reason why the exception object did not obey the default rule, and perhaps an indication that this reason applies to the object-in-question also. For example, in the Hi Similarity version of the experiment, a problem specified where computer manuals are typically found: subjects were told that both the IBM and the Mac manuals were being reviewed by support staff because new software had been purchased. This assertion could be interpreted as a *reason why* the exception object (the Mac manual) was not where manuals typically were, *and also could be construed as giving a reason* to believe the object-in-question (the IBM manual) might also violate the rule. We enshrine this view as:

Explanation-based Exceptions: When the given information provides both a relevant explanation of why the exception-object violates the default rule and also provides a reason to believe that the object-in-question is similar enough in this respect that it will also violate the rule, then infer that the object *does* violate the rule.

So, it is not similarity alone, but rather the availability of information which explains why the exception is an exception, and which hints that the object-in-question might fall under that explanation.

Another interesting finding of this experiment was about Benchmark Problem #3, where it appeared that if an object violates one default rule, subjects will view it as being likely to violate other default rules as well. We enshrine this observation as:

Second-Order Default Reasoning: If the available information is that the object-in-question violates other default rules, then infer that it will violate the present rule also.

(Others might prefer to call this the "Guilt by Past Association" rule, or maybe the "Bad Egg" principle – "once a bad egg, always a bad egg").

6. Some concluding remarks

First we would like to emphasize the preliminary nature of these results. There are many further directions in which experimentation in this area could be taken, such as following up on the observation that subjects seem to be influenced by the specificity of the information they are given, by the amount of information available about the two objects, by the "similarity" that subjects perceive between the two objects, and by the number of previously-violated default rules. Each of these claims deserves more, and more careful, experimentation.

But turning our attention to the matters discussed in this paper, there are two types of conclusions that we would like to urge.

First type of conclusion: Unlike most other reasoning formalisms, nonmonotonic or default reasoning is "psychologistic" – that is, it is defined *only* as what people do in circumstances where they are engaged in "commonsense reasoning". It follows from this that such fundamental issues as "what are the good nonmonotonic inferences?" or "what counts as 'relevance' to a default rule?", etc., are only discoverable by *looking at people and how they behave.* It is *not* a formal exercise to be discovered by looking at mathematical systems, nor is it to be decided by such formal considerations as "simplicity" or "computability", etc.

Second type of conclusion: The results of the experiments reported here point to some particular considerations which seem to be critical to non-monotonic theories. First, we have some evidence that apparent similarity between the exception object and an object-in-question impacts whether or not the default rule would apply. The results for the basic default reasoning problems indicate that apparent similarity between two objects, and the specificity of information about how a rule is violated, impact the application of the default rule. How can relevance and specificity be worked into a logic of non-monotonic reasoning? One idea that we mentioned was called "explanation-based default reasoning", in the same sense used in the machine learning literature. That is, a reasoner attempts to explain why some default rule does not apply to one object and then sees if that explanation can fit another object. This would shift the investigation of non-monotonic reasoning away from a specification of a logic per se and towards the specification of some process that (perhaps deductively) determines whether an explanation constructed for one case fits a second case. We also drew attention to the notion of a "second-order default", where an object is seen as violating many other default rules and is therefore seen as likely to violate the current rule also. The idea is that a default reasoning system should countenance "generally exceptionable objects". And finally we drew attention to a feature that merits attention at some time: that people in general want to allow their artificial agents to draw fewer default conclusions then they themselves are willing to draw. This last is perhaps not an issue for default reasoning systems themselves to address, but is certainly a topic that needs to be addressed in the wider realm of intelligent artificial agents and their role in society.

Acknowledgements

The authors would like to thank research assistants Siobhan Nearey and Matt Duncan, and the Department of Psychology at the University of Alberta for use of its research pool. They would also like to acknowledge the assistance of NSERC individual research grants, a joint NSERC equipment grant, and the University of Alberta Central Research Fund.

References

Bacchus, F.: (1991), *Representing and reasoning with probabilistic knowledge*, The MIT Press, Cambridge.

Baron, J.: (1994), *Thinking and Deciding*, Cambridge Univ. Press, NY.

Belnap, N.: (1978), 'A Useful Four-Valued Logic', in J. Dunn, G. Epstein (eds.), *Modern Uses of Multiple-Valued Logic*, Reidel, Dordrecht, pp. 8–37.

Cohen, L.J.: (1981), 'Can Human Irrationality be Experimentally Demonstrated?', *Behavioral and Brain Sciences* **4**, pp. 317–370.

Evans, J.: (1987), *Bias in Human Reasoning: Causes and Consequences*, Lawrence Erlbaum, Hillsdale, NJ.

Gigerenzer, G.: (1991), 'How to make Cognitive Illusions Disappear: Beyond "Heuristics and Biases"', *European Review of Social Psychology* **2**, pp. 83–115.

Gigerenzer, G.: (1994), 'Why the Distinction between Single-event Probabilities and Frequencies is Important for Psychology (and Vice Versa)', in G. Wright, F. Ayton (eds.), *Subjective Probability*, John Wiley & Sons, NY, pp. 129–161.

Gigerenzer, G.: (1998) 'Ecological Intelligence: An Adaptation for Frequencies' in D. Cummins, C. Allen (eds.), *The Evolution of Mind*, Oxford University Press, NY, pp. 9–29.

Kahneman, D., Slovic, P., Tversky, A.: (1982), *Judgment Under Uncertainty: Heuristics and Biases*, Cambridge UP, Cambridge.

Lifschitz, V.: (1989), 'Benchmark problems for formal nonmonotonic reasoning, version 2.00', in M. Reinfrank, M.J. de Kleer, M. Ginsberg (eds.), *Nonmonotonic Reasoning*, Springer, Berlin, pp. 202–219.

McCarthy, J.: (1980), 'Circumscription – a Form of Non-Monotonic Logic', *Artificial Intelligence* **13**, pp. 27–39.

McDermott, D., Doyle, J.: (1980), 'Non-monotonic Logic, I', *Artificial Intelligence* **13**, pp. 41–72.

Moore, R.: (1985), 'Semantical Considerations on Nonmonotonic Logic', *Artificial Intelligence* **25**, pp. 75–94.

Pearl, J.: (1988), *Probabilistic Reasoning in Intelligent Systems: Networks of Plausible Inference*, Morgan Kaufmann, San Mateo.

Pelletier, F.J., Elio, R.: (1997), 'What Should Default Reasoning Be, By Default?', *Computational Intelligence* **17**, pp. 165–187.

Pollock, J.: (1987), 'Defeasible Reasoning', *Cognitive Science* **11**, pp. 481–518.

Reiter, R.: (1978), 'On Reasoning by Default', *Proceedings of TINLAP-2*, Univ. Illinois: Assoc. for Computational Linguistics, pp. 210–218.

Reiter, R.: (1980), 'A Logic for Default Reasoning', *Artificial Intelligence* **13**, pp. 81–132.

Selman, B., Kautz, H.: (1989), 'The Complexity of Model-Preference Default Theories', in M. Reinfrank, M.J. de Kleer, M. Ginsberg (eds.), *Nonmonotonic Reasoning*, Springer, Berlin, pp. 115–130.

Wason, P.: (1968), 'Reasoning About a Rule', *Quarterly Jour. of Experimental Psychology* **20**, pp. 273–281.

Wason, P., Johnson-Laird, P.: (1972), *The Psychology of Reasoning: Structure and Content*, Harvard Univ. Press, Cambridge.

Yates, J.F.: (1990), *Judgment and Decision Making*, Prentice Hall, Englewood Cliffs.

VII

METHODOLOGY

EXPERIMENT AND THE GROWTH OF EXPERIMENTAL KNOWLEDGE

Alan Chalmers
Unit for History and Philosophy of Science
University of Sydney, NSW 2006, Australia

1. Introduction

Attempts to address the problems facing positivism by stressing the role of theory in science, that became popular in the nineteen sixties, ran into problems of their own. The more recent 'new experimentalism' has arisen as an attempt to meet those latter problems. In this paper I offer an appraisal and critique of the new experimentalism, focusing especially on the most sophisticated exposition and philosophical defense of it, the account given by Deborah Mayo in her book *Error and the Growth of Experimental Knowledge*.

From the point of view of positivist-style philosophies of science, scientific knowledge is based on a secure foundation of facts established by observation. As far as its mode of argumentation is concerned, science is seen as proceeding from the bottom up, deriving scientific knowledge from facts by some kind of inductive inference. Karl Popper challenged both the idea that science is based on secure observational foundations and that there is available to science a mode of induction that is at once clearly formulated, ampliative and free from a range of telling objections. For Popper and many that followed in his footsteps observation is theory – or paradigm – dependent to the extent that a change in what is considered to be the best theory or paradigm will occasion some changes in what are considered to be the observable facts. (It took an understanding of Newton's law of inertia to appreciate that the claim that the earth is stationary is false in spite of the extent to which it seemed to have been borne out by experience.) From a Popperian point of view, argument in science is top-down rather than bottom-up. The deductive consequences of theories are tested against observation, with those theories that fail the test being rejected. The best theories are those that best withstand tests.

For those, whom I take to be in the majority, that take it as given that science progresses in some unambiguous sense and that it offers reliable guidance to technology, the top-down approach fails because it cannot adequately grasp those aspects of science. It is too fallibilist and too negative. The gen-

P. Gärdenfors, J. Woleński and K. Kijania-Placek (eds.), In the Scope of Logic, Methodology and Philosophy of Science, Vol. I, 157-169.

eralizations that constitute scientific knowledge can never be confirmed to any significant degree. Not even the empirical base of science represents secure ground for a Popperian, insofar as observation is theory dependent and fallible. Many doubt that the fact that a theory has withstood Popperian tests is sufficient to capture its scientific credentials within the top-downers scheme of things and appeal to the 'underdetermination of theory by evidence'. The ability of a theory to stand up to tests does not pick it out as superior to an infinity of other possible theories able to withstand the same tests. Not even falsifications are decisive because of the Duhem/Quine problem. Any lack of match between theory and observation indicates that there is a problem somewhere in the maze of theory plus auxiliary assumptions plus observational claims, but top-down reasoning, the logic of falsification, is not able to locate the cause of the trouble in the maze.

Only if the above problems are solved or dissolved can the position of the top-downers be taken as capturing what most take to be the distinctive feature of science, the progressive and reliable way in which it learns from experience. Popper picked as the theme for his collection of essays *Conjectures and Refutations* (1969, p. vii) the thesis that 'we can learn from our mistakes' where it transpires that a mistake is a theory that turns out to be refuted by experience. But, as Mayo has stressed, Popper's top-down methodology does not tell us what we can in fact learn from this kind of mistake other than that the falsified theory is mistaken.

> I can therefore gladly admit [wrote Popper (1969, p. 231)] that falsificationists like myself much prefer an attempt to solve an interesting problem by a bold conjecture, *even (and especially) if it soon turns out to be false*, to any recital of irrelevant truisms. We prefer this because we believe that this is the way in which we learn from our mistakes; and that in finding that our conjecture was false, we shall have learnt much about the truth, and shall have got nearer to the truth.

I argued, two and a half decades ago, in my (1973), when my academic career was in its infancy, that this is wrong headed. We learn in science when we find that there is a white spot as predicted by the wave theory of light, that there are radio waves, that light rays are indeed bent when they pass close to the sun and so on. We learn most from our successes rather than our failures. The top-downers have a hard time explaining how this can be the case. Popper himself stressed that the degree to which a theory has passed tests "says nothing whatever about future performance, or about the 'reliability' of a theory" (1979, p.18, italicized in the original). To accept this is to acknowledge that science has nothing positive to offer technology, a counter-intuitive conclusion to say the least.

A tendency amongst the top-downers since the 1960's has been to strengthen the efficacy of arguments in science by increasing the input coming in from the top, by developing notions of research programs, paradigms, research traditions and the like which somehow accommodate and sanction high level theory. Lakatos's research programs contain conventionally accepted hard-cores, Kuhn's paradigms contain community sanctioned rules and so on. A general problem run into by these theory-dominant positions, apart from the status of some of their components, involves the comparison of one framework with another. If methods for theory generation and theory appraisal are program or paradigm dependent, in what sense can a framework be said to be better than a rival? In what sense can scientific revolutions be said to be progressive? The greater the role attributed to theoretical frameworks in the evaluation of data and the justification of theory with respect to data, the greater the problem of incommensurability becomes, it would seem. By stressing more and more the importance of input from the top, the top-downers have lost their grip on the importance of input from the bottom. They have lost an appropriate grip on the sense in which scientific revolutions result in increased reliability and experimental control. It would seem that the problems that beset the top-downers from Popper onwards need to be confronted by a healthy re-introduction of some bottom-up reasoning. At least part of the motivation of the new experimentalism is to do just that. Let us see to what extent they have been successful.

2. A secure experimental base for science

The new experimentalists turn to experiment, as opposed to mere observation, to seek a secure basis for science that can be appealed to to solve some of the problems that the radical theory dependence of the top-downers leads to. If there is a fund of experimental results that can be securely established independently of high level theory then those results can be drawn on as a reliable guide to technology, whilst the steady accumulation of such results afford at least one sense in which science unambiguously progresses.

The production of controllable and demonstrable experimental effects give the new experimentalists part of what they want. Let us consider, for example, the first, primitive electric motors constructed, following two months of laborious trial and error, by Michael Faraday (Gooding 1990). These motors involved a wire, suspended in such a way that its lower end, dipping into a pool of mercury surrounding one end of a magnetized iron rod, was free to rotate around that rod, and did indeed rotate once a current was passed through the wire. Faraday's understanding of his motor involving magnetic lines of force circulating the wire, a mode of theorizing that was at the time novel and shared by few. However, an appreciation of the demonstrable reality of the electric motor effect did not depend on an understanding or acceptance of Faraday's theory. This is brought home by the way in which Faraday quickly capitalized on his discovery and secured priority over his rivals both in Britain and in

Continental Europe. Each of those rivals received a version of Faraday's motor in the mail, together with instructions on how to make it work. They were instructed to add mercury to the specified level, connect the terminals to a battery and hold one end of a bar magnet against the iron rod. Once they did so, they would find that the wire would rotate, Faraday confidently assured them. He also pointed out that they could reverse the sense of the rotation by reversing the battery connections or the magnet. Rivals such as Ampére accommodated the motor effect into a theoretical framework that was far different from Faraday's and had no place for the latter's mysterious lines of force. But that did not alter the fact that Faraday's motors worked as much for his rivals as for him. An appreciation of that experimental fact simply did not require an appeal to high level theory. If it had, Faraday's use of the mail to secure his claims to priority and advertise his prowess would not have worked in the straightforward way that it did.

One can list a wide range of experimental effects whose reality can be demonstrated in a way that does not require an appeal to high level theory. Hertz produced radio waves in such a way that the proponents of all the competing theories of electromagnetism at the time had no option but to acknowledge their existence and attempt to accommodate them into their theories. Whatever the high level theory that explains them there is no doubting that there are such things as lasers. In all these cases, experimental effects, once discovered, are permanent. There is no way in which acceptable changes in high level theory could have the consequence that the discovered effects don't exist after all. Here is once sense, at least, in which science can be said to progress through revolutionary changes in theory.

Not all experiments are as transparent as those displaying experimental effects in the unambiguous way in which a laser beam can be made to cut its way through a steel plate. There is a sense in which the appreciation of many experimental results are theory dependent insofar as they require an appreciation of how various instruments work and an understanding of how various extraneous effects have been guarded against and so on. Hertz's erroneous conclusion that cathode rays are not deflected by an electric field was at least partly due to his failure to appreciate the extent to which ionization of the residual gas in his discharge tubes could give rise to unwanted electrostatic fields. But even here new experimentalists, notably Ian Hacking, have pointed to a range of techniques experimentalists have for distinguishing between effects and artifacts in a way that bypasses an explicit appeal to high-level theory. Dense bodies are seen in red blood cells in exactly the same location on the microscopic grid using both an electron microscope and a fluorescence microscope. This is strong evidence that the bodies are structures in the blood cells rather than artifacts. How else could one account for the appearance of the same bodies in microscopes working on quite different physical principles? This argument from coincidence goes through without appealing to detailed theory of the workings of either microscope. Another strategy is to practically intervene to see if a questionable observation changes in the anticipated way. Do the lines on an electron micrograph represent details of crystal structure or not? If those lines are distorted in the expected way when the specimen crystal as a whole is distorted, then the verdict that they do is supported (Menter 1956, pp. 123–124). Experimentalists have a range of strategies involving probing and intervening to test for the reality of observed effects that fall far short of an explicit appeal to theory.

Deborah Mayo has provided us with a more quantitative and thoroughly worked out position on what can be established by experimental reasoning and how. She spells out the ways in which experimental regularities can be established, often without appeal to high level theory. Experimental knowledge is knowledge of the frequency with which specified experimental procedures produce results of a specified kind. Establishing experimental knowledge is by no means straightforward. Much skilled work and many tests are involved in producing an experiment of a specified kind and checking that it is indeed one of that kind. For instance, Jean Perrin's experiments on Brownian Motion required suspensions of particles of uniform size. It took him months of centrifuging to produce such suspensions, and a number of independent tests to establish that the process had been successful. Likewise, testing the efficacy of a drug involves ensuring, checking and counter-checking that the treated and control group of subjects are alike in relevant respects whilst any experiment in physics requires that extraneous effects, due to stray cosmic rays, vibrations due to passing trains, thermal noise etc. be eliminated or allowed for. Experimental results are a far cry from the positivists' protocol sentences of the 'I have a red patch in my field of view now' variety.

Once it has been established that an experiment is indeed of a specified kind then the results of such experiments support some claim insofar as it is unlikely that the results would have eventuated were the claim false. Establishing an experimental claim involves probing the possible ways in which it could be mistaken and ruling them out. Experimental knowledge of the low level kind delineated by Mayo is perhaps not quite as problem-free as she would like. After all, we can only probe for the sources of error that we know about. But leaving such quibbles aside, it is undoubtedly the case that many experimental claims can be vindicated by experimental techniques that do not draw on high level theory. Perrin's experimental demonstration that Brownian Motion is random involved tracking the path of particles, observed every thirty seconds with a camera lucida, analyzing the successive displacements in a variety of ways and showing them to be in accord with what is to be expected of random walks. As Mayo (1996, pp. 228–229) observes, 'the number of tests, checks and rechecks Perrin performed amounted to statistical overkill' But note that this small result, established by fairly direct means involving no appeal to high level theory, goes a long way towards establishing a major consequence, namely, that the second law of thermodynamics is at best only statistically true. So experimental knowledge can be construed as reliable and established independent of high level theory, and can, on occasions, have some highly important consequences for high level theory. Nevertheless, it is important to be clear of the limited scope of experimental knowledge as we have characterized it thus far. As Mayo herself observes (1996, p. 187), experimental knowledge is 'knowledge of the results to be expected were certain experiments carried out'. The probabilities attached to such expectations are probabilities in frequentist statistics, and as such they cannot be detached from the experimental processes they characterize. We can note that even this restricted sense of experimental knowledge gives us much of what is needed to understand science as a guide to technology. However, few will be content to leave it at that. In science it is intended that claims supported in experimental situations can confirm laws and theories applicable outside of those situations. Can bottom-up reasoning sanctioned by the new experimentalists be employed for this purpose? Mayo, for one, attempts to show what is involved, although she insists that this is a 'separate step' from that involving the justification of the experimental claims themselves.

3. Confirming laws and theories by experiment

The problem to be confronted is how experimental knowledge can be used to vindicate scientific laws and theories. Mayo approaches the problem by proposing a plausible and useful notion of a severe test which is a generalization of what is already involved in her error statistics. The production of evidence, *e*, by test procedure, *T*, counts as a severe test of hypothesis, *H*, to the extent that *T* would be unlikely to yield *e* were *H* false. The point here is to be stringent and demanding about the conditions under which an hypothesis can be borne out by evidence. Mere compatibility of an hypothesis with the evidence is not enough. The test procedure must be sensitive enough to enable the identification and elimination of ways in which *e* might eventuate were *H* false. Mayo aims to exploit this notion of a severe test to show how scientific knowledge can be vindicated in a way that avoids the underdetermination problem. Whilst it may be the case that there are many, perhaps an infinity, of laws or theories compatible with the available evidence, the aim is to show that there is no more than one law or theory that is severely tested by the procedures that produced the evidence.

Mayo's account of severe testing is designed to confront the problem of underdetermination in a way that Popper failed to do. For Popper, the confirmation of any novel prediction of a theory is an instance of that theory passing a severe test. From Mayo's point of view such tests are often not severe at all. What Newton's theory has to say about comets survived a severe test when Halley's comet was first observed to return as predicted, but the fact that Newton's theory gets it right about comets is no indication that it gets it right about everything. (As a matter of fact, we know that it doesn't get it right about fast moving electrons, for example.) There are many possible theories, differing from Newton's, that can get it right about comets, and the choice between them is certainly underdetermined by observations of comets. Popper freely admitted such circumstances, of course. For him, the fact that a theory successfully makes a novel prediction might be a good reason to prefer it over its predecessor, but it is not a good reason to accept it as true or probable. A Popperian encourages further work on a theory to put its consequences to the test, perhaps thereby achieving further successful novel predictions but eventually revealing its shortcomings. This is too fallibilistic for Mayo, who aims to grasp a sense in which scientific knowledge, once severely tested by experiment stays severely tested. That is what her more demanding notion of a severe test is meant to achieve.

Let us focus on a key example of Mayo's that illustrates her position. It concerns the relationship between Einstein's General Theory of Relativity and the eclipse expeditions led by Eddington to determine the deflection of starlight by the sun. Here I omit the fascinating detail involved in measuring the deflection of starlight intriguingly discussed by Mayo, accept to point out that this involved the comparison of images of stars on photographic plates and did not involve appeal to high level theory. Taking that experimental knowledge as given, we face the question of what theoretical claims are supported by it. Mayo's claim is that whilst Einstein's law of gravity, a consequence of his General Theory of Relativity, is supported by the results of the eclipse experiments, the General Theory itself is not. To argue this she appeals to her notion of a severe test. She asks, would the observed deflection have been possible if Einstein's law of gravity were false and some alternative true. Mayo (1996, pp. 288–292) justifies a negative answer by pointing out how all the Newtonian alternatives to Einstein's law were found wanting and eliminated, either because the effect predicted by those alternatives was of the wrong order of magnitude to explain the observed deflection of starlight or because they had other implications that were false or contradictory. Mayo (1996, p. 291) approvingly quotes a passage from an article in *Nature* written at the time. 'Hence we seem to be driven by exhaustion to the Einstein law as the only satisfactory explanation'.

But there is no principle of exhaustion in Mayo's error statistics. What the argument pinpointed by Mayo shows is that Einstein's law of gravity was shown to be superior to available alternatives by virtue of the fact that it could survive tests that those alternatives could not survive. But that is not sufficient to show that there is no possible alternative, Newtonian or otherwise, not yet thought of. That is, Mayo's case does not solve the problem of underdetermination that she argues is such a shortcoming of Popper's position. Both Popper and Mayo argue for laws and theories on the grounds that they have survived severe tests. Mayo's attempt to strengthen Popper's notion of severity is laudable, and necessary for the kinds of reasons she identifies. However, if a test of a general hypothesis, H, is severe only if it is able to rule out all possible alternatives to H, then severe tests of general laws will be very rare indeed. Mayo's stringent demands cannot be met. That is, Mayo's error statistics has not enabled her to pull a rabbit out of the urn, so to speak, and solve the problem of induction.

It would be a mistake to dismiss Mayo wholesale in the light of the foregoing comments. She has identified a major weakness in the Popperian position and her attempts to address the problem are certainly in the right direction. This is brought out by the further details of her discussion of what the eclipse experiments established and did not establish. Having argued that the Eddington eclipse experiments constituted support for Einstein's law of gravity, Mayo goes on to caution that it should not be concluded from this that those experiments confirmed Einstein's General Theory of Relativity. This is so because there are a whole class of theories of space-time that predict Einstein's law of gravity of which Einstein's theory is only one. Mayo's question, could Einstein's law of gravity be true and the General Theory of Relativity false has to be answered in the affirmative, whereas a negative answer is necessary if Einstein's General Theory of Relativity is to be regarded as severely tested by the eclipse experiments. If the General Theory of Relativity has been experimentally tested only in the areas of light deflection, Mercury's perihelion, the red shift and the Mossbauer effect, then all we are entitled to say is that that theory gets it right for those effects, not that it gets it right about everything.

It might be thought that this kind of problem can be overcome by invoking an argument from coincidence that parallels the one we have already encountered at the experimental level. If a theory successfully predicts a wide variety of phenomena surely it would be a remarkable coincidence if it were able to do so if false. In rejecting this kind of argument Mayo has history on her side. For instance, Fresnel's wave theory of light successfully predicted a wide range of optical phenomena in the decades that followed its first proposal. That is why it was preferred to Newton's particle theory. But, in spite of this, Fresnel's theory proved to be false in significant respects. In particular, the elastic ether that was understood to be an integral part of it proved to be non-existent. On Mayo's behalf we can say that whilst some consequences of Fresnel's theory governing interference, diffraction and transmission phenomena did survive severe tests, the theory in its generality did not.

Suppose a theory, T, has as consequences hypotheses H_1 and H_2 that are largely independent of each other. Then the survival of a severe test by H_1 does not automatically have the consequence that H_2 has survived a severe test also. But this is, in effect, what is involved if it is assumed that T has survived a severe test just because H_1 has. Insofar as this reflects Mayo's reasoning it is surely sound. This way of looking at confirmation calls for what Mayo calls 'partitioning'. This involves clarifying the entailments within a theory to indicate which parts of the theory are necessary for which predictions, in the way we have already seen Mayo do in the case of light bending and the General Theory of Relativity. Whilst I acknowledge the importance of this way of looking at things, I point out that it has consequences that to some extent tell against Mayo's attempt to distinguish her position from Popper's. The way in which theories can be partitioned is not something that is immediately apparent.

The theoretical task involved in the partitioning of a theory constitutes work to be done just as the empirical testing of it does. It was not immediately apparent that electromagnetic theory could dispense with the ether, nor that the General Theory of Relativity could accommodate a whole class of theories of space-time. Once this point is acknowledged then Mayo's position has the consequence that verdicts concerning how severely tested a claim is with respect to some specified evidence produced in some specified way are subject to change in the light of theoretical work on partitioning. At the time of Hertz's first production of radio waves, physicists took them to be confirmation of the existence of the electromagnetic ether. After Einstein those waves are not taken as a severe test of the existence of the ether at all. The non-transparency of partitioning renders judgements about the severity of tests revisable.

4. Finding a home for high level theory

An implication of the foregoing paragraphs would seem to be that, from Mayo's perspective, high level theories can never be severely tested in their generality and hence can never be vindicated. One response to this, reminiscent of Pierre Duhem, would be to identify scientific knowledge with the stock of well-tested experimental knowledge and to treat high level theories merely as heuristic aids to the establishment of experimental knowledge. Any such view would need to take the heuristic role very seriously. After all, it was a prediction of the General Theory of Relativity that led to the Eddington eclipse expeditions and the discovery of the deflection of starlight in the first place! But can the role of theory in science be confined to a heuristic one, however strongly that heuristic role is construed. There are reasons to doubt that it can.

There are cases where theory and the results of observation and experiment stand in a reciprocal relationship and reinforce each other in a way that renders the observational or experimental knowledge theory-dependent. As a consequence, claims that low level knowledge can be established independently of high level theory are undercut. I will illustrate by reference to the relationship between the information contained in micrographs of crystal structure, showing the all-important dislocations in such structure, produced by an electron microscope and theory, both of crystal structure and the workings of the electron microscope. Interpretation of the micrographs, and the decision that structures in them were indeed images of crystal structures and their dislocations, were no straightforward matters. Some modes of argument were arguments from coincidence of the kind highlighted by Ian Hacking and other new experimentalists. For instance, if the spacing of crystal planes calculated on the assumption that lines on a micrograph were indeed images of those planes agreed closely with the spacing as measured by X-ray diffraction techniques, then this lent some support to the interpretation of the micrographs as images of those planes. But other arguments lent more heavily on theory. Certainly the scientists involved appealed to the match between the predictions based on the quantum mechanical detail of the interaction between electron microscope and specimen, on the one hand, and the structure discernible on the micrographs, on the other, as evidence, both for the success of their theoretical treatment and for their interpretation of the observable evidence. Two researchers M.J. Whelan and P.B. Hirsch (1956, p. 1303) who worked on the first quantum mechanical account of the formation of electron microscopic images of dislocations in metals make this point quite explicitly. In their own words, '[The] general agreement obtained between theory and experiment confirms that the fringes [observable on the micrographs] are due to stacking faults and that the theory developed – is basically correct.' The interdependence of theory and experiment, evident here at the outset of the use of electron microscopy in crystallography, became even more important as the electron microscope was pushed towards its theoretical limit.

Peter Hawkes is an electron microscopist at the National Scientific Research Center in Toulouse, France, who, incidentally, was involved as a research student in some of the early work I have already reported. He writes, 'There was a very instructive polemic a few years ago concerning the reality or not of certain butterfly-like patterns, and theory was able to demonstrate that such patterns would be expected if the illuminating beam was tilted (accidentally or deliberately). The point of this example is that observation unassisted by theory was positively misleading and it was an application in which the e.m. [electron microscope] was the only tool that could provide information.' (Private correspondence, 1998)

I do not know how representative of observation in contemporary physics these examples are, but I see no reason to believe that such cases are uncommon. What they involve is an interdependence of observation and theory that seems to demand the inclusion of theory, and fairly high-level theory at that, in judgements of what counts as experimental knowledge. They tell against a separation of merely heuristic theory and securely established experimental knowledge. Interestingly, whilst this may be a problem for some aspects of Mayo's position, examples of the kind I have invoked can be accommodated by her notion of a severe test, informally at least. We can ask, could the match between theory and interpretation of observation have been achieved if both were not indeed correct (at least in the relevant respects) and the answer would seem to be that the match would be highly unlikely if this were not the case.

Another key use of theory in science involves 'deductions from the phenomena' in something like the Newtonian sense. Something like this is involved, for example, when various conservation laws are employed to deduce the properties of microparticles that do not leave tracks in particle detectors from observations of those that do. It seems to me that the conservation of energy, for example, cannot be severely tested according to the strict requirements that Mayo sets for the establishment of experimental knowledge any more than the General Theory of Relativity can and this makes it impossible to accommodate many of the key uses of the conservation of energy in science.

A close identification of scientific knowledge and experimental knowledge makes it difficult to justify or comprehend the application of science outside of experimental situations. At best, it would seem that experimental knowledge can be employed in technologically contrived situations where the controlled conditions are sufficiently like the experimental ones in which the presumed knowledge has been tested. But what of the use of the laws of electromagnetism by meteorologists or, even more removed from experimental test situations, the use of conservation of energy or General Relativity in Big Bang Cosmology? The more one confines scientific knowledge to what can be instantiated in controlled experimental situations, which Mayo tends to do, the more difficult it becomes to justify the application of science to the world outside of those contrived situations. Assuming theories to be applicable generally, and beyond the experimental situations in which they are vindicated, requires standards for confirmation that fall short of Mayo's strict demands.

5. Is underdetermination a problem?

Mayo singles out underdetermination as a problem that Popper did not solve and aims to do better. Her idea, as we have seen, is to show that whilst there may be an infinity of theories just as compatible with the available evidence as those that constitute our scientific knowledge, there is not an infinity of theories that is severely tested by that evidence. For her a claim is severely tested by a procedure yielding some evidence to the extent that that evidence would be unlikely to be produced were the claim false. Mayo's position can be largely sustained if the claims at issue are about the likelihood of experimental results eventuating from well-defined experimental set-ups. But once we progress beyond that to the support of laws and theories, which Mayo herself signals to be a separate step, then Mayo's notion of a severe test is not sufficiently strong to avoid underdetermination. It may be that the evidence can pick out a law as superior to all available rivals, but it is not sufficient to rule out all possible rivals, a point I have illustrated with reference to Mayo's treatment of Einstein's law of gravity. Is this a failing of Mayo's position (or Popper's for that matter)? Is underdetermination a problem? I shall argue that it is not.

We have seen that the fact that a theory withstands tests in the domain of a variety of phenomena that it predicts is not sufficient to establish it in its generality. Arguments from coincidence cannot show that a general theory is all right, but surely they do show a strong sense in which the theory in question cannot be all wrong. Fresnel was wrong about the existence of an elastic ether, but he was right that light has a wave character in a wide variety of circumstances. How else can we account for the wide range of successful predictions made by his theory? What is more, the equations derived from contemporary electromagnetic theory describing the relative intensities of light beams reflected and refracted in various circumstances are identical to those derived by Fresnel provided we do not read too much into what is meant by the 'amplitude' of the waves involved in each case. This is the circumstance that John Worrall (1989) exploits in his defense of "Structural Realism". It is not only well established experimental knowledge, but also some limiting case of theoretical knowledge that lives on through major scientific change. A native of our host country, Władysław Krajewski (1977), has defended this notion under the name of the correspondence principle, as has my own former supervisor Heinz Post (1971) and former postgraduate colleague Noretta Koertge (1969). It is because of the extent to which experimental knowledge and also some limiting case of well confirmed theoretical knowledge is retained through radical theory change that underdetermination does not have drastic consequences for the reliability and progressive character of science. Relativity theory does not undermine the wide applicability of the medieval science of weight, although that latter theory would not be of much use in a space ship. The medieval science of weights has been superseded but was, and is, practically efficacious in a wide variety of circumstances nevertheless. I think it plausible to say that scientists are more confident of applying Newton's laws of motion to the trajectory of a spacecraft in the light of the refutation of those laws by relativistic effects than they would have been otherwise. We can recognise the limits of Newtonian theory in the light of Einstein's and can recognise that spacecraft do not travel at sufficient speed to transcend those limits. It is true that underdetermination introduces fallibility into science insofar as we do not know in advance the circumstances under which our theories will break down. But who would wish to defend a methodology of science that denied that?

To sum up, the new experimentalists, especially Deborah Mayo, have demonstrated a sense in which experimental science has at its disposal techniques for establishing experimental knowledge that does not involve an appeal to high level theory. Such experimental knowledge, once well-confirmed stays well confirmed, and the accumulation of that knowledge provides one sense in which science progresses reliably and cumulatively. Mayo has highlighted how to exploit partitioning so that we can be discerning in our claims about the extent to which theoretical knowledge is confirmed by experiment, but her methodological strictures does not enable her to overcome underdetermination in a way that she would like to so her position does not differ so much from Popper's as she implies. I have argued that the kind of fallibility that underdetermination brings with it is a kind that science can live with.

The high level theories unavoidably utilised in science, but which cannot be confirmed in a way that lives up to Mayo's high demands, will almost certainly turn out to be false in some respects, but not so wide of the mark that appealing to them will lead us too far astray in a wide variety of circumstances. The top-down reasoning of the theory-dominant tradition can and does fruitfully co-exist with the bottom-up reasoning of the new experimentalists. Anyway, that's how it looks from down-under.

References

Chalmers, A.F.: (1973), 'On learning from our mistakes', *British Journal for the Philosophy of Science* **24**, pp. 164–173.

Gooding, D.: (1990), *Experiment and the Making of Meaning*, Reidel, Dordrecht.

Koertge, N.: (1969), *A Study of Relations Between Scientific Theories: a Test of the General Correspondence Principle*, Unpublished Ph.D. Thesis, University of London.

Krajewski, W.: (1977), *Correspondence Principle and Growth of Science*, Reidel, Dordrecht.

Mayo, D.: (1996), *Error and the Growth of Experimental Knowledge*, University of Chicago Press, Chicago.

Menter, J.W.: (1956), 'The direct study by electron microscopy of crystal lattices and their imperfections', *Proceedings of the Royal Society of London, Series A* **126**, pp. 119–135.

Popper, K.R.: (1969), *Conjectures and Refutations*, Routledge, London.

Popper, K.R.: (1979), *Objective Knowledge*, Oxford University Press, Oxford.

Post, H.R.: (1971), 'Correspondence, invariance and heuristics: in praise of conservative induction', *Studies in History and Philosophy of Science* **2**, pp. 213–255.

Whelan, M.J., Hirsch, P.B.: (1956), 'Electron diffraction from crystals containing stacking faults', *Philosophical Magazine* **1**, pp. 1121–1142 and 1303–1324

Worrall, J.: (1989) 'Structural realism: the best of both worlds?', *Dialectic* **43**, pp. 99–124.

THEORY TESTING, STATISTICAL METHODOLOGY, AND THE GROWTH OF EXPERIMENTAL KNOWLEDGE*

Deborah G. Mayo
Department of Philosophy,
Virginia Tech

Abstract The idea that scientific progress is a matter of the growth of experimental knowledge has several implications for an adequate methodology of evidence, inference, and testing. Far from desiring an account for rationally accepting, preferring, or "probabilifying" large-scale theories with limited evidence, we need an account that shows us how to learn more about the phenomena of interest by probing individual hypotheses severely. I illustrate with experimental general relativity from 1960–1980. Progress hinged on systematically considering how discrepancies in key parameters could have failed to be detected with existing data. An understanding of statistical methodology shows how we learn about theoretical parameters and hypotheses by means of experimental parameters and hypotheses that are identical, or are good approximations, to the theoretical ones.

1. Introduction

My idea of the growth of experimental knowledge reflects several insights that have emerged from the increased interest in experiment among philosophers and historians of science over the past fifteen years. Rather than look to high-level theory as the seat of scientific progress, these "new experimentalists" urge, we should look to local aspects of experiment. Learning from evidence, in this new experimentalist philosophy, depends not on appraising large-scale theories but on local experimental tasks of estimating backgrounds, modeling data, distinguishing experimental effects, and discriminating signals from noise.

*This paper was written under tenure of a National Science Foundation, Science and Technology Studies Award. I gratefully acknowledge that support. For several useful comments on earlier drafts of this paper, I thank Alan Chalmers.

P. Gärdenfors, J. Woleński and K. Kijania-Placek (eds.), In the Scope of Logic, Methodology and Philosophy of Science, Vol. I, 171-190.

I am happy to chime in with this experimentalist call to move away from so-called "theory-dominated" philosophies of science, and toward the local, nitty-gritty details of "the life of experiment". However, my work stands in contrast to those who take the complexities and context-dependencies of experimental practice as grounds for giving up on the goal of developing a general account of scientific inference and testing. I propose instead to take what we have learned from focusing on experiment to develop a more adequate account of scientific inference and the growth of knowledge. The growth of knowledge, in the account I favor, has to do not with replacing or confirming or "probabilifying" or "rationally accepting" large-scale theories, but with testing specific hypotheses in such a way that there is a good chance of learning something-whatever theory that information becomes part of. This learning proceeds by testing experimental hypotheses and inferring those hypotheses that pass probative or severe tests. Progress is not in terms of appraising high level theories – which may not even be available – but in terms of the growth of this experimental knowledge.

Some philosophers of science, even while describing themselves as "theory-dominated", seem willing to go with me at least part of the way in the account of experimental inference I have proposed. They grant it offers a "useful corrective" (Chalmers 1999) to the tendency to focus too much on theory, and they allow that it is preferable to traditional logics of confirmation, as well as (for most) to contemporary Bayesian confirmation accounts (Laudan 1997). They grant further that experimental knowledge has many of the assets I claim for it: that it is highly robust through changes in higher-level theories, that low level testing can be the basis for scientific revolutions, that it gives us a handle on, or at least greatly mitigates, recalcitrant problems (theory-ladeness of data, Duhem's problem). *But still they are not happy!* They want more! They demand to know the relationship between experimental inference and experimental knowledge on the one hand, and testing, appraising, accepting full-blown theories, on the other. Can these methods of experimental testing ever allow us to say that a full theory has passed a probative or severe test? And if not, do they not require some supplementary account of large-scale testing (perhaps on the order of a comparative account, that I will be describing)?

> I regard it as an outstanding and pressing problem in the philosophy of the natural sciences to augment the insights of the new experimentalists with a correspondingly updated account of the role or roles of theory in the experimental sciences, substantiated by detailed case studies (Chalmers 1999, p. 251).

Although first and foremost my previous work has been directed toward developing an updated account of evidence and inference, I agree with this call. Today's discussion springs out of my recent attempts to consider what a correspondingly updated account of the roles of theory would look like, within the experimentalist account of inference that I favor.

2. Statistical methodology: the error-statistical account

To begin I must sketch the main ingredients of the testing account I recommend. Perhaps the best way to do so is by contrasting my account with traditional philosophical accounts of evidence and inference. These traditional, logical empiricist, accounts consist of rules relating statements of evidence and hypotheses by means of either an *inductive logic* of confirmation (e.g., Carnap, Reichenbach) or a *deductive logic* of falsification (e.g., Popper). All such accounts seek to provide logical relations between evidence and hypothesis whether it be one of confirmation, support, or corroboration. We can call such accounts *logics of evidential relationship*, or *E-R logics*.

Both the inductive and falsificationist variants were plagued with numerous philosophical problems and paradoxes, and fell short of answering those (e.g., Kuhn) who questioned whether scientific practice could be captured by logics of the sort that philosophers sought. Methodological rules for testing are so vague, some claimed, that the same data could be taken to "justify" hypothesis H just as well as H's rivals on the basis of the same rules – i.e., data underdetermines hypotheses. As a result, many proposed that hypotheses could only be tested in a "holistic" manner, relative to entire *theoretical systems* (Duhem-Quine) or *paradigms* (Kuhn). For post-Kuhnian holists testing hypotheses does not occur apart from appraising larger units – an entire *research program* (Lakatos) or *tradition* (Laudan). If philosophers are not to abandon the goal of justifying science, many thought, our task would become that of showing what is "rational" about large-scale theory change especially in scientific "revolutions". So, we should remember that at least part of the reason recent philosophers have emphasized large-scale theory appraisal stems from the failures of the logical empiricist accounts of evidence and testing. The question that arises is whether it is possible that possessing a more adequate account of empirical inference would bring with it a redirection of our goals back to the more localized testing of earlier accounts of evidence, albeit with a more adequate account of inference. We will have to see.

The framework I suggest moves away from the two key features of E-R logics: first, rather than start its work with given evidence, or accepted evidence statements, the account I favor includes the task of generating, modeling, and analyzing data; and second, the question of how well evidence warrants hypotheses in my account turns not on a logical computation, but on empirical information about the *overall reliability* or *severity* of the experimental testing procedures. To determine if test data provide strong evidence for a hypothesis requires considering how reliably a test is able to detect a given type of error; that is, it requires knowing the test's *error probabilities*. I call all accounts that embrace this principle *error probability accounts*.

Under standard error probability methods I include Neyman and Pearson tests and estimation procedures, as well as Fisherian and related methods. Nevertheless, I advocate reinterpreting standard statistical accounts so as to avoid ever-present criticisms and misuses, and to move away from the traditional characterization of Neyman-Pearson tests as tools for good long run decisions. I interpret them instead as tools for obtaining reliable experimental knowledge. Because I retain what I think is fundamental in these methods – the use of error probabilities in discerning the import of data – I call my account of testing *error statistics*.

A main way to contrast the two kinds of approaches is by means of their quantitative measures. The quantities in evidential-relationship, E-R, approaches are probabilities or other measures (of support or credibility) assigned to hypotheses. The error statistician, in contrast, employs probabilities to characterize the properties of methods, e.g., of testing or of estimation procedures. An example of an error probability would be the probability that a given procedure of testing would reject a null hypothesis erroneously – the probability of a *Type I error*. Such notions, from standard statistical practice, form the backbone of the error statistical account and, although I urge the relevance of these ideas beyond formal statistical contexts, I wish to emphasize that it is a mistake to completely divorce this account of testing from its formal statistical source, as many seem inclined to do. The statistical methods do real work for us – and it is only by understanding them properly that one can take advantage of the astute insights they offer for more informal contexts.

Most notably, thinking of formal error statistical testing alerts us at once that it is impossible to assess reliability or severity with just statements of data and hypotheses divorced from the experimental context in which they were generated, modeled, and selected for testing. We may implement this statistical lesson by identifying three main elements for experimental inquiry in general, which we can represent as three types of models: *models of primary hypotheses, models of data*, and *models of experiment* that link the others by means of test procedures.[1] A substantive scientific inquiry is to be broken down into one or more local hypotheses that make up the *primary questions* or *problems* of separate inquiries. These local problems, I propose, can be framed in terms of one or more standard or *canonical errors*: about the values of quantities (or parameters), about causes, about accidental or spurious effects, and about assumptions involved in testing other errors. The experimental models serve as the key linkage models connecting the primary model to the data models. Because of the errors of observation, and the inability to directly connect primary questions to observable quantities, the experimental and data models are often statistical in nature. I should be clear that my notion of experiment is not limited to cases where there is literal control or manipulation – indeed, statistical considerations have their greatest role precisely when we cannot literally control or manipulate factors but at best can model *what it would be like (statistically)* if we could control or manipulate them.

Such counterfactual information is given by what is officially called the *sampling distribution* of a statistic – I call it the experimental distribution[2] – and much of my work has been to show how this statistical information may be used to arrive at experimental knowledge. The reasoning used in arriving at this knowledge follows an informal pattern of argument that I call an *argument from error* or *learning from error*. The argument is guided by the following thesis:

> *Arguing From Error*: It is learned that an error is absent when (and only to the extent that) a procedure of inquiry (which may include several tests) with a high probability of detecting (i.e., signaling the presence of) the error if and only if it is present, nevertheless detects no error.

Such a procedure of inquiry may be called a *reliable* or highly *severe* error probe. We can argue that an error is absent if it fails to be signaled by a highly severe error probe. (Analogous arguments are used to infer the presence of an error.) Formal error statistical tests serve to provide tools which allow us to ensure ahead of time that errors will be correctly detected (i.e., signaled) with high probabilities.

Alternatively, the argument from error can be described in terms of a test of a hypothesis H, that a given error is absent. The evidence indicates the correctness of H when H passes a *severe test* – one with a very high probability of failing H, if (and only if) H is false.

More formally, I define, for an experimental test procedure, ET:

> *Hypothesis H passes a severe test with e* if (a) e fits H^* and (b) the experimental test procedure ET had a high probability of producing a result that accords *less well* with H than e does, if H were false or incorrect.

* for a suitable notion of fit or "distance".

We are familiar with the principle underlying the severity requirement from day to day reasoning. For an analogy, consider a tool for a medical diagnosis such as an ultrasound probe. If the diagnostic tool has little or no chance of detecting a given disease – say a cyst of given size – even if it is present, then a passing result, a "clean bill of health", fails to provide grounds for thinking the cyst is absent. That is because the tool had very low severity. To connect such a tool with a test, hypothesis H may be the assertion: there is no cyst (perhaps of a given size). The error that is absent according H is supposing there is no cyst when in fact there is one (perhaps of a given size). Alternatively, if the tool has an extremely high chance of detecting the disease or cyst, just in case it is present, then a clean bill of health is a strong indication that the cyst (at least of the given size) is absent, that is, it is strong evidence that hypothesis *H* is correct.

Since my main purpose today is not to discuss the severity notion, I will not delve into several details and qualifications that its full understanding requires (but see Mayo 1996, 1997a,b).[3] Two, however, are too important to overlook:

First, the severity of a test must always be understood as the severity of a test that passes (or infers, or finds evidence for) some hypothesis *H* with specified data. A test can be severe for one inference and not for another.[4] The second thing to emphasize is that the severity condition can not be identified with one of various likelihood measures; in particular condition (b) of the severity definition is *not* the same as saying that *e* is very improbable given not-*H*, i.e., that $P(e \mid H \text{ is false})$ is low. For example *H* might assert a given effect is real, not due to chance, and a result *e* might be very improbable under not-*H*, where not-*H* asserts that the observed effect is just a coincidence, or due to chance – but, depending upon the experimental sampling rule, the test may have a very high probability of unearthing some such statistically unusual result *or other*, even by chance alone. Such a result would not severely pass the hypothesis *H*, that the effect is real (see, for example, Mayo (1996), chapter 10, and Mayo and Kruse (2001)).

One need not start from considering a hypothesis and then seeing if data provide evidence for it; one can also start with data and work one's "way up" as it were. This is what goes on in an estimation problem, where parameters are estimated from data. It is this version of severity that is especially relevant for us today:

> Data *e* (generated by experimental test procedure ET) is good evidence for the estimate (i.e., set of parameter values) that ET severely passes with *e*.

In this way, a confidence interval estimation procedure rules out, with severity, parameter values improbably far from the experimental result, licensing the inference that the resulting interval estimate contains the correct value of the parameter of interest.

An important challenge in setting out this notion of severity is to show how it avoids the obstacles that vitiate other, broadly analogous, notions of severity (e.g., that of Popper). The most serious obstacle is this: How can one satisfy the severity requirement – that there be a high probability of detecting *H*'s errors if *H* is false – when there are always alternatives to *H* that have not even been thought of? One thing that enables my account of severity to answer this objection is that it is a *piecemeal* account of testing.

The thrust of experimental design is to deliberately create contexts which enable questions to be asked one at a time. Within an experimental testing model, "*H* is false" refers to the denial of a specific primary hypothesis *H*: if

H states a given effect is systematic – of the sort brought about more often than by chance – then not-H states it is due to chance; if H states a parameter value deviates from 0 by no more than a certain margin of error, not-H states it deviates by more than this amount, if H states that factor F is responsible for at least p% of an effect, not-H states it is responsible for less than p%, and so on. It is the particular experimental model, in the series of models of an inquiry, that designates the hypothesis and space of alternatives under test.

3. The error-statistical account and large-scale theory testing

This piecemeal idea still holds when we are probing a large-scale theory, and now we start to move into new territory. Even large-scale theories, when we have them (in this account), are applied and probed only by a piecemeal testing of local experimental hypotheses. Rival theories T_1 and T_2 of a given phenomenon or domain, even while corresponding to very different primary models (or rather, very different answers to primary questions), need to be applicable to the same data models, especially if T_2 is to be a possible replacement for T_1. In particular, T_2 must be able to *describe and analyze the significance of the experimental outcomes that its rival T_1 can.* This in turn suggests ways they may be linked to shared experimental models which allows severely probing and discriminating them.

3.1. Implications of the piecemeal account for large-scale testing

Although the question about large-scale testing goes beyond the focus of my previous discussions of (localized) testing (e.g., in Mayo 1996), there are several implications or groups of theses that emerge fairly directly from that account, and I shall begin by listing them:

1. *Large-scale theories are not severely tested all at once*: To say that a given experiment E is a test of theory T is an equivocal way of saying that E probes what T says about a particular phenomenon or experimental effect, i.e., E attempts to discriminate the answers to a specific question H. We can abbreviate what T says about H as $T(H)$, and we can write: $T_n(H) = h_n$. *This piecemeal idea of what is passed lets us warrant the severity assessment, and say "what is so good" about a hypothesis that has passed a severe test.*

2. *A severity assessment is not threatened by alternatives at "higher levels"*: If two rival theories, T_1 and T_2, say the same thing with respect to the effect or hypothesis H being tested by experimental test ET, i.e., $T_1(H) = T_2(H)$,

then T_1 and T_2 *are not rivals* with respect to experiment E. *Thus, a severity assessment can remain stable through changes in "higher level" theories.*

3. *Severity discriminates between theories that "fit" the data equally well.* T_1 is discriminated from T_2 (whether known, or a "beast lurking in the bush"[5]) by identifying and testing experimental hypotheses on which they disagree i.e., where $T_1(H) \neq T_2(H)$, that is, $h_1 \neq h_2$. Even though *two rival hypotheses might "fit" the data equally well, they will not generally be equally severely tested by experimental test ET.*

The above points may be seen to concern themselves with: *reliability, stability, and avoidance of serious underdetermination*, respectively. They are the basis for a further claim I endorse, namely:

4. Learning from piece-meal tests is the central engine for learning about higher level theories.

As I see it, theories, i.e., theoretical models, serve analogous roles that are to experimental models in the tasks of learning from empirical data. Just as experimental models serve to describe and analyze the relevance of any of the experimental data for the experimental phenomenon, theoretical models serve *to analyze the relevance of any of the experimental inferences (estimates and tests) for the theoretical phenomenon.*

I shall explain and elucidate these theses in the midst of tackling the claims of my "theory-dominated" critics (or "rescuers"). First, let me sketch what they say.

3.2. The holist-comparativist rescue:

One can see from my first point in 3.1 why philosophers who view progress in terms on large-scale theory change advocate a comparative testing account as a needed supplement to my account of local experimental testing. Since a large-scale theory may, at any given time, contain hypotheses and predictions that have not been probed at all, and thus hypotheses which could not be said to have passed severely (in my sense), it would seem impossible to say that such a large-scale theory had severely passed a test as a whole. A comparative testing account, however, would allow us to say that the theory is best tested so far, or we should prefer it so far. This position is set out clearly in recent comments by Larry Laudan, in a section entitled, "Comparativism to the Rescue".

> When we ask whether GTR can be rationally accepted, we are not asking whether it has passed tests which it would almost certainly fail if it were false. As Mayo herself acknowledges, we can rarely if ever make such judgments about most of the general theories of the sciences. But we can ask 'Has GTR passed tests

> which none of its known rivals have passed, while failing none which those rivals have passed?'
> Within such a comparativist perspective, we can say that a theory has been severely tested [as a whole] provided that it has survived tests its known rivals have failed to pass (and not vice versa). (Laudan 1997, p. 314)

Alan Chalmers, to whom I am also indebted for turning my attention to this problem, maintains that I already must be, implicitly, appealing to something like Laudan's comparativist account of testing high level theories:

> [Mayo's] argument for scientific laws and theories boils down to the claim that they have withstood severe tests better than any available competitor. The only difference between Mayo and the Popperians is that she has a superior version of what counts as a severe test. (Chalmers 1999, p. 208)

Amalgamating Chalmers' and Laudan's suggestions, we can identify what they have in mind for a comparativist account as follows:

> [3.2] *Comparativist Testing*: A theory has been well or severely tested provided that it has survived (severe?) tests its known rivals have failed to pass (and not vice versa).[6]

While I am grateful to Laudan and Chalmers for turning my attention to this issue, I shall argue that the comparativist-holist move that they advocate is no rescue, but rather conflicts with the main goals of the severity account. But this critique bears positive fruits: It points the way to understanding the view of testing and learning about large-scale theories that emerges from, and is at home within, the error-statistical account of experiment. I shall proceed by discussing a cluster of issues that arise with respect to the points delineated in 3.1.

4. Critique and contrast

4.1 Point 1: On the comparativist appraisal, we are stuck with this problem: *one cannot say about the comparatively best tested theory what severity requires; that the ways the theory or claim can be in error have been well-probed and found to be absent (to within the various error margins of the test).* It seems disingenuous to say all of theory T is well tested when one knows there are ways T can be wrong that have not been probed that there are regions of implication not checked at all. One is back to a problem that beset Popper's account – namely being unable to *say what is so good about the theory that (by historical accident) happens to be the best tested so far?*

Whereas we *can* give guarantees about the reliability of the piecemeal experimental test, we cannot give guarantees about the reliability of the procedure advocated by the comparativist-tester (understood along the lines of [3.2]). Their procedure is basically: Go from passing hypothesis H to passing all of T – but this is a highly *un*reliable method – anyway, it is entirely unclear how one could ever assess its reliability. By contrast, we can apply the severity idea because the condition "given H is false" (even within a larger theory) always means "given H is false with respect to what it says about *this particular* effect or phenomenon". If a hypothesis $T(H)$ passes a severe test we can infer something positive: that the theory T gets it right about the specific claim, H, that severely passes. Alternatively, we may infer that the theory is correct with respect to what it says about the given aspect that severely passes: that a given error has been reliably ruled out.

Granted, it may often be shown that ruling out a given error is connected to, and hence provides evidence for, ruling out others. I am certainly not denying that severity can license going from one severely tested hypotheses to others – quite the contrary. The ability to do so is a very valuable and powerful way of crosschecking and building on results. Sometimes establishing the connections needed for crosschecking are given by theoretical background knowledge, other times sufficient experimental knowledge will do. But, whether these connections are warranted is an empirical issue that has to be looked into on a case by case basis – whereas the comparativist of [3.2] is saying we are licensed to regard theory T as well tested, so long as T is the best tested so far.

Rational Acceptability: It is not that the error-statistical account cannot countenance finding a theory "rationally acceptable", preferred, or worthy of pursuit – locutions often used by comparativists – upon reaching a point where its key experimental predictions have been severely probed and found to pass. One could even announce at this point that the theory had solved a set of key experimental problems, and take this as grounds for "deciding to pursue" it further. But these decisions are distinct from testing, and as it is an account of testing which I seek to supply, such decisions would call for a supplement to what I am offering.[7]

We should be concerned, too by the threat to the *stability* of severity assessments that the comparativist account would yield – the second point in 3.1:

4.2. Point 2. Suppose an experimental test ET is probing answers to a question: what is the value of a given parameter μ? Then, if a particular answer or hypothesis severely passes, this assessment is not altered by the existence of a theory which gives the same answer to this question. More generally, our account lets us say that severely passing $T(H)$ (i.e., what T says about H), gives us experimental knowledge about this aspect of T, and this assessment remains even though experimental knowledge may undergo repeated improvements, revisions, and even reconceptions. By contrast, as soon as an alternative theory comes to light that does as well as T does on this (and other tests), the comparativist would regard T as no longer severely tested. The comparativist would be forced to change the assessment of how well theory T had been tested. (Theory T has to "give back the crown" as it were – simply because another full-blown theory can pass the same tests.)

On the error-statistical testing account I favor, if a rival theory T_2 agrees with T_1 with respect to the effect or prediction under test, then the two theories are not rivals *so far as this experimental test is concerned* – no matter how much they may differ from each other in their full theoretical frameworks or in prediction ranges not probed by the experimental test ET. Let me qualify this: My claim is not that two theories fail to be rivals just because the test is insufficiently sensitive to discriminate what they say about the phenomenon under test; my claim is that they fail to be rivals when the two say exactly the same thing with respect to the effect or hypothesis under test.

My severity assessment reflects this. If theory T_1 says exactly the same thing about H as T_2 does $(T_1(H) = T_2(H))$, then T_2 cannot alter the severity with which the test passes H.[8] The shifting assessment advocated by the comparative-holist account goes against a key rationale for seeking severe tests: to identify experimental knowledge that *remains* stable through theory change, even if theories at higher levels may bring reinterpretations of this knowledge. Take the example that both Laudan and Chalmers raise in this connection: the General Theory of Relativity (GTR).

Consider tests of what GTR says about a particular hypothesis H such as the *Equivalence Principle* – roughly the claim that bodies of different compositions fall with the same accelerations in a gravitational field. Having severely passed this principle – by passing a series of null hypotheses – experimenters were warranted in saying, as they did, that any viable gravity theory would have to be one that affirmed the equivalence principle (in the domain of earth bound objects). They were warranted in saying, to quote leading experimental relativist Clifford Will, that: "gravity is a phenomenon of curved spacetime, that is, it must be described by a 'metric theory' of gravity" (Will 1993, p. 10).

Now the comparativist allows that corroborating a part of GTR corroborates (defeasibly) GTR as a whole. In fact, however, corroborating the Equivalence Principle is regarded as only distinguishing metric vs. non-metric theories – it is not regarded as testing GTR against rival metric theories.[9] How could it, when they are saying the same thing with respect to the Equivalence Principle? The error statistical account reflects this.

Granted, this depends upon the way that gravitational theories were partitioned, leading Chalmers to remark:

> Mayo's position has the consequence that verdicts concerning how severely tested a theory is with respect to some specified evidence are subject to change in the light of theoretical work on partitioning [the space of theories] (Chalmers, this volume, section 3).

As I see it, what changes in light of theoretical and experimental work on partitioning is not the severity with which a given hypothesis has been tested, but rather the assessment of which alternatives to the theory we have managed to rule out, by dint of the severely passed hypothesis. However, Chalmers' remark underscores an important task that philosophical discussions of theory testing overlook; namely, that of asking, once a given experimental effect *H* has severely passed: What more does *H* teach me about the phenomena (e.g., of gravitation)? What kinds of alternatives could affirming *H* fail to discriminate? I will return to this task.

4.3 Point 3: This refers to a key principle of error statistics, which is also the basis for solving a number of philosophical problems. For example, it is often argued that data underdetermines hypotheses because data may equally well warrant conflicting hypotheses according to logical measures of evidential relationship. It is true that these logical measures underdetermine hypotheses, but we can often distinguish, on grounds of severity, the well-testedness of two hypotheses, and thereby get around underdetermination charges. I take this up elsewhere (e.g., 1996, 1997a, 1997b). Here my interest is mainly in how the principle in point 3 bears on our question of how to go from low level experimental tests to higher level theories.

I had alleged that: "When enough is learned from piecemeal studies, severe tests of higher level theories are possible" (Mayo 1996, p. 191), to which Laudan responds with severe doubt:

> There are, so far as we know, indefinitely many general theories all of which 'fit' whatever lower-level hypotheses have actually passed severe tests. If that is true, the prospects for severity to flow upwards by a process of gradual concatenation appear doomed to failure. (Laudan 1997, p. 314)

Chalmers makes similar claims.

What is being overlooked here is that just because a theory fits (or is compatible with) the data that severely pass lower-level hypotheses, it does not follow that the theory has been tested at all, much less severely tested. The assumption is that I would have to give any such theory just as much credit, but this is false. We can grant, for example, that a rival theory could always be erected to accommodate the data, but a key asset of the error statistical account is its ability to distinguish the well-testedness of hypotheses and theories by the reliability or severity of the accommodation method. Not all fits are the same. Thus we may be able to show, building on individual hypotheses, that one theory, *at some level* (in the series or hierarchy of models) or close variants to this theory, severely passes. In so doing, we can show that no rival to this theory can also pass severely. Although a rival theory, at that level, can be engineered to fit the data, doing so results in the rival theory passing with very low severity.[10]

Admittedly, showing this requires looking at the actual detailed features of the data recorded at the level of the data models, not just at the inferred experimental effect or phenomenon. Ironically, many who claim to be champions of the new experimentalist focus on the nitty-gritty details of data do not avail themselves of this information when it comes to relating experimental phenomena to theory, and they find themselves stuck with an underdetermination that is avoidable. It sounds plausible to say there can always be some rival – when all that rival would have to do is fit already known experimental effects – at the level of experimental models. Things are very different if one takes seriously the constraints imposed by the information in the detailed data.

> Once we progress beyond [well-defined experimental set-ups], to the support of laws and theories, which Mayo herself signals to be a separate step, then Mayo's notion of a severe test is not sufficiently strong to avoid underdetermination (Chalmers 2002, this volume, section 5).

While I grant there is a separate step in moving from experimental inferences to full blown theories (just as there is a distinct step from raw data to data models), I did not intend this to signal that it takes one beyond "well-defined experimental set-ups" amenable to the error statistical methodology.

Let us see how: Since my critics or "rescuers" consider GTR, I will too. I agree with John Earman that "gravitational research in this century does not conform to any of the currently available models of macromethodology, [...] Kuhn, Lakatos, Laudan, and others." (Earman 1993, p. 173) But I think it conforms well with the error statistical model.

5. Experimental gravitation

Experimental testing of GTR is generally divided into three periods: from 1920–60; from 60–80, and from 80 onward. The very labels scientists have given these periods already suggest deep contrasts with the best known philosophical accounts (the models of macromethodology).

From the comparativist's perspective, 1920–60 would plainly be an era in which GTR enjoyed the title of "best tested" theory of gravity: it had passed the "classical" tests to which it had been put and no rival that had a superior testing record existed to knock it off its pedestal. By contrast, from 1960–1980, a veritable 'zoo' of rivals to GTR had been erected, all of which could be made to fit these classical tests. So in this later period GTR, from the comparativist's perspective, would have fallen from its pedestal, and the period might be regarded as one of crisis, threatened progress, or the like. But in fact, the earlier period is widely regarded (by experimental gravitation physicists) as the period of "stagnation" or at least "hibernation"; whereas the later period, though marked by the zoo of alternatives, is widely hailed as "the golden era" or "renaissance" of GTR. Yet these labels make good sense for our severe tester.[11]

The golden era came about thanks to events of 59–60 which set the stage for new confrontations between GTR's predictions and experiments. Nevertheless, the goals of this testing were, not to decide if GTR was correct in all its implications, but rather, in the first place, to learn more about GTR (what does it really imply about experiments we can perform?); and in the second place, to build models for phenomena that involve relativistic gravity: quasars, pulsars, gravity waves and such. In short, the goal was to learn more about gravitational phenomena.

Comparativist testing accounts (in the sense described in 3.2), eager as they are to license the entire theory, ignore what for our severe tester is the central engine for making progress, for getting ideas for fruitful things to do next to learn more. Rather than asking, given our evidence and theories, which theory of this domain is the best? The severe tester asks, given our evidence and theories, what do we know about this phenomenon? Far from arguing for GTR on the grounds that it has survived tests that existing alternatives could not, as our comparativist recommends, our severe tester would set about exploring just *why* we are *not* allowed to say that GTR is severely probed as a whole-in all the arenas in which gravitational effects may occur. Even without having full-blown alternative theories of gravity in hand we can ask (as they did in 1960): *how could it be a mistake to regard the existing evidence as good evidence for GTR*? Certainly we could be wrong with respect to predictions and domains not probed at all. But how could we be wrong even with respect to what GTR says about the probed regions, in particular solar system tests? One must begin where one is.

Far from desiring to accept all of GTR before its time, experimental relativists wished to delineate a framework that would not bias them toward accepting GTR prematurely (Will 1993, p. 10). Given the difficulty in testing gravitational theories, scientists were especially self-conscious about delineating such a framework in the 60s – hence the case is a good one for our purposes. They needed a framework that would let them describe violations to GTR's hypotheses – discrepancies with what it said about specific gravitational phenomena in the solar system.

To this end, a set of related experiments was modeled within what was called the *parametrized post-Newtonian* or PPN formalism. The PPN framework sets out a list of parameters that allowed for a systematic way of describing violations of GTR's hypotheses. These alternatives, by the physicist's own admission, were set up largely as straw men with which to set firmer constraints on these parameters. Whereas it is not even clear, from the comparativist point of view (as given in [3.2]), what motivation there would be for deliberately erecting rival theories to GTR in 1960 – after all, GTR was not facing anomalies – it is motivated from the point of view of getting more experimental knowledge about gravity: *for this was the only way to extend the regions that could be said to have been severely probed.* It was the only way to learn more about gravity.

Researchers quite deliberately limited the PPN framework to probing a portion or variant of GTR:

> The PPN framework takes the slow motion, weak field, or post-Newtonian limit of metric theories of gravity, and characterizes that limit by a set of 10 real-valued parameters. Each metric theory of gravity has particular values for the PPN parameters. (Will 1993, p. 10)

The PPN framework went beyond known rivals to GTR, and was instead the basis for inventing new classes of rivals. True it did not include all logically possible gravity theories, but only those that passed the criteria for viable gravity theories.[12]

Viable theories had to account for experimental results already severely passed, but they also had to satisfy another important role for a theory of experimental gravity, (one which relates to my fourth thesis, in 3.1); namely, the ability to show the relevance and importance of the experimental data for gravitational phenomena. Viable theories would have to be able to analyze and explore experiments about as well as GTR.

The PPN framework permitted researchers to compare the relative merits of various experiments ahead of time in probing the solar system approximation or solar system variant of GTR. Appropriately modeled astronomical data supply the "observed", i.e., estimated, values of the PPN parameters which could then be compared with the different values hypothesized by the diverse theories of gravity. This permitted the same PPN models of experiments to serve as intermediate links between the data and several alternative primary models based on GTR and its rival metric theories.

This mediation was a matter of measuring, or more correctly, *inferring*, the values of PPN parameters by means of standard statistical (estimation) techniques. What is most deserving of greater attention, I propose, are the strategies by which a primary theoretical parameter about the given post-Newtonian parameter (e.g., the deflection of light by gravity λ) is probed by *turning it into* a claim or hypothesis about a statistical parameter (in the experimental model). Then, inferences from (statistically modeled) data to these statistical distributions are used to learn answers to the primary questions. These inferences consisted of multiparameter least squares fits to parameters in models of data. Each estimation interval exemplifies the second statement of severity: *e* is good evidence that the correct value of the parameter is *not* among those excluded from the estimation interval. Researchers could then "go up" from the data, as it were, to infer ranges for discrepancies from GTR. Some elements of this piecemeal strategy, for the case of λ, can be sketched by means of the hierarchy of models as in Figure 1.

The standard strategy in linking the data model and the experimental model is to find a function of the parameter and the data such that this function follows a known statistical distribution (e.g., the Chi-square distribution). One can then use (Chi-square) statistics to estimate the various parameters in the model via such methods as confidence interval estimates. Putting together the interval

Figure 1.

PRIMARY: Testing the Post-Newtonian Approximation of GTR: Parametrized Post-Newtonian (PPN) formalism Delineate and test predictions of the metric theories using the PPN parameters: Use estimates to set new limits on PPN parameters and on adjustable parameters in alternatives to GTR e.g., λ How much spatial curvature does mass produce?
EXPERIMENTAL MODELS: PPN parameters are modeled as parameters of statistical distributions: they are statistical hypotheses (relating to models of the experimental source) Form interval estimates: hypotheses about parameter values that pass (and do not pas) severely e.g., hypotheses about λ in optical and radio deflection experiments
DATA: Models of the experimental source (eclipses, quasar, moon, earth-moon system, pulsars) Least squares fits of several parameters, one of which is a function of the observed statistic and the PPN of interest (the function having known distribution) e.g., least squares estimates of λ from "raw" data in eclipse and radiointerferometry experiments.
DATA GENERATION & ANALYSIS, EXPERIMENTAL DESIGN

estimates, it is possible to constrain the values of the PPN parameters and thus "squeeze" the space of theories into smaller and smaller volumes, as depicted in Figure 2. In this way entire chunks of theories could be ruled out at a time (i.e., all theories that predict the values of the parameter outside the interval estimates). By getting increasingly accurate estimates, more severe constraints are placed on how far theories can differ from GTR, in the respects probed. By 1980 it could be reported that "one can now regard solar-system tests of post-Newtonian effects as measurements of the 'correct' values of these parameters" (Will 1980, p. 309)

In this manner, scientists do accumulate and "flow up" (as Laudan put it) – up to a given level (e.g., solar system variants of GTR). This experimental knowledge does not merely permit us to infer that we have a correct parameter value but also to correctly understand gravity or how gravity behaves in a given domain.

It might be claimed that the appraisal of GTR based upon gravitational experiments can also be explained by the large-scale theory testers, in particular, the comparativists. Even if we were to grant that they, as well as other accounts, may be able to reconstruct after-the-fact what went on, it is important to see that the current account is interested in forward-looking methods, not after-the fact reconstructions. We can understand the strategies and inferences – the

Figure 2.

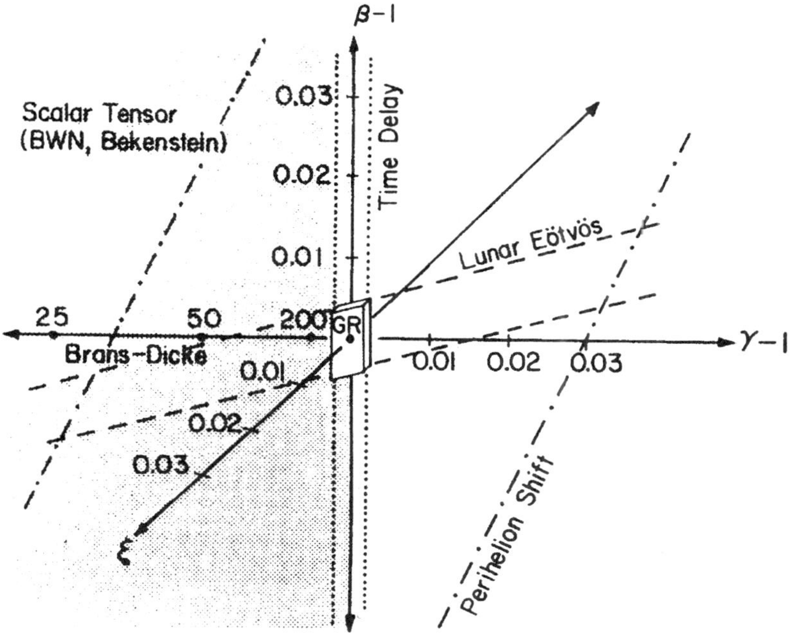

specific ones actually made in given experiments (not all tidied up and made to fit one's favorite macromethodology)-by means of the goals of severe testing and the statistical methods for obtaining experimental knowledge. We can also motivate what happens next in this episode:

Going Beyond Solar System Tests. After this point, progress is again made by recognizing the errors still not ruled out.

> All tests of general relativity within the Solar System [...] have an important qualitative weakness: they say nothing about how the "correct" theory of gravity might behave when gravitational forces are very strong, such as near a neutron star [...]. (Taylor et al 1992, p. 132)

The discovery (in 1974) of the binary pulsar 1913+ 16 opened up the possibility of probing new aspects of gravitational theory: the effects of gravitational radiation. This is a fascinating example in its own right which I cannot take up here[13] (see Damour and Taylor 1991; Lobo 1996, pp. 212–215; Will 1986).

6. Concluding remarks

We began by asking: How do low-level experimental tests allow inferences about theories (e.g., GTR)? Our answer is: by allowing us to learn about theoretical parameters and hypotheses by means of experimental parameters and hypotheses that are identical to, or are good approximations of, the theoretical ones. One might regard theories, i.e., theoretical models, as serving an analogous role to experimental models in the tasks of learning from empirical data. Experimental models serve to describe and analyze the relevance and significance of the experimental data for the experimental phenomenon, and theoretical models serve *to analyze the relevance and significance of any of the experimental inferences (estimates and tests) for the theoretical phenomenon.*

Were one to pursue the error statistical account at the level of large-scale theories, one would be interested to ask not, How can we severely pass high level theories? But rather ask: *How do scientists break down their questions about high level theories into piecemeal questions that admit of severe testing?* And how do the answers to these questions enable squeezing (if not exhausting) the space of predictions of a theory or of a restricted variant of a theory? We are not inductively eliminating one theory at a time but rather classes of theories, classes defined by giving a specified answer to a specific (experimental) question.

Note, too, that what is sought is not some way to talk about a measure of the degree of support or confirmation to one theory compared with another, but rather ways to measure how far off what a given theory says about a phenomenon is from what a "correct" theory would need to say about it. Although we may not have a clue what the final correct theory of the domain in question will look like, the value of the experimental knowledge we can obtain now might be seen as giving us a glimpse of what a "correct" theory would say as regards to the question of current interest, no matter how different the full theory might otherwise be.

Notes

1. This idea of a series of models is adapted from the account of Patrick Suppes 1969.

2. Because of the tendency to equate "experimental" with cases of manipulation, and because this account is relevant also for "observational" studies, I now prefer the term inferential distribution, as it is the distribution that is central for reaching and criticizing inferences (tests or estimations).

3. For a systematic formal articulation of the severity concept and how it serves as a basis for reinterpreting Neyman and Pearson statistical testing, see Mayo and Spanos, forthcoming.

4. To illustrate, consider a diagnostic tool with an extremely high chance of detecting a disease. Finding no disease (a clean bill of health) may be seen as passing hypothesis H_1: no disease is present. If H_1 passes with so sensitive a probe, then H_1 passes a severe test. However, the probe may be so sensitive as to have a high probability of declaring the presence of the disease, even if no disease exists. Declaring the presence of the disease may be seen as passing hypothesis H_2: the disease is present. If H_2 passes a test with such a highly sensitive probe, then H_2 has not passed a severe test. That is because there is a very low probability of not passing H_2 (not declaring the presence of the disease) even when H_2 is false (and the disease is absent). The severity of the test that hypothesis H_2 passes is very low.

5. I allude to a phrase in Earman 1992.

6. I place "severity" in parenthesis here with a question-mark because at least in Laudan's recommended definition, it is not required that the individual tests passed themselves be severe.

7. After all, Laudan (e.g., Laudan 1977) himself has stressed that we should distinguish theory pursuit from other stances one might take toward theories.

8. In denying that higher-level theories need to be ruled out to severely pass a given effect, I do not mean to deny that one can make a mistake in supposing a hypothesis severely passes – but such challenges will come, not from higher levels, but rather from lower ones. A key set of challenges come from what I group under "experimental assumptions". Mistakes can happen because the actual experimental data do not satisfy the assumptions of the experimental model, or because the experimental test was not sufficiently accurate or not sufficiently precise to reliably inform about the primary hypothesis or question.

9. At least as of 1960.

10. For example, although it was fine to erect alternatives to GTR (say in the 1960's), that were intentionally assured of agreeing with the predictions that already passed fairly severely, the alternative theory could not just postulate these known effects. See Note 12. Such a theory, even though it could "fit" the observed effects, would have been very poorly tested.

11. For excellent discussions, see Earman (1993), Will (1986), (1993), (1996).

12. Criteria for a Viable Gravity Theory (during the "Golden Era"):

(i) *It must be complete*, i.e., it must be capable of analyzing from 'first principles' the outcome of any experiment of interest. It is not enough for the theory to *postulate* that bodies made of different material fall with the same acceleration [...]
[This does not preclude "arbitrary parameters" being required for gravitational theories to accord with experimental results.]

(ii) *It must be self-consistent*, i.e., its prediction for the outcome of every experiment must be unique, i.e., when one calculates the predictions by two different, though equivalent methods, one always gets the same results [...]

(iii) *It must be relativistic*, i.e., in the limit as gravity is 'turned off' [...] the nongravitational laws of physics must reduce to the laws of special relativity [...]

(iv) *It must have the correct Newtonian limit*, i.e., in the limit of weak gravitational fields and slow motions, it must reproduce Newton's laws [...] (Will 1993, pp. 18–21).

13. For a brief discussion of how the hierarchy of models would be applied to the binary pulsar analysis, see Mayo (2000).

References

Chalmers, A.: (1999), *What is This Thing Called Science? Third Edition*, University of Queensland Press, Australia.

Chalmers, A.: 200x, 'Experiment and the Growth of Experimental Knowledge', This volume.

Damour, T., Taylor, T.H.: (1991), 'On the Orbital Period change of the Binary Pulsar PSR 1913+16', *Astrophysical Journal* **366**, pp. 501–511.

Earman, J.: (1992), *Bayes or Bust: A Critical Examination of Bayesian Confirmation Theory*, MIT, Cambridge.

Ehlers, J., Perry, J. and Walker, W. (eds.): (1980), *9th Texas Symposium on relativistic Astrophysics*, New York Academy of Sciences, New York.

Hall, G.S., Pulham, J.R.: (1996), *General Relativity: Proceedings of the Forty Sixth Scottish Universities Summer School in Physics*, SUSSP Publications, Edinburgh and The Institute of Physics, London.

Laudan, L.: (1977), *Progress and Its Problems*, University of California Press, Berkeley.

Laudan, L.: (1997), 'How About Bust? Factoring Explanatory Power Back into Theory Evaluation', *Philosophy of Science* **64**, pp. 303–316.

Lobo, J.: (1996), 'Sources of Gravitational Waves', in G.S. Hall, J.R. Pulham (1996), pp. 203–222.

Mayo, D.: (1996), *Error and the Growth of Experimental Knowledge*, The University of Chicago Press, Chicago.

Mayo, D.: (1997a), 'Duhem's Problem, the Bayesian Way, and Error Statistics, or 'What's Belief Got to Do With It?' ' and 'Response to Howson and Laudan', *Philosophy of Science* **64**, pp. 222–244, 323–333.

Mayo, D.: (1997b), 'Severe Tests, Arguing from Error, and Methodological Underdetermination', *Philosophical Studies* **86**, pp. 243–266.

Mayo, D.: (2000), 'Experimental Practice and an Error-Statistical Account of Evidence', *Philosophy of Science* **67** (Part II Symposia Papers, ed. by D. Howard), pp. S193–S207.

Mayo, D., Kruse, M.: (2001), 'Principles of Inference and Their Consequences', in D. Corfield and J. Williamson (eds.) *Foundations of Bayesianism*, Kluwer Academic Publishers, Dordrecht, pp. 381–403.

Mayo, D., Spanos, A.: forthcoming, 'A Post-data Interpetation of Neyman-Pearson Methods Based on a Conception of Severe Testing'.

Suppes, P.: (1969), 'Models of Data', in P. Suppes (1969), *Studies in the Methodology and Foundations of Science*, D. Reidel, Dordrecht, pp. 24–35.

Taylor, J., Wolszczan, A., Damour, T. and Weisberg, J.: (1992), 'Experimental constraints on strong-field relativistic gravity', *Nature* **355**, pp. 132–136.

Will, W.C.: (1980), 'The Confrontation Between General Relativity and Experiment, in J. Ehlers, J. Perry, and W. Walker (1980), pp. 307–321.

Will, W.C.: (1986), *Was Einstein Right?*, Basic Books, New York.

Will, C.W.: (1993), *Theory and Experiment in Gravitational Physics*, Cambridge University Press, Cambridge.

Will, C.W.: (1996), 'The Confrontation Between General Relativity and Experiment. A 1995 Update', in G.S. Hall, J.R. Pulham (1996), pp. 239–281.

NEW EVIDENCE FOR OLD

John Worrall
London School of Economics

Abstract If the development of science is, *pace* Kuhn and Co, to be explained as a rational process on essentially empiricist criteria, then there must be some sort of justified epistemic confirmatory premium on predictions of certain kinds, as opposed to "mere accommodations". Only then could the history of mature sciences be explained as consisting of the replacement of well-confirmed theories by even better confirmed theories. There is, of course, a long-running dispute in philosophy of science over exactly this issue. This dispute formed an important part of the debate between Mill, who believed that only "the ignorant vulgar" could think that predictions "count more", and Whewell who claimed that predictive success gave the theory that enjoyed it "the stamp of truth beyond the power of ingenuity to counterfeit". And the issue has surfaced in various forms since then – most recently in the much-discussed "old evidence problem" that allegedly afflicts the Bayesian view of confirmation.

I have argued in earlier papers that there is, in fact, no confirmatory premium for new evidence *as such*: It is true that a crucial distinction needs to be made in confirmation theory between two types of evidence; and true that the distinction coincides *to some extent* with the old/new distinction; however the important difference is not that between new and old evidence but between evidence that has not, and evidence that has, been "used in the construction" of the theory whose confirmation is at issue.

This claim has itself been criticised on various grounds, by Bayesians (like Colin Howson) and non-Bayesians (like Deborah Mayo) alike. Mayo proposes a different view aimed at capturing the same intuitions that underlie my earlier account; while Howson dismisses as "entirely bogus" the whole claim that using evidence in the construction of a theory diminishes that evidence's power to confirm that theory.

In this paper, I shall, show that both Howson's and Mayo's analyses are flawed and hence defend and clarify my original claim. This defence puts the rationale for the claim on a much clearer and firmer footing than hitherto; and thereby, I believe, significantly clarifies this whole important and longstanding debate.

P. Gärdenfors, J. Woleński and K. Kijania-Placek (eds.), In the Scope of Logic, Methodology and Philosophy of Science, Vol. I, 191-209.

1.

The fact that there continues to be a lively debate in the philosophy of science literature about "novel facts" would surely strike the disinterested observer as astounding. After all, the idea that a theory is in general better supported by hitherto unknown evidence that it correctly predicts than it is by already known evidence that it entails seems beyond the pale of credibility. Why should anyone ever have defended the outlandish idea that the time-order of theory and evidence is of epistemological-methodological importance? Let me start then by reminding you why some perfectly sensible people defended this apparently senseless idea.

The clue lies in the surprising complexity of the deductive structure of empirical tests of scientific theories. As many commentators had clearly seen intuitively, but as Duhem was the first to spell out in some detail, the sort of claim that we tend to think of as a "single" scientific theory – Newton's theory, the classical wave theory of light, Maxwell's theory of electromagnetism – has no empirical consequences when considered "in isolation". In order to have a full deduction of a directly empirically testable statement from such a theory, other theoretical assumptions and hypotheses are invariably necessary as extra premises. In order to test Newton's theory through observations of apparent planetary positions, for example, particular – though clearly theoretical – claims about the positions and masses of the main objects in the solar system are required, along with a set of optical assumptions underpinning the working of the telescopes involved and allowing for such hypothesised phenomena as atmospheric refraction. Call these extra assumptions "auxiliaries".

Duhem also in effect pointed out that for most "single" scientific theories, there is a further level of complexity – the "single" theory itself is naturally thought of as consisting of a *core* claim together with some set of *more specific* theories. Consider, for example, the classic case of the two theories vying for acceptance in early 19th century optics. "The" wave theory was made up of the core claim that light consists of some sort of periodic disturbance transmitted through some sort of elastic material medium, together with a set of more specific assumptions (ones which evolved through time) about the type of periodic disturbance and the exact mechanical constitution of the medium. Similarly "the" emission theory consisted of the core claim that light consists of some sort of particle fired off by luminous bodies, together with a set of more specific assumptions (ones which again evolved through time) about the type of material particles involved (what, for example, differentiates those particles that produce red light from those that produce blue light) and the forces to which they are subjected in various circumstances (when reflected from a plane mirror, refracted into glass, or "diffracted" when passing by the edge of opaque objects).

So the full deductive structure of an empirical test is always

a. CENTRAL THEORY

b. AUXILIARY THEORIES

Hence, empirically checkable consequence.

Here the set *b* is, despite some misleading remarks by Quine and even by Duhem himself, *of course* finite, though invariably large (larger than you might initially think); while *a* – at least in many cases – itself breaks down naturally into a core component and a set of more specific claims. Call the conjunction of *a* and *b* a "theoretical system based on the core idea in the central theory *a*."

But now let *e* be some piece of evidence – the result of Young's two slit experiment, say – that seems intuitively to tell against a particular "isolated" theory, in this case the emission theory of light. The above analysis straightforwardly entails that there must be theoretical systems based on the core emission idea that entail that piece of evidence *e*. And indeed optical scientists of the early 19th century considered such systems – *either* attributing the pattern of dark and light bands to a physiological effect (thus questioning the original auxiliary assumptions underlying the instruments used – in this case our own eyes) *or* attributing those bands to particular (and peculiar) forces that emanate from the double-slit screen and affect the paths of the light-particles (thus questioning one of the original more specific assumptions involved in the central theory).

Since the above is true for *any* piece of evidence and *any* "isolated" theory, it threatens to rule out any account of the rationality of theory-change in science in terms of the winning theory having support from more evidence than the losing theory. *If* having the right empirical consequences is the only criterion, then since any core idea can be incorporated into a theoretical system that has the right empirical consequences, there is no empirically-based rational preference for any such core idea over any other.

2.

So this was what made the initially implausible idea that predictions "count more" appealing: if true it would solve this problem in a crisp and clear way, and hence neutralise the threat to an empiricist account of rational theory-change. "Novel facts count more" would imply that, since interference effects had been predicted by the wave theory, those effects continued to count more for that theory even once they had been accommodated, after the event, by the rival emission theory.[1]

There are, however, at least two problems. One is that although the novelty claim might capture well-established intuitive judgments of support in certain particular cases from the history of science, it clearly contradicts well-established intuitive judgments of support in other cases. One such much-discussed case concerns Einstein's account of the precession of Mercury's perihelion (compared to his prediction of the star-shift data)[2]; another (as I showed elsewhere)[3] concerns Fresnel's 1819 wave account of the straightedge diffraction results first produced by Grimaldi in 1665; and yet another concerns the account by Copernicus's theory of the long known phenomena of planetary stations and retrogressions and of the bounded elongation of Mercury and Venus.[4] In all these cases, the fact that a new theory entailed correctly some already well-known phenomenon counted strongly in its support – just as strongly as any hitherto unknown phenomenon whose existence the corresponding theory predicted.

The second problem with the "novel facts count more" view is that despite its appeal as a solution of some cases of the Duhem problem (and even if, contrary to fact as we just saw, it agreed with the intuitive solution in all particular cases), the idea just remains outlandish. Why on earth *should* the apparently purely contingent historical issue of whether or not a theory was first developed before some particular piece of evidence became available matter at all in an account of the *rational* support that that evidence lends to theory? (Let me say in parentheses that, while the novelty view has been almost universally abandoned there is an interesting exception: Patrick Maher has defended the temporal view in a novel way.[5] Although I do not have the time to argue it here, I believe that Maher's arguments in so far as they are successful at all in fact underwrite the non-outlandish heuristic view that I continue to defend as explained below.)

3.

Zahar and I developed a different view which

(i) aimed, like the novel facts view, to neutralise the Duhem threat to rational theory-change;

(ii) agreed with *all* well-established intuitive judgements of support in particular cases rather than just some of them; and

(iii) unlike the novel facts view has, we argued, a clear epistemic justification.[6]

The view is sometimes called, following Nickles, Mayo and others,[7] the "Use Novelty" (UN) account – though, since the UN charter is that novelty is not the issue at all, the name is somewhat misleading. The account directs attention to what seems really methodologically suspect in cases like the emission theory of light and interference effects. The problem in these cases is not that the evidence involved was already known when the theory was first articulated, but rather that the evidence was *used in the construction of the theory* – in the paradigm case, this means that the evidence was used to fix the value of some parameter (or parameters) in the theory under consideration.

In such cases, some general theory T involves, let's say for simplicity, a single parameter left free by any consideration of a theoretical kind. T itself does not entail e. But e can be used to fix the value of the parameter, hence producing the more specific theory T' which *does* entail e (*modulo* auxiliaries and initial conditions).

To anticipate a point I shall need to labour below, the claim never was that there is anything suspicious about this procedure *per se* (indeed it plays a regular and indispensable role in the scientific enterprise). But Zahar and I did claim that it means that e cannot be taken fully to support T' despite the fact that T' entails e. Real support for T' must be sought through *independent* evidence – that is, evidence aside from e that T' entails but which its more general predecessor T fails to entail. So, in the example just discussed, Young's two slit result could indeed be accommodated within the emission theory – essentially by giving oneself an expression for the diffracting force produced by the double-slit apparatus with a large number of free parameters and working out *from the details of the known experimental result* what those parameter-values have to be. Since the details of the experimental outcome follow from the rival wave theory without the involvement of any such adjustable parameters – two coherent series of waves just do alternately constructively and destructively interfere depending on the phase relations at particular points – these results support the wave theory over the emission theory, despite the adaptability of the latter theory to them.

The crucial positive feature of Fresnel's theory in respect of these interference or diffraction results was clearly expressed some years after the invention of the theory by the British scientist George Gabriel Stokes:

> The theoretical distances of the several fringes [...] were a matter of pure prediction; for the only unknown quantity involved in the theoretical expression, the length of the wave, had been determined by Fresnel by independent methods [...] so that *not a single arbitrary constant was left to be determined by some measurement of a fringe in some one particular case, whereby an [...] accordance between theory and observation might have been brought about.* (Stokes 1884, p. 63; emphasis supplied)

Although the notion of parameter-fixing is a precise mathematical one, there are clear informal analogues. For example, the central "creation science" claim that god created the universe in 4004 BC essentially as it now is, provides itself with an indefinite number of "free parameters" waiting to be filled in with details of how exactly the universe is. Creationists simply use whatever observations are available to fix the details of their more specific theories. This gives rise to what might be called the "Gosse dodge" – the fossil record (or rather *so-called* fossil record) allegedly provides no evidence for Darwinian evolution over creation, since all aspects of this record can be accounted for by supposing that they are just features of the way god chose to create the world: he just decided to plant bone-*like* structures in tar pits and desert sands and just decided to make various rocks complete with pictures that look awfully like the imprints of skeletons of extinct creatures. The UN view straightforwardly captures the intuitively obvious judgment that no amount of Gosse-dodging equilibrates the evidential scales between Darwinism and "scientific" creationism-Darwinism with plausible geological assumptions (pretty well) predicts that there will be fossils around, creationism only accommodates the "fossil" record after the event. But the important fact from the point of view of what really counts as evidence for the relevant theoretical claim is not simply that science already had a fossil record when Creationists developed their theories, but rather that those particular theories were based on, constructed in the light of, tailored to, that record.

So what then exactly does the UN view claim? One purpose of the present paper is to get much clearer on exactly this issue. However for the moment, let's suppose that it is captured – at any rate approximately – by what Nickles called the "no double use rule". This is the rule that "you can't use the same fact twice": once in the construction of a theory and then again in its support. Or alternatively, and more precisely though also more problematically: the bits of evidence that support a theory are those which (i) follow from the theory (*modulo* initial conditions and perhaps other bits of "background knowledge") but (ii) were not used in the construction of that theory – by for example fixing the value of a parameter within it.[8]

4.

Much of this may seem "old hat", but I thought it important to rehearse the general background before coming to the main aim of the present paper. There have been, in the past 15 years or so, a number of important criticisms of the UN charter which call for a response they have not yet received – allegedly destructive criticisms from Bayesians such as my colleague Colin Howson and Allan Franklin and more sympathetic, constructive but no less powerful criticisms by non-Bayesians such as Thomas Nickles and particularly Deborah Mayo.[9] These turn out to raise not just local issues, but a range of central issues in philosophy of science. I cannot attempt to respond adequately to them all here. Instead I shall concentrate on one – which I nonetheless believe to be the most central. (It is certainly one common to all critics.) I shall outline the clarification (or amendment, you can take your pick) of the position that this criticism forces and must leave for another occasion the question of how this more elaborate position ties in with alternative accounts developed by the Bayesians and by Deborah Mayo.

Beforehand, however, I need to emphasise a couple of details which were explicit, or at any rate clearly implicit (or so I claim) in the original position. I emphasise them, not for the boring reason of claiming priority for what are in fact some fairly obvious points, but because they turn out to be significant for what comes later.

First, there never was any suggestion that using phenomena in the construction of theories, fitting theories to known evidence, is a "bad thing".[10] Adams and Leverrier clearly used the initially anomalous data on Uranus's orbit in the construction of the Newtonian theory that predicted the existence of the hitherto unrecognised planet Neptune; and this development is universally and surely correctly regarded as one of the greatest successes of the Newtonian framework.[11] The important aspect of this development, however, was exactly the predictive success – Neptune was found more or less where and with more or less the properties that the modified Newtonian theory entailed. What happened, then, in this case was that there was a specific theoretical system T based on the core Newtonian tenets that entailed the wrong results about Uranus's orbit; scientists like Adams and Leverrier then *retreated* by freeing up certain aspects of that initial system (basically the commitment to a particular number of other planets), and then used the erstwhile refuting evidence from Uranus to refill the parameter values; in this way they created a new theoretical system T' which *both* got Uranus's orbit right *and* predicted the new planet. Given the predictive success in this case, it seems to matter little what we say about the support of the initially refuting data from Uranus's orbit for the eventual system T' – though good bookkeeping suggests that we say here too that the Uranian data (as opposed to that from Neptune) supplies no real support for T'. But clearly, since it led to a theoretical system that did enjoy impressive new support, the use of that data in the construction of that theoretical system cannot be a "bad thing".

The complaint against creationism, then, that is underwritten by the UN position is not that it uses data about fossils to construct particular theoretical systems based on its core idea, but rather that the *only* data entailed by those systems is the data thus fed into it: unlike the Newtonian case, there is no *independent* evidence for the theoretical system thus constructed. Using data in the construction of a theory is no bad thing, but the data used do not support (or do not fully support) the theory thus constructed, in particular if two theoretical frameworks based on different core ideas are vying for acceptance and if the first framework produces a theory that straightforwardly entails some piece of evidence (already known or not) while the second framework then uses that evidence to construct a particular theory that entails it, then this does not balance the evidential scales – so far as this particular piece of evidence goes at any rate the first theoretical framework remains better supported.

The second important detail that I claim was in the UN account from the start is this. Not only does the UN view not condemn using facts to fix theories in general, it is quite consistent with the fact that sometimes the *best* theoretical explanation of some data is one that has been constructed using those data. For example, the best explanation for the observed lack of stellar parallax in, say, the 18th Century was surely the Copernican/Newtonian one that there is indeed an apparent parallactic motion produced by the orbit of the earth around the sun but, because of the enormous distance to the nearest star compared to the distance to the sun, the extent of that parallactic motion placed it beneath the limits of observability permitted by the then available telescopes. At least at the time this account was first proposed, there were no independent theoretical reasons for adopting any such view about the comparative distances to the sun and nearest (other) star – instead a view about this ratio was "read off" the failure to observe any parallax. The account nonetheless provided the best available explanation – but not because of anything to do with the data about apparent stellar motions, but rather because the account enjoyed empirical success, quite unmatched by any other available account, *in other quite distinct areas*. Given that it was best empirically supported through other data, the framework wins the right to give the best explanation of some phenomenon, even when that explanation is itself entirely *ad hoc* – in the sense that that explanation was constructed from the data without that particular construction being subject to any then available independent test.

5.

After these clarifications of the view, I now consider the direct criticism that has been brought against the UN account. It has been claimed that, just like its temporal novelty predecessor, the UN account is in outright contradiction with well-established intuitive judgements about empirical support in particular cases. Although they make the point in somewhat different ways, Tom Nickles, Colin Howson and Deborah Mayo all in effect claim that the "double use rule" conflicts with scarcely deniable intuitions about particular cases – there are lots of cases in which data used in the construction of a theory also support it (in many cases maximally so). The rule, as my friend Colin Howson puts it, "makes nonsense of quite basic and eminently reasonable scientific appraisals" and is on that and other accounts "entirely bogus".[12]

Colin Howson likes to emphasise standard statistical examples such as the following. We are given that an urn contains only black and white balls though in an unknown proportion; we are prevented from looking inside the urn but can draw balls one at a time from it. Suppose that a sample of size n has been taken (with replacement) of which k have been found to be white. Standard statistical estimation theory then recommends the hypothesis that the proportion of white balls in the urn is $k/n \pm \varepsilon$, where ε is calculated as a function of n by standard confidence interval techniques. The sample evidence is the basis here of the particular hypothesis constructed and surely also supports it at least to some (good) degree – the evidence for the hypothesis just is that a proportion k/n of the balls drawn were white.

Deborah Mayo cites and analyses in more detail the same case and also cites the following "trivial but instructive example" (p. 271). Suppose one wanted to arrive at what she describes as "a hypothesis H" about the average SAT score of the students in her logic class. She points out that the obvious – indeed uniquely sensible – way to arrive at H is by summing all the individual scores of the n students in the class and dividing by n. The "hypothesis" arrived at in this way would clearly be "use-constructed". Suppose it is that the average SAT score for these students is 1121. It would be absurd to suppose that the data used in the construction of the "hypothesis" that the average SAT score is 1121 fails to support that hypothesis. On the contrary, as she writes:

> Surely the data on my students are excellent grounds for my hypothesis about their average SAT scores. It would be absurd to suppose that further tests would give better support.[13]

Exactly so: the data provide not just excellent, but, short of some trivial error, entirely *conclusive* grounds for the "hypothesis" – further tests are entirely irrelevant. This is exactly why it seems odd to talk of a "hypothesis" at all in these circumstances.

Despite this oddness, Mayo's example and the statistical examples cited both by her and by Howson, do point in the direction of an important consideration. Exactly what that consideration is is best seen through the sort of case highlighted by Tom Nickles in his (1987) – that of theories "deduced from the phenomena".

6.

During the heyday of hypothetico-deductivism, Isaac Newton was often regarded as providing classic support for Einstein's dictum that anyone seeking to understand science should concentrate on what scientists do and ignore entirely what they *say* they do. Newton was patently a great scientist, probably the greatest ever, and yet his claims about how to do good science – that hypotheses should be eschewed and properly accredited scientific theories must be "deduced from the phenomena" – seemed to be logical nonsense. Theories transcend any available data – *both* "horizontally" in generalising on necessarily finite data *and* "vertically" in attributing the data to underlying unobserved mechanisms. Even if we allow inductive generalisation, Kepler's and Galileo's "data" tell us how planets and terrestrial bodies move, Newton's theory tells us that they move in the way they do because of the underlying, non-directly observable force of gravity. On the other hand, deductive logic is non-content-increasing. The conclusion of any inference that is deductively valid was already contained in the premises (even though a deduction may of course surprise us psychologically speaking in that we may not have *realised* that the conclusion was indeed implicit in the premises). Since Newton's theory clearly transcends, has content not shared by, Galileo's and Kepler's "data", his – often repeated – claim to have deduced his theory from the data must be logical nonsense.

More recent years, however, have seen a rehabilitation of Newton's method. Dorling, Glymour, Zahar, Harper and others[14] have all shown that Newton meant what he said – he really did provide deductions of theories from the phenomena –, and that his method is, at any rate on the surface, quite different from anything hypothetico-deductive. Of course not even Newton can defy the laws of logic and what "deduction from the phenomena" really means is deduction from the phenomena *plus some usually very general "background" principles*. Scientific theories are shown to follow deductively not from data alone, nor even from generalised data alone – both claims would indeed be logical nonsense – but from generalised data *plus* general principles that are arguably part of "background knowledge".[15]

Once the real idea of this method is grasped, it is easy to see how ubiquitous its application in science has been. John Norton has shown for example, in some recent papers, how the basic principles of even as "revolutionary" a theory as the quantum theory were argued for by "demonstrative induction" (aka "deduction from the phenomena").[16] Norton argues, I believe fundamentally correctly, that it is the fact that theories are so often deducible in this way that makes the phenomenon of underdetermination of theory by data, so beloved by philosophers, invisible to working scientists. Very often a theory is deducible from, and so uniquely determined by, general principles that scientists already accept plus some – usually new – experimental data.

The details in cases of "revolutionary" science tend to be complex, but "normal" or intra-research programme cases can be very straightforward.

Suppose a scientist already accepts the general wave theory of light – the theory that light from any particular source consists of waves of some wavelength or other transmitted through the luminiferous aether. This general theory does not specify the wavelength of any particular kind of monochromatic light – say light from a sodium arc. The scientist would like a more detailed theory that does specify that wavelength. Does she lay on her couch and make a Popperian bold conjecture? Such an attempt to find a needle in an infinite haystack would be entirely ill-advised. Instead she would deduce the specific theory, involving the specific value of the wavelength, from the phenomena. She would look for some consequence, e, of her general theory T, where e characterises some observable magnitude (fringe separation in some particular experiment, say) as a one-to-one function of the wavelength. She would perform the experiment using light from a sodium arc, measure the magnitude at issue – here, the fringe separation (call the result of this measurement e') – and infer to a more specific theory T'. So for example, subject to a couple of idealisations, one can infer from the general wave theory that, in the case of the famous two-slit experiment, the (observable) distance X from the fringe at the centre of the pattern to the first fringe on either side is related to (theoretical) wavelength λ, via the equation $X/(X^2+D^2)^{1/2} = \lambda/d$ (where d is the distance between the two slits and D the distance from the two-slit screen to the observation screen – both of course observable quantities). It follows analytically, of course, that $\lambda = dX/(X^2+D^2)^{1/2}$. But all the terms on the right hand side of this last equation are measurable. Hence particular observed values will determine the wavelength and so the more specific theory T'.

The relevance of this to the use-novelty issue will be obvious: in cases of theories deduced from the phenomena, the phenomena at issue are clearly used in the construction of the theory (and used in the clearest possible sense, they are essential premises in its derivation) and yet those phenomena not only support the specific theory arrived at, they do so in a near maximal way. When asked why she accepts T' – the particular version of the wave theory involving the particular assignment of a value to the wavelength of sodium light – the answer a 19th century physicist would give is exactly the evidence about the fringe distances from which she derived that particular theory. *Given that she already accepts the general theory*, this evidence is indeed not just support but conclusive support for that particular theory. What better support for a theory from data could there be than that the data deductively entail the theory – given other assumptions that one already accepts?

Deborah Mayo's example cited earlier can be thought of as a special case of deduction from the phenomena: in the SAT score case we just have straight deduction from the data, essentially no background assumptions are necessary. In Colin Howson's statistical case the underlying ideas are, I think, generalisations of those underlying deduction from the phenomena – though obviously the details are different, since in these probabilistic cases the evidence neither deductively implies the constructed hypothesis, nor does the constructed hypothesis deductively imply the evidence. Nonetheless here too the general model – that the process of drawing balls from an urn is a Bernouilli process with unknown parameter k – is part of background knowledge, indeed it is just "given" as an unquestioned assumption. And the evidence (of the relative frequency of white balls in some sample of draws from the urn) is used, not deductively but in accordance with again accepted statistical techniques, to arrive at a specific theory (about the proportion of white balls in the urn as a whole) which specific theory then makes highly probable (and so is intuitively supported by) the evidence that was used in its construction.[17]

7.

So, end of story? Worrall discovers deduction from the phenomena and has to admit that the critics of the UN account of support were right all along – not only are there cases where evidence used in the construction of a theory also provides support for that theory, those cases are exactly the clearest-cut cases of maximal, or near-maximal support. To repeat, what better support could there be than deducibility (or its near relative in Colin Howson's statistical case)?

Well, hold on to that essential phrase that in these cases the data entail (or "quasi-entail") the theory *given other general assumptions one already accepts* and recall the original problem.

Although the argument for support from the interference fringe data for the specific theory T' in our example seems intuitively impeccable, the argument is formally entirely analogous to exactly the sort of most methodologically-suspect cases discussed earlier. Creationists too can deduce their Gossefied theory from the phenomena about the alleged fossils – *given* the general creationist theory that they already accept, then the specific version with the "fossils" as aspects of the creation is deducible from the data.[18]

Or consider another much-discussed example. Immanuel Velikovsky developed a theory about a giant comet that broke away from Jupiter and made a series of orbits around the earth before settling down to a quieter life as the planet Venus. The "close encounters" between the comet and the earth were, according to the theory, responsible for such remarkable (alleged) phenomena as the falling of the walls of Jericho and the parting of the Red Sea. Velikovsky accepted that such cataclysms could hardly have been restricted to selected parts of the Middle East and looked for records of similar natural pyrotechnics in other contemporary record-keeping cultures. The search revealed some embarrassing gaps. But Velikovsky was equal to the task: he postulated that, for the scribes in *some* cultures, the events associated with the close encounter with the comet had proved so traumatic that "collective amnesia" had set in. And he was of course able to deduce the detailed version of his theory that specified which cultures did and which did not suffer from collective amnesia from the phenomena – the phenomena of the existence of (more or less) suitable records from some cultures and their non-existence from others: collective amnesia afflicted precisely those cultures $C_1, \ldots, C_n$ for which no suitable records of cataclysms exist.

8.

Given this formulation of the problem, its solution surely jumps out at you. The judgement of support in the "deduction from the phenomena" cases is ineliminably *conditional*, an ineliminably *intra*-paradigm or *intra*-research programme judgement. *Given* that a general framework, or research programme, is already accepted, then the data give – in the case of a genuine deduction – not just *some* support for the specific theory, but conclusive support. Given that the general form of the wave theory was already accepted, then the fringe data *entail*, and therefore obviously support, the specific form of the wave theory with the particular values for the wavelength-parameters. But equally, if you already accepted the general Velikovskian theory then the cataclysmic record data would entail (or almost entail), and hence give maximal (or near maximal) support for, the specific form of the Velikovskian theory complete with particular values (in that case 0 or 1) for the "collective amnesia parameters".

The correct judgement seems to be, then, that, in all cases of this kind, support is for a particular theory *given* the already accepted general framework or programme, where the conditional remains undischarged. The support, speaking loosely but suggestively, does not spill over to the general framework. The fringe data give you conclusive reason for holding one particular form of the wave theory, if you are going to hold the wave theory at all, but no reason at all for holding the wave theory at all. The "cataclysm" record data (including lack of records) give you (near) conclusive reason for holding one particular form of the Velikovskian theory, if you are going to hold the Velikovskian theory at all; but no reason at all for holding the Velikovskian theory at all.

But is there any other sort of empirical support? Are there cases of support for a specific theory where the support *does* spill over to the general underlying framework idea? The answer is that there are two sorts of case both exemplified in the wave theory example, but not, exactly as one would expect, in the case of Velikovsky's theory.

First in the wave theory there are phenomena that "fall naturally out of the core idea" – this was exactly the case, for example, with straightedge diffraction (remember the passage from Stokes that I quoted earlier). And *secondly*, the more specific version of the wave theory we mentioned, complete with values for the theoretical wavelengths, having been constructed using certain fringe data, then implied *other* independent data – about, for example, fringe separations in different experiments. (Indeed the standard situation in these genuinely scientific cases is that the free parameters are *overdetermined* – there is a range of experimental results $e_1, \ldots, e_n$ *any one of which* would determine the value of the initially free parameters to produce a theory that would then entail, and so obtain independent support from, the other $n-1$ results.)

Neither type of case of the support for the general theory is exemplified in either Velikovsky's or the creationist's theory. Neither of these theories enjoyed any straightforward, non-use-constructed empirical success nor, having produced specific versions in the light of particular evidence, did any of these specific versions go on to make successful independent predictions.

The UN account applies exactly to this kind of support that is not conditional on prior acceptance of the underlying general theory. And it is, moreover these types of case of support – the ones to which the UN rule *does* apply – that carry the more force. The intuition that the fringe data supply powerful support for the fixed-parameter wave theory – because they entail that theory given the general wave theory as premise – relies essentially on the fact that there are other data, not used in its construction that support that general wave theory. The intuition that Kepler's and Galileo's data supply powerful support for Newton's theory, relies exactly on the fact that there is plenty of UN-obeying support for the general principles that underlie Newton's deduction of his theory from that data. The intuition that the lack of suitable records does *not* support Velikovsky's collective amnesia hypothesis and that the alleged fossil record does not support Gossified creationism is based exactly on the following fact. Although, given the general theory, the data in each case entail (or at least come close to entailing) those specific versions of the theory, there is no UN-satisfactory support for either general theory in the first place. The fossil record gives you conclusive reason for holding the Gossified version of Creationism, if you hold any version of Creationism at all, but it gives you no reason at all for holding any version of Creationism. The lack of suitable records gives you (near-)conclusive reason for holding Velikovksy's theory complete with collective amnesia, if you hold Velikovsky's general theory, but it gives no reason at all to hold Velikovsky's general theory.

9.

Some brief concluding remarks:

i. This distinction between two types of support is not itself *ad hoc* within the UN programme. Once you realise that this programme was never committed to the idea that "fitting your theories to the data is a bad thing", the distinction is just an elaboration of the reasons why it need be no bad thing.

ii. Unfortunately – methodological life would be altogether simpler otherwise – the judgment in the UN-satisfying cases that the support "spills over" from the specific to the general underlying theory is *not* a simple reflection of the fact that the general idea entails the phenomena at issue without the need of *any* auxiliaries. The core wave idea on its own does *not* entail the straightedge diffraction phenomena. This would be inconsistent with the Duhem thesis which in fact holds universally. Instead, the straightedge diffraction phenomena were entailed by that core idea together with the "natural" auxiliaries – those motivated, though not entailed, by the general heuristic ideas underlying the wave theory.

iii. This is why the – difficult – notion of naturalness within a framework cannot be avoided. It is also exactly why the UN rule, despite perhaps first appearances, does have an epistemic rationale. Early critics, like Alan Musgrave,[19] were quick to suggest that the UN view was no better off than the temporal novelty view in terms of possession of a cogent epistemic rationale. I have asked why on earth the temporal-order of theory and evidence should in itself have any significance for rational support. But it might seem that there is at least equal justification for asking why on earth a theory's empirical merits, once it is "on the table", should depend at all on how it was constructed – on how it got onto the table in the first place.

But the answer is now easy to see. In science we are interested – indeed, I would argue, principally interested – in how much support a specific theory obtains *as representative of the general theory that underlies it.* After all, if we were just interested in codifying, correctly dealing with, some range of empirical phenomena, say fringe phenomena, it would not matter at all, except perhaps pragmatically, which of two rival theories we used if they both in some way accommodated all the phenomena. Scientists in the early 19th Century were principally interested in how much support the facts gave to Fresnel's *specific* theory of diffraction (consisting of the general wave theory plus specific assumptions tieing it to the case of diffraction) *as representative of that general wave theory*. The judgment that principally required justification was that the sensible way forward in *other* areas of optics – such as crystal optics – was to build different specific theories on that same general idea.

But given Duhem's point that the general theory has no empirical consequences of its own, this judgment about "the sensible way forward" involves considerations *both* of how the specific theory based on it stands deductively to relevant facts *and* of how the specific assumptions relate to the general idea. And this latter relationship is exactly what heuristics is about – the complex, but I claim objective, relationships between general motivating idea and particular theories built around it.[20]

iv. Although I have given what I hope is a clear account of the basic idea underlying the heuristic account of support, much remains to be done – especially in terms of detailed comparison of the view with other approaches. It will, for example, no doubt be suggested that the basic idea of the account can be given a much more precise formulation in Bayesian terms. I have no doubt that the heuristic account can indeed be given a Bayesian formulation – but this is exactly the problem with Bayesianism. The fact that every conceivable position in the prediction vs accommodation debate has been defended on the basis of some Bayesian position is a perfect illustration of the fact that "the" position can explain everything and so really explains nothing. However, this clearly requires careful articulation which I am not able to provide here.

A fuller treatment also requires a detailed comparison of the heuristic view I have defended with the accounts of Nickles and Mayo. I have argued that their chief criticism of what they took to be my view misses the mark. But they each present positive accounts that can plausibly be argued to be equivalent to the clarified (or modified?) view I have explained here. Indeed, it might be argued that Deborah Mayo's views about 'severe tests' can be used to develop a deeper account, one that yields my own heuristic view but which operates from an intuitively more compelling starting point. Surely much of what I have said can be reexpressed intuitively using the notion of a severe test – the "fossils" do not provide a severe test of the Gossified version of creationism, planetary stations and retrogressions do provide a severe test of Copernicanism despite having been known for centuries, and so on. However, I would argue that, when it comes to which notion is the deeper, or more explanatory, it is in fact heuristic considerations that explain our intuitions about severe tests rather than the other way round (despite the relative familiarity – due to Popper's influence? – of the latter notion). Again this requires careful argumentation which I cannot provide here.

One final important omission is this. Except for a brief reaction to the criticism of the UN view by Colin Howson, I have been considering throughout the case of deterministic theories where, *modulo* auxiliary assumptions and initial conditions, we do have direct deducibility of the evidence concerned from the theories concerned. In the case of statistical-probabilistic theories, matters may well be more complicated. There, the issue of how well some specific theory fits the data is in general a question of degree. Suppose that we have two separate general classes of statistical theory G_1 and G_2, and specific theories T_1 and T_2 developed from each general class using some data e. It may well be that the extent to which the two specific theories T_1 and T_2 are supported by e, also has consequences for the extent to which the general classes of theory, G_1 and G_2, are supported by e. But this is because, speaking intuitively, the best fit with the data achievable with a specific theory from one of the general classes, G_1, say, is better than the best fit achievable with a specific theory from the other general class G_2. So, for example, the fact that some particular parabola fits the data (of simultaneous values for some variables x and y) better than any particular straightline, gives some reason to think that the general theory that the relationship between x and y is parabolic (plus an "error term") is better supported than the general theory that the relationship is linear (plus an error term). Suppose in contrast, that we have two general deterministic frameworks (say Ptolemaic and Copernican astronomy) and two specific theories P_n and C_m produced within those frameworks on the basis of some evidence e – so both specific theories are "use-constructed" with respect to e. Here the degree of fit between e and P_n and C_m will be maximal – each deductively entails e (*modulo* accepted auxiliaries). And so it does seem intuitively clear in this case that *e on its own*, while it may give conclusive reason to someone who already accepts the Copernican framework to accept C_m (and similarly give conclusive reason to someone who already accepts the Ptolemaic framework to accept P_n), cannot give any reason to prefer one *general* framework over the other.[21]

Notes

1. There is a separate potential argument from within the Bayesian framework. Since, whenever e is already known, it is already part of background knowledge, for any such e, $p(e) = 1$. The view – natural for a Bayesian – that e supports T if and only if $p(T,e) > p(T)$, then entails that no known e ever supports any T. ($p(T,e) = p(T\&e)/p(e) = p(T\&e)$ (if $p(e) = 1$); and $p(T\&e) \leqslant p(e)$.) This would then imply that novelty of evidence is a necessary condition for it to support any theory. However, this was, I think, always regarded as a *problem* for Bayesianism (the "old evidence problem" – see Glymour (1980)) – it seemed forced to say that temporal novelty mattered in this strong sense and yet no one really believed that it could. (Later there were Bayesians – one is Patrick Maher – who at least *appear* to hold that novelty is important.)
2. See, for example, Earman and Glymour (1980).
3. See my (1985).
4. See Lakatos and Zahar (1976).
5. See his (1988).
6. See Lakatos and Zahar (1976) and Worrall (1978) and (1985).
7. See Nickles (1987), and Mayo (1996).
8. I am not, of course, claiming that these are necessary conditions for T to support e – for a start probabilistic theories can be confirmed by evidence which they do not entail. They are however sufficient (whereas the temporal prediction view is neither necessary nor sufficient).
9. Nickles (1987) and particularly Mayo (1996).
10. Contrary to Colin Howson's (1990).
11. Though there have been suggestions that even this case is not as straightforward as has normally been assumed (see Bamford 1996).
12. Nickles (1987), Howson (1990) and Mayo (1996).
13. Mayo (1996).
14. Dorling (1973) and (1974), Glymour (1980), Zahar (1981), Harper (1991) and (1993)
15. For a detailed account of Newton's method, its strengths and the problems it raises see my (2000).
16. Norton (1993) and (1995).
17. But there are special difficulties in extending this account to statistical cases. The present account should be regarded as applying only to the case of deterministic theories.
18. The basis Creation hypothesis is (essentially) that the world was created as it now is observed to be; hence plugging in observations of the way the world is leads deductively to specific versions of the theory.
19. Musgrave (1978).
20. My friend Elie Zahar has a lot to answer for in having made the suggestion (in his (1973)) that heuristics involves consulting private diaries and correspondence.
21. There are no doubt many other complexities in the statistical case that need attention too. In particular the arguments in Foster and Sober (1994). Much of Deborah Mayo's analysis too is, of course, directed at statistical cases.

References

Bamford, G.: (1996), 'Popper and his commentators on the discovery of Neptune: A close shave for the law of gravitation?', *Studies in History and Philosophy of Science* **27**, No.2, pp. 207–232.

Dorling, J.: (1973), 'Demonstrative Induction: Its Significant Role in the History of Physics', *Philosophy of Science* **49**, pp. 360–372.

Dorling, J.: (1974), 'Henry Cavendish's Deduction of the Electrostatic Inverse Square Law from the Result of a Single Experiment', *Studies in the History and Philosophy of Science* **4**, pp. 327–348.

Earman, J., Glymour, C.: (1980), 'Relativity and Eclipses: The British eclipse expeditions of 1919 and their predecessors', *Historical Studies in the Physical Sciences* **11**, pp. 49–85.

Foster, M., Sober, E.: (1994), 'How to tell when Simpler, More Unified, or less Ad-Hoc, Theories will Provide more Accurate Predictions', *British Journal for the Philosophy of Science* **45**, No. 1, pp. 1–35.

Glymour, C.: (1980), *Theory and Evidence*, Princeton University Press, Princeton.

Harper, W.: (1991), 'Newton's Classic Deductions from the Phenomena', *PSA 1990*, vol. 2, pp. 183–96.

Harper, W.: (1993), 'Reasoning from Phenomena: Newton's Argument for Universal Gravitation and the Practice of Science', in P. Theerman, A.F. Seef (eds.), *Action and Reaction*, University of Delaware Press, Delaware City, pp. 30–78.

Howson, C.: (1990), 'Fitting theory to the facts: Probably not such a bad idea after all', in C.W. Savage (ed.), *Scientific Theories*, Minnesota Studies in the Philosophy of Science, vol. 14, University of Minnesota Press, Minneapolis, pp. 224–244.

Lakatos, I., Zahar, E.: (1976), 'Why did Copernicus's Research Programme Succeed Ptolemy's?', in Westman, R. (ed.), *The Copernican Achievement*, University of California Press, Los Angeles.

Maher, P.: (1988), 'Prediction, Accommodation, and the Logic of Discovery', in A. Fine, J. Leplin (eds.), *PSA 1988*, vol. 2, Philosophy of Science Association, East Lansing, pp. 273–285.

Mayo, D.: (1996), *Error and Growth of Experimental Knowledge*, University of Chicago Press, Chicago.

Musgrave, A.E.: (1978), 'Evidential Support, Falsification, Heuristics and Anarchism', in G. Radnitzky, G. Andersson (eds.), *Progress and Rationality in Science*, Reidel, Dordrecht, pp. 118–136.

Nickles, T.: (1987), 'Lakatosian heuristics and epistemic support', *British Journal for the Philosophy of Science* **38**, pp. 181–205.

Norton, J.: (1993), 'The Determination of Theory by Evidence: The Case for Quantum Discontinuity, 1900–1915', *Synthese* **97**, pp. 1–31.

Norton, J.: (1995), 'Eliminative Induction as a Method of Discovery: How Einstein Discovered General Relativity', in J. Leplin (ed.) *The Creation of Ideas in Physics*, Kluwer, Dordrecht, pp. 38–74.

Stokes, G.G.: (1884), *Burnett Lectures. On Light*, Macmillan, London.

Worrall, J.: (1978), 'The Ways in which the Methodology of Scientific Research Programmes improves on Popper's Methodology', in G. Radnitzky, G. Andersson (eds.), *Progress and Rationality in Science*, Reidel, Dordrecht, pp. 18–39.

Worrall, J.: (1985), 'Scientific discovery and theory-confirmation', in J.C Pitt (ed.). *Chance and progress in modern science: Papers related to and arising from the Fourth International Conference on History and Philosophy of Science*, D. Reidel, Dordrecht, pp. 301–332.

Worrall, J.: (2000), 'The Scope, Limits and Distinctiveness of Newton's Method of 'Deduction from the Phenomena' ', *British Journal for the Philosophy of Science* **51**, March 2000, pp. 45–80.

Zahar, E.: (1973), 'Why did Einstein's program supersede Lorentz?', Parts 1 and 2, *British Journal for the Philosophy of Science* **24**, pp. 95–125, 223–262.

Zahar, E.: (1989), *Einstein's Revolution: A Study in Heuristic*, Open Court, La Salle.

VIII

PROBABILITY, INDUCTION, AND DECISION THEORY

PREFERENCE LOGIC AND RADICAL INTERPRETATION

Kanger Meets Davidson[1]

Wlodek Rabinowicz
Department of Philosophy
Lund University

The primary purpose of this paper is to trace the intellectual effects of an encounter between two very different philosophers working in two seemingly unconnected areas. As will be seen, Stig Kanger's meeting with Donald Davidson led the latter to modify his influential theory of radical interpretation and gave the former an inspiration for a rather striking paradox in preference logic. While the paradox can be dissolved, radical interpretation confronts some serious difficulties.

1. A paradox in preference logic

Stig Kanger's paradox can be found in "A Note on Preference Logic". This characteristically short two-page note was his contribution to a Festschrift for Thorild Dahlquist, published in Uppsala in March 1980. To introduce the paradox, let us suppose that $\succcurlyeq$ is a preference relation on a set of states of affairs (propositions) that is assumed to be closed under Boolean operations. I.e., $\succcurlyeq$ is the type of relation that is studied in preference logic. We may read $A \succcurlyeq B$ either evaluatively, as "State A is at least as good as state B", or descriptively, as "State A is at least as preferred as state B". We use $\succ$ to refer to strict preference: $A \succ B$ is to be read as "A is strictly preferred to (or, better than) B". Now, consider two conditions that one might want to impose on $\succcurlyeq$:

Interpolation of Exclusive Disjunction (IED): For all states A, B,

$$\text{if } A \succ B, \text{ then } A \succcurlyeq (A \div B) \succcurlyeq B.$$

Here, $A \div B$ stands for the symmetrical difference of A and B, respectively, i.e., for the Boolean analogue of exclusive disjunction ("either A or B, but not both").

P. Gärdenfors, J. Woleński and K. Kijania-Placek (eds.), In the Scope of Logic, Methodology and Philosophy of Science, Vol. I, 213-233.

Four Levels (4L): There are some states A, B such that

$$A, B, -A \text{ and } -B \text{ all occupy different levels in } \succcurlyeq .$$

That is, either $A \succ B$ or $B \succ A$, and similarly for all the other pairwise comparisons between the four states. $-A$ and $-B$ stand for the complements of A and B, respectively. Strict preference $\succ$ is immediately definable in terms of weak preference $\succcurlyeq$:

$$A \succ B =_{\text{df}} A \succcurlyeq B \text{ and it is not the case that } B \succcurlyeq A.$$

Similarly, we can define indifference as follows:

$$A \approx B =_{\text{df}} A \succcurlyeq B \text{ and } B \succcurlyeq A.$$

The two conditions above "seem evident", says Kanger, but they cannot be upheld in tandem:

Kanger's impossibility result: There is no weak (= transitive & complete) ordering $\succcurlyeq$ that satisfies both the Four Levels and the Interpolation of Exclusive Disjunction.

The assumption of completeness for $\succcurlyeq$ is very strong. In many cases, we would like to allow for gaps in the preference ordering. If states A and B are significantly different from each other, we may well insist that neither $A \succcurlyeq B$ nor $B \succcurlyeq A$ are the case. Can we avoid to assume completeness and still prove that Kanger's two conditions are mutually incompatible?

Fortunately, we can. It is enough to assume that $\succcurlyeq$ is transitive.[2] In fact, as may be seen from the proof below, it would be enough just to assume that $\approx$ is a transitive relation. While the transitivity of indifference is relatively uncontroversial when indifference is interpreted evaluatively, the transitivity requirement is more problematic on the descriptive interpretation. As is well known, if preferences of a subject are determined by multidimensional pairwise comparisons, then intransitivities of indifference are to be expected. Still, these well-known phenomena might also be interpreted in a different way – as showing that a subject's preference ordering should be determined holistically rather than by a series of independently conducted pairwise comparisons. Transitivity will then function as an apriori constraint on any adequate holistic determination of preference.

Here, then, is the paradox in its final version:

Strengthened impossibility result: There is no transitive relation $\succcurlyeq$ that satisfies both the Four Levels and the Interpolation of Exclusive Disjunction.

The proof that follows slightly simplifies Kanger's original version. By the transitivity of $\succcurlyeq$, $\succ$ is transitive.[3] Consequently, (4L) implies the existence of states $A, B, -A$ and $-B$ that are linearly ordered by strict preference, in one of the twenty-four possible ways. In Kanger's somewhat sadistic version of the proof, we are supposed to go through all these possible cases, one by one, and for each case demonstrate that (IED) is violated. Here, we shall be more economical.

Proof: Assume (IED). We first show that, given (IED), the exclusive disjunction of unequally valued states must be equal in value to one of its disjuncts:

Lemma: If $A \succ B$, then $A \approx (A \div B)$ or $B \approx (A \div B)$.

Proof of Lemma: Assume that $A \succ B$. By (IED),

(i) $A \succcurlyeq (A \div B) \succcurlyeq B$.

Thus, in particular, $A \succcurlyeq (A \div B)$. We consider two cases:
Case 1: $(A \div B) \succcurlyeq A$. Then $A \approx (A \div B)$, by the definition of $\approx$, and we are home.
Case 2: It is not the case that $(A \div B) \succcurlyeq A$. Then $A \succ (A \div B)$, by the definition of $\succ$. Thus, applying (IED) once again, this time to A and $A \div B$,

(ii) $A \succcurlyeq (A \div (A \div B)) \succcurlyeq (A \div B)$.

Since

$$A \div (A \div B) = (A \cap -(A \div B)) \cup ((A \div B) \cap -A) = (A \cap B) \cup (B \cap -A) = B,$$

it follows from (ii) that $B \succcurlyeq (A \div B)$. Since we already know from (i) that $(A \div B) \succcurlyeq B$, it follows that $B \approx (A \div B)$, by the definition of $\approx$. And so we are home again.

We next show that:

Lemma + the transitivity of $\approx \Rightarrow$ (4L) is violated.

Proof by reductio: Suppose (4L) holds for some A and B. Then we have either $A \succ B$ or $B \succ A$. In each case, Lemma implies:

(i) $A \approx (A \div B)$ or $B \approx (A \div B)$.

We also have either $-A \succ -B$, or $-B \succ -A$. In each case, Lemma implies:

(ii) $-A \approx (-A \div -B)$ or $-B \approx (-A \div -B)$.

Since $A \div B = -A \div -B$, (ii) is logically equivalent to:

(iii) $-A \approx A \div B$ or $-B \approx A \div B$.

(i) and (iii) imply that two of the four states $A, B, -A, -B$ are on the same level as $A \div B$. But given the transitivity of $\approx$, this means that these two states occupy the same level, contrary to the hypothesis. □

2. The paradox dissolved

How are we to deal with this paradox? To begin with, one might argue that the two conditions are not as evident as Kanger suggests. That (4L) is not uncontroversial will be seen below. But what is especially important is that the seemingly strong intuitive appeal of (IED) is deceptive. In the first place, the value of the exclusive disjunction may simply be *incomparable* with the values of the disjuncts. The evaluation of disjunctive states is notoriously contested. But even if we ignore this possibility, we might still want to reject (IED). If the disjuncts are *mutually compatible* with each other, the value of their exclusive distinction need not lie somewhere in-between the values of the disjuncts. After all, in the exclusive disjunction of A and B, the alternatives envisaged are not simply A and B but A-and-not-B and B-and-not-A. Thus, the value of such a disjunction should lie somewhere in-between the values of the mutually incompatible alternatives. These alternatives coincide with A and B only if A and B themselves are incompatible with each other. This suggests that (IED) is intuitive only when it is weakened:

Interpolation of the Exclusive Disjunction of Incompatibles (IEDI):
For all *mutually incompatible* states A and B, if $A \succ B$, then $A \succcurlyeq (A \div B) \succcurlyeq B$.[4]

Unlike (IED), (IEDI) is fully compatible with the Four Levels, as will be seen below.

Does it mean, then, that (IED) is a condition without any appeal whatsoever? I do not think so. The Interpolation of Exclusive Disjunction does have some independent plausibility. Even when two unequally valued states are mutually compatible, it is not easy to see how their exclusive disjunction could be preferred or dispreferred to *both* of them at the same time. In order to finally dissolve the paradox, therefore, we should explain how it is possible that the Four Levels and the Interpolation of Exclusive Disjunction, taken separately, appear to be rather plausible (if not quite "evident"), even though, as we have seen, they cannot be upheld together. Such an explanation is not hard to come by. (4L) and (IED) are both plausible, because each of them is separately satisfied by some relatively plausible interpretations of $\succcurlyeq$. Examples of such interpretations will be presented below.

Case 1: The Four Levels is satisfied while the Interpolation of Exclusive Disjunction is violated. Example: the *expected utility* interpretation.

On this interpretation, $A \succcurlyeq B$ iff $U(A) \geqslant U(B)$, where U measures the expected value of a state. Following Jeffrey (1983), the expected utility of a state is here taken to be the weighted sum of the expected utilities of its different possible realisations, with weights being the *conditional* probabilities of the realisations in question. Thus, if P is the underlying probability function, we assume that the following holds for all states A and B:

(EU) If $P(A \cap B) = 0$ and $P(A \cup B) > 0$,
$U(A \cup B) = P(A/A \cup B)U(A) + P(B/A \cup B)U(B)$

It is easy to see that (EU) immediately implies (IEDI): If A and B are incompatible, then $A \div B = A \cup B$. Consequently, $U(A \div B) = U(A \cup B)$ equals the weighted sum of the utilities of the disjuncts. Thus, the utility of $A \div B$ must lie somewhere between the utilities of A and B.[5]

(IED), on the other hand, may well be violated by this expected utility interpretation, as is shown by the following example. Let C, D and E be three equiprobable states that are mutually exclusive and jointly exhaustive. Thus, $P(C) = P(D) = P(E) = 1/3$, and $P(C \cap D) = P(D \cap E) = P(C \cap E) = 0$. Let the U-values for C, D and E be $0, 2$ and 3, respectively. Consider $A = C \cup D$ and $B = C \cup E$. Note that $A \div B = D \cup E$. Using (EU), we can calculate the utilities of A, B and $A \div B$:

$$U(A) = P(C/A)U(C) + P(D/A)U(D) = 1/2 \times 0 + 1/2 \times 2 = 1.$$

Similarly, $\quad U(B) = 1/2 \times 0 + 1/2 \times 3 = 1,5,$

while $\quad U(A \div B) = 1/2 \times 2 + 1/2 \times 3 = 2,5.$

Thus (IED) is violated but the Four Levels holds; states A, B and their both complements have all different utilities:

$$U(-A) = U(E) = 3, \quad U(-B) = U(D) = 2.$$[6]

Case 2: The Interpolation of Exclusive Disjunction is satisfied while the 4-Level Condition is violated. Examples: *extremal* preference (maximin or maximax), or *ceteris paribus* preference.

Suppose that preferences between states are derived from preferences between *possible worlds*. The preference between states is defined as the preference between selected world-representatives of the states in question. To be more precise, assume a weak preference ordering $\succsim$ on the set U of possible worlds. We identify states with sets of worlds: a state is identified with the subset of U that consists of all worlds in which that state obtains. Let c be a *choice function* from states to worlds, such that $c(A)$ is defined iff $A \neq \emptyset$, and for each such non-empty A, $c(A) \in A$. Then, define $A \succcurlyeq B$ as $c(A) \succsim c(B)$. (The relation $\succcurlyeq$ does not obtain between A or B, if either of them is empty.)

Suppose that the choice function c is based on some underlying linear ordering $\nmid$ of U: $c(A)$ picks out that world in A that comes highest in $\nmid$. (If such a world is to exist, for any non-empty A, whether finite or not, it is not enough that $\nmid$ is linear; *every* subset of U, and not just the finite ones, must contain the maximal $\wr$-element. In other words, U must be *well-ordered* by the converse of $\wr$.)

Possible interpretations of $\wr$:

(i) $\wr$ is a “linearisation” of $\succsim$ on U (with equally good worlds coming in an arbitrary order). That is, $\wr$ is any linear ordering such that for all A and B, if A comes above B in $\succsim$, then $A \wr B$. Then $\succcurlyeq$ is *maximax*.

(ii) $\wr$ is a “linearisation” of the converse of $\succsim$ (with equally good worlds again coming in an arbitrary order). Then $\succcurlyeq$ is *maximin*.[7]

(iii) $\wr$ is the ordering of worlds with respect to their similarity to the actual world (= the status quo, the reference world). For any worlds x and y, $x \wr y$ iff x is more similar to the actual world than y is. Then $A \succcurlyeq B$ stands for: *Ceteris paribus*, it would be better that A than that B. In other words: It would be better that A rather than that B, other things being equal (to what they actually are).[8]

It can now be shown that if $\succcurlyeq$ is based on a preference relation $\succsim$ between world-representatives of states, selected by means of a choice function c that is based on a linear ordering $\wr$ of U, then (IED) is satisfied and (4L) is violated.

Proof: Clearly, since $\succsim$ is transitive on worlds, the derived ordering $\succcurlyeq$ is transitive on states. Therefore, in view of the impossibility result, it is enough to prove (IED). Suppose that $A \succ B$, i.e., $c(A)$ comes higher than $c(B)$ in the $\succsim$-ordering. Let x be that element of $\{c(A), c(B)\}$ that comes highest in the $\wr$-ordering. Clearly, $x \notin A \cap B$, since otherwise we would have $c(A) = c(B)$ and so it would not be the case that $c(A)$ comes higher than $c(B)$ in ordering $\succsim$. Since $x \in A \cup B$, it follows that $x \in A \div B$. Consequently, $c(A \div B) = x$. But then $c(A \div B) = c(A)$ or $c(A \div B) = c(B)$. In each case, $A \succcurlyeq (A \div B) \succcurlyeq B$.[9]

3. What’s this got to do with Davidson’s radical interpretation?

Here is Davidson’s problem of *radical interpretation*: What does it take to understand another person, more or less from scratch? Davidson’s goal is to elicit what a subject *means* by what he says, what he *believes* and what he *prefers*. The elicitation should be based on some relatively unproblematic set of empirical data.

While in his earlier work, collected in *Truth and Interpretation* (Davidson 1986), the goal of radical interpretation was just to disentangle meaning from belief,[10] the new project involves an additional task: determination of the subject’s desires. In this way, understanding a person allows us to understand not just what he thinks but also what he does, given that what he does is a function of what he believes and desires.

This project of a three-way elicitation – of meaning, belief and preference – has been pursued by Davidson over two decades in a series of publications starting with his 1978 lecture "Towards A Unified Theory of Meaning and Action" (published two years later, see Davidson (1980)) and continuing with such papers as "Expressing Evaluations" (Davidson 1982), "A New Basis for Decision Theory" (Davidson 1985), "The Structure and Content of Truth" (Davidson 1990), and "The Folly of Trying to Define Truth" (Davidson 1996).

In order to understand Davidson's project, it is instructive to compare it with Quine's well-known conception of radical translation (cf. Quine 1960). Here is how Davidson himself describes Quine's undertaking:

> Noting that, while there is no direct way to observe what speakers mean, all the evidence required to implement communication must be publicly available, Quine surveys the relevant available evidence, and asks how it could be used to elicit meanings. [...] For Quine, the key observables are acts of *assent* and *dissent*, as caused by events within the ambit of the speaker. From such acts it is possible to infer that the speaker is caused by certain kinds of events to hold a sentence true. [Added in footnote: The step from observed assents to the inferred attitude of holding true is not, I think, explicit, in Quine.] Just here a basic challenge arises. A speaker holds a sentence true as a result of two considerations: what he takes the sentence to mean, and what he believes to be the case. [...] How can the roles of these two explanatory factors be distinguished and extracted from the evidence? [...] Quine's key idea is that the correct interpretation of an agent by another cannot intelligibly admit certain kinds and degrees of difference between interpreter and interpreted with respect to belief. As a constraint on interpretation, this is often called by the name Neil Wilson gave it [Wilson (1959)], the *principle of charity*. (Davidson 1990, pp. 318f)

Quine's problem is then that different hypotheses about the speaker's meaning can be defended by adjustments in the hypotheses about his beliefs. Quine's solution of this problem is that the speaker's beliefs must obey *constraints* imposed by the principle of charity that requires a far-reaching consensus in beliefs between the speaker and us. Given these constraints on beliefs, the available data about the speaker's assents to and dissents from sentences (including the external circumstances of such assents and dissents), we can fix the speaker's meaning (up to remaining indeterminacies).

The differences between Quine and Davidson are at least sixfold:

(i) While Quine pursues a project of (radical) *translation*, Davidson is interested in *interpretation*.

> While Quine is concerned with the conditions of successful translation from a speaker's language into the interpreter's, I emphasise that the speaker needs to know of the semantics of the speaker's language, that is, what is conveyed by the T-sentences entailed by a theory of truth [for the speaker's language]. (Davidson 1990, p. 319).

(ii) Quine takes the circumstances that prompt assents to (or dissents from) observation sentences to be patterns of stimulation of nerve endings rather than external objects or events. These proximal stimuli are taken to determine the empirical content of such sentences in the speaker's language, which implies that their translation is relatively unproblematic: they are translated into the sentences in the interpreter's language that are correlated with the similar stimulation patterns. (*Relatively* unproblematic, that is. For some second thoughts on this issue, cf. Quine (1990), sections 15 and 16.) Davidson's view on this issue is more in line with common sense:

> [...] interpretation depends (in the simplest and most basic situations) on the external objects and events salient to both speaker and interpreter, the very objects and events the speaker's words are then taken by the interpreter to have as subject matter. (Davidson 1990, p. 321)

(iii) For Quine, charity is much less inevitable than it is for Davidson. While we normally assume that the speaker shares our beliefs to a large extent, this assumption of consensus is not inescapable: it might be overturned by considerations of *simplicity*.

> The linguist assumes that the native's attitudes and ways of thinking are like his own, up to the point where there is contrary evidence. He accordingly imposes his own ontology and linguistic patterns on the native wherever compatible with the native's speech and other behavior, *unless a contrary course offers striking simplifications*. (Quine 1990, pp. 48f., my italics.)

(iv) Furthermore, Quine's version of charity requires us to assign to the speaker not so much the beliefs we actually hold but rather the beliefs we imagine we *would* have held in the speaker's shoes.

> [The translator] will favor translations that ascribe beliefs to the native that stand to reason or are consonant with the native's observed form of life. [...] Practical psychology is what sustains our radical translator all along the way, the method of his psychology is empathy: he imagines himself in the native's situation as best he can. (Quine 1990, p. 46.)

Davidson appears to be less prepared to make such allowances for expected divergences and, when he does make them, he takes empathy to uncover divergences in needs and valuations rather than divergences in beliefs. The interpreter is counselled

> to interpret agents he would understand as having, in important respects, beliefs that are mostly true and needs and values the interpreter shares or imagines himself sharing if he had the history of the agent and were in compatible circumstances. (Davidson 1985, p. 93)

(v) While Quine's project is strictly behaviourist – the basic data concern an outward assent and dissent behaviour of the speaker – Davidson is fully prepared to allow as empirical data the speaker's *mental attitudes* to sentences, such as holding a sentence to be true. To be sure, such a mental attitude is manifested in an assent behaviour, but the attitude and the behaviour are not the same thing.

(vi) However, Davidson thinks that knowing which sentences are assented to or held true by a subject under which external circumstances would be *insufficient* as the data basis for elicitation. To fix the meaning, and especially to fix the meaning of theoretical sentences, we need to determine *evidentiary relations* between the different sentences in the speaker's language: what he counts as evidence for what, what sorts of evidence would make him consider a given sentence as more probable. We need *probability assignments* rather than simple yes-or-no attitudes of holding true. As Davidson illustrates this point:

> [...] a sentence [tentatively] interpreted as meaning that there is a patter on the roof, if held true (given a high probability), ought to increase the probability of the sentence [tentatively] interpreted as meaning that it is raining. In this way, by marking what the speaker takes as evidence for the truth of a sentence, it is possible to interpret sentences and words of an increasingly abstract and theoretical nature. (Davidson 1982, p. 15)

But how are we to access the subject's probabilities? Here it is time to introduce another important source of inspiration for Davidson: Frank Ramsey's program for decision theory. The task Ramsey put himself was to simultaneously determine a subject's probability assignments (quantitative degrees of belief) *and* his quantitative degrees of preference for different outcomes. (Cf. his paper "Truth and Probability", posthumously published in Ramsey (1931).) The data for this elicitation were to be the subject's ordinal preferences over various gambles, as revealed by his (actual and hypothetical) choices. Being prepared to gamble on an event shows something about an agent's probability for the event in question and about his valuation of the possible outcomes.

As Quine, Ramsey encountered the problem of compensatory adjustments: changes in the hypotheses about the subject's probabilities could be compensated for by adjustments in the hypotheses about his degrees of preference for various outcomes. He solved that problem by imposing a number of *constraints* on the subject's preferences over gambles (including such constraints as transitivity, completeness, etc.). Given the constraints, the preferences could be seen as going by the expected utility of gambles and the constraints made it possible to uniquely determine the probability assignments that underlied these expected utilities. When event probabilities were determined, it was then easy to determine the degrees of preference for outcomes (up to positive linear transformations).

Davidson accepts Ramsey's idea of taking ordinal preferences as basic data, but rejects using preferences over *gambles* as basic. As he points out, when the subject is given a choice between gambles, each gamble is presented to a subject as a proposition of the form:

$$A \text{ if } C, B \text{ if not } C.$$

Thus, preference over gambles is an *intensional* attitude – such preferences are attitudes towards propositional objects. As such, they are unfit to function as empirical data for radical interpretation.

The same objection immediately applies to Ethan Bolker's and Richard Jeffrey's approach to expected utility (cf. Jeffrey 1983): their base for the elicitation of probabilities and degrees of preference are a subject's ordinal preferences over propositions:

that A is more desirable than that B.

Again, such a base is unfit to function as an empirical point of departure for radical interpretation.

But from Davidson's point of view, Bolker-Jeffrey constraints on preferences over propositions have one important advantage as compared with Ramsey's: they are formulated in such a way as to allow the objects of preference to be any set of entities whatsoever, as long as the Boolean operations are definable on that set. In view of the close connection between Boolean operations and truth-logical sentential connectives, it thus becomes possible to replace propositions as objects of preference with *linguistic entities*.

This leads to Davidson's own proposal. *Provided* we can identify the truth-functional (Boolean) connectives in a subject's language, why not replace Bolker-Jeffrey preferences over propositions with preferences over (otherwise) uninterpreted *sentences* as basic data for radical interpretation?

The subject prefers the truth of p to the truth of q (in symbols, $p \succcurlyeq q$).

This is a mental attitude towards linguistic objects. As such, it still is an *intentional* attitude, but no longer an *intensional* one. The objects of the attitude are not propositions, but sentences that obey well-defined identity conditions. As Davidson puts it:

> [...] the objective was not to avoid intentional states; it was to avoid *individuative* intentional states, *intensional* states, states with (as one says) a propositional object. A preference for the truth of one sentence over another is an extensional relation that relates an agent and two sentences (and a time). Because it can be detected without knowing what the sentences mean, a theory of interpretation

> based on it can hope to make the crucial step from the nonpropositional to the propositional. (Davidson 1990, p. 323)

Constraints on the preferences among sentences are the same as Bolker-Jeffrey conditions on preferences among propositions, with one extra constraint added:

$$\text{If } p \leftrightarrow q \text{ is a truth-functional tautology, then } p \approx q.$$

For the subject, his sentences are meaningful. But their meaning is originally not given to the interpreter. To replicate Bolker-Jeffrey project of elicitation,

(i) we first need to identify the *truth-functional connectives* in the subject's language.

Otherwise, simple sentences may be treated as black boxes. Then, using Bolker-Jeffrey methods,

(ii) we elicit the subject's *degrees of preference* and *probabilities* for sentences.

Unlike in Ramsey's case and in the case of other standard expected utility theories, utilities in Bolker-Jeffrey approach are elicitable only up to so-called fractional linear transformations (with four free parameters instead of the usual two), and the probability assignment is only determined within certain limits.[11] But this remaining indeterminacy is a cost that Davidson is quite willing to pay.

In the last step, given the subject's probabilities for uninterpreted sentences (plus our knowledge of the external circumstances), and making use of the Principle of Charity,

(iii) we determine the subject's intensional attitudes: his *degrees of (intensional) belief* and the *meaning* of the sentences in his language.

With the meaning in place, the subject's degrees of preference for sentences allow us to determine the third factor in the subject's mental life, his *degrees of intensional preference*. Which completes the task of radical interpretation.

4. In search of truth-functional connectives

But how are we to achieve task (i)? How can we identify truth-functional connectives, in an otherwise uninterpreted language, just on the basis of its user's preferences over sentences? This presupposes, of course, that we have already managed to identify the sentences of the language in question, and that we have found some way to test, for each pair of such sentences, whether the subject prefers one sentence to the other, or vice versa, or is indifferent between them. Davidson does not discuss how this can be done. Nor does he discuss the obvious objection that Bolker-Jeffrey constraints on preferences may be much too exacting for any subject to be able to obey them in full. But let us suppose that we can ignore these problems. What then? How are the truth-functional connectives to be identified?

The first try was made by Davidson in the summer of 1978, at the Wittgenstein seminar in Kirchberg. In the published version of this lecture (Davidson 1980), he suggests that we first should find the connective **?** that stands for *neither ... nor* in the subject's language. We can do it, he claims, by examining the subject's preferences for sentences. Since, as is well known, all the other truth-functional connectives are definable in terms of that one, the rest is easy. For example:

A connective $\boldsymbol{n}$ stands for negation iff for all sentences p, $\boldsymbol{n}p \approx p\boldsymbol{?}p$.[12]

And then,

A connective $\boldsymbol{c}$ stands for conjunction, iff for all sentences p and q,
$p\boldsymbol{c}q \approx \boldsymbol{n}p\boldsymbol{?}\boldsymbol{n}q$.

And so on.

But how can we determine that **?** is what we have been looking for? Well, Davidson suggests the following test: **?** stands for *neither ... nor* iff:

$$\text{For all sentences } p \text{ and } q,\ p \succ q \text{ iff } q\boldsymbol{?}q \succ p\boldsymbol{?}p.$$

Note that, if the hypothesised interpretation of **?** is correct, $q\boldsymbol{?}q$ and $p\boldsymbol{?}p$ stand for $\neg q$ and $\neg p$, respectively. Thus, what Davidson relies on here is the preference principle according to which p is preferred to q iff $\neg q$ is preferred to $\neg p$.

As it stands, this proposal is unsatisfactory, for at least two reasons, one of which is partly recognised by Davidson himself:
(1) The principle $p \succ q$ iff $\neg q \succ \neg p$ does not generally hold for the intended expected utility interpretation of the preference relation. On this interpretation, the equivalence only holds for *probabilistically independent* p and q. This means, that "we must devise a way of telling, from preferences among sentences, that two sentences are independent" (ibid., p. 11). While Davidson thinks it likely that such a test for probabilistic independence can be developed, he does not provide it in the paper.

In fact, it can be shown that the needed test may be impossible to obtain. As we have seen, in Bolker-Jeffrey theory, the subject's probability assignments are not uniquely determinable. In particular, as Levi (1999) has proved, if p and q are probabilistically independent on one of the probability assignments that are compatible with the evidence, then there will be infinitely many other such admissible assignments on which p and q are *not* independent. This means that the test of probabilistic independence is unavailable on the Bolker-Jeffrey approach!

Levi (*ibid.*) suggests that such a test could still be available to *Davidson*, who takes the interpreter to rely on more information than the mere preference data about the subject. The interpreter can access his own beliefs and values and thus might be able to reduce the indeterminacies in his interpretation by invoking the Principle of Charity. I am not convinced, however, that this solution can be of help at the present early stage of interpretation. Charity comes into play at stage (iii), when the interpreter tries to determine the subject's intensional attitudes and in this process is supposed to maximise the consensus between himself and the subject. At present, we are still at the very beginning of the interpretation process, when the interpreter has not yet managed to fully determine the subject's *extensional* attitudes (= attitudes to sentences). He still has to find out the subject's degrees of preference and probability for *uninterpreted* sentences. At this preliminary stage, there is no room for charity.
(2) Still, I may be too pessimistic.[13] But even if a test of probabilistic independence could be developed, Davidson would not yet be in the clear. There are other truth-functional connectives that **?** might stand for and $p?p$ will still correspond to negation. An example is the Sheffer stroke (= *not both*). So even if we found **?** such that, for all probabilistically independent p and q, $p \succ q$ iff $q?q \succ p?p$, we would still be unable to tell whether **?** stands for *neither-nor* or for *not both*.

Davidson made a second try, which he never published, in his lecture in Oslo, in the fall of 1979. This time, instead of *neither-nor*, he proposed to start with a search for *exclusive disjunction* in the subject's language. In fact, the latter looks like a right connective to start with as far as preference data are concerned. Note that Ramsey's gambles may be understood as such disjunctions: If it is read truth-functionally, a gamble description "A if C, B if not C" is equivalent with the exclusive disjunction "Either (C and A) or (not-C and B)". Admittedly, this truth-functional reading of "if" in the gamble description is quite unsatisfactory: it ignores the subjunctive connection that is supposed to obtain between the gamble event (C) and the prizes (A and B). What is even more important, unlike *neither-nor*, exclusive disjunction does not suffice for the definition of all the remaining truth-functional connectives. However, if the exclusive disjunction could be identified in the subject's language, we would at least make *some* progress in the process of interpretation.

Davidson took it for granted that exclusive disjunction satisfies the interpolation principle (IED), possibly because he did not clearly distinguish it from the (relatively) innocuous condition (IEDI).[14] Therefore, he suggested that connective **?** stands for *either ... or* iff it satisfies the interpolation condition:

$$\text{For all } p \text{ and } q, \text{ if } p \succ q, \text{ then } p \succcurlyeq (p?q) \succcurlyeq q.$$

It is here that Kanger comes in: In Oslo 1979, Kanger showed Davidson that, on the intended expected utility interpretation of $\succcurlyeq$, (IED) does not generally hold.[15] Kanger's paradox from 1980 is, as far as I can see, an indirect result of this exchange, even though Davidson's name is never mentioned in that paper. As Davidson conjectures (personal communication), this polite silence might have been partly motivated by Kanger's feeling that, whatever might be said against it, (IED) is still an internally plausible principle, worth considering in its own right.

Kanger had also been helpful in connection with Davidson's last and final try to solve the connective problem.[16] This time, Davidson proposed to start with the *Sheffer stroke*. He presented this idea in his Hägerström lectures in Uppsala, with Kanger as a host, in the Spring of 1980, and he kept to it in all his later work on the subject (cf. Davidson 1982, 1985, 1990 and 1996).

Thus, the idea is to first identify **?** that stands for *not both* and then to identify all the remaining truth-functional connectives in terms of **?**. The latter task is easy since all such connectives are definable by means of the Sheffer stroke. (The identification method to be used is thus the same as the one sketched above in connection with *neither-nor*.) But what about the Sheffer stroke itself?

The expected utility interpretation implies that, for all p,

(a) if $p \succ T$, then $T \not\succcurlyeq \neg p$,

where T is an arbitrary tautology. We also have,

(b) if $T \succ p$, then $\neg p \not\succcurlyeq T$.

We note first that if **?** stands for the Sheffer stroke, then

$p\mathbf{?}p$ stands for $\neg p$, while $q\mathbf{?}(q\mathbf{?}q)$ stands for a tautology: $\neg q \vee \neg\neg q$.

Consequently, Davidson asserts, **?** stands for the Sheffer stroke iff,

(i) for all p and q,

(α) if $p \succ q\mathbf{?}(q\mathbf{?}q)$, then $q\mathbf{?}(q\mathbf{?}q) \not\succcurlyeq p\mathbf{?}p$,
(β) if $q\mathbf{?}(q\mathbf{?}q) \succ p$, then $p\mathbf{?}p \not\succcurlyeq q\mathbf{?}(q\mathbf{?}q)$;

and

(ii) for some p and q, it is not the case that $p\mathbf{?}p \approx q\mathbf{?}q$.

According to Davidson, no truth-functional connective apart from the Sheffer stroke satisfies both (i) and (ii).
(I have somewhat simplified Davidson's condition (i). Instead of $q\mathbf{?}(q\mathbf{?}q)$, he

makes use of a more complicated expression in which he substitutes $q?r$ for q in $q?(q?q)$. This is unnecessary, since the simpler $q?(q?q)$ already stands for a tautology if **?** is the Sheffer stroke.)

Condition (ii) is first added in Davidson (1990), probably in order to avoid the obvious objection that (i) by itself would be insufficient to pick out the Sheffer stroke. It is easy to check that (i) is satisfied even if **?** stands for the *material implication*. It is also satisfied if **?** stands for the "*tautological*" connective that for any two sentences yields truth as value. In both cases, (i) translates into the trivially valid:

$$(\alpha)\quad \text{if } p \succ T, \text{ then } T \not\succeq T, \qquad \text{and} \qquad (\beta)\quad \text{if } T \succ p, \text{ then } T \not\succeq T.$$

In fact, *pace* Davidson, conditions (i) and (ii) are still insufficient to pick out the Sheffer stroke. Given just (i) and (ii), **?** might stand for the "*contradictory*" connective that for any two sentences yields falsity as value. Since on the intended interpretation, the expected utility of contradiction is undefined, the contradictory sentences do not belong to the field of $\succeq$. Consequently, if **?** were such a connective, (i) and (ii) would be vacuously satisfied.

As a matter of fact, given just (i) and (ii),**?** might also stand for *neither... nor*. Since on this interpretation $q?(q?q)$ is again contradictory, (i) would be vacuously satisfied and (ii) would be true.

To exclude these remaining interpretive possibilities, we should add an extra condition, such as:

(iii) for some p and $q, p \not\succeq q?(q?q)$.

But are we now in the clear? Not quite. Some objections, not mentioned by Davidson, still remain:

Objection 1: What if **?** is not a *truth-functional* connective at all? For example: What if it stands for *it is impossible that both*? Or *it is improbable that both*?

The Sheffer stroke is the only truth-functional connective that satisfies conditions (1)–(iii). This much can be shown. But the subject's language may contain several non-truth-functional connectives that also satisfy these conditions. This possibility has not been excluded by Davidson; it is a potentiality he never even considers!

The non-truth-functional interpretation could be excluded if Davidson's interpreter had access to some additional information about the subject's sentential attitudes, apart from the evidence about the subject's preference ordering on sentences. Suppose the interpreter also knows which sentences are *held to be true* and which are *held to be false* by the subject. (While it is a requirement of rationality that the two sets of sentences be disjoint, they will normally not be jointly exhaustive.) Then the task of identifying the truth-functional connectives is easy. Thus, **?** stands for the Sheffer stroke iff:

(α) Whenever the subject holds p or q to be false, he holds $p?q$ to be true;

(β) Whenever the subject holds $p?q$ to be true, he does not hold both p and q to be true;

(γ) The subject holds $p?q$ to be false iff he holds both p and q to be true.[17]

But, as we have seen, Davidson wants to make do without the information about which sentences are held to be true/false by the subject. In this respect, Davidson's approach to radical interpretation has changed in recent years, as compared with his earlier work collected in *Truth and Interpretation*. The reasons for this austerity are not quite clear: After all, the attitudes of holding true/false, if directed to sentences, are just as extensional as the attitudes of preferring one sentence being true rather than another. Possibly, Davidson's motivation was aesthetic: it is clearly more elegant to use only one kind of data (preference data) instead of two. But just as possibly, under Jeffrey's influence, he might have come to suspect that the simple yes-or-no attitudes of holding true/false are not easily ascribable to a person. In particular, their relationship to a person's probability assignments is notoriously unclear. As is well known, holding true cannot be identified with assigning high probability: the former, unlike the latter, is supposed to be closed under conjunction.

Objection 2: What is there to guarantee that the preference data reflect the subject's *expected utility* comparisons?

As we have seen in the first section, there are other plausible interpretations of $\succcurlyeq$. To be sure, if the interpreter knew that the subject's preferences satisfied Bolker-Jeffrey constraints, then he would know that they can be interpreted as in expected utility terms. But in order to know that they satisfy the constraints, he must first identify the truth-functional connectives in terms of which these constraints are formulated. At the same time, the interpreter's procedure for the identification of the connectives assumes that the subject's preferences do obey the relevant constraints, which the interpreter cannot yet know at that stage.

I do not think that the two objections mentioned above are unanswerable. Given Davidson's general *holistic* approach to theorising, they might both be met by a standard recipe: Why not try it out and see how it works? Thus, suppose we find a connective **?** in the subject's language that obeys conditions (i)–(iii). We can then start the process of interpretation with the hypothesis that **?** does indeed stand for the Sheffer stroke. *If* given this hypothesis, the subject's preferences over sentences do turn out to satisfy the Bolker-Jeffrey constraints, then we may conclude that **?** was the Sheffer stroke *and* that the preference data in fact reflected the subject's expected utility comparisons. On the other hand, if the hypothesis turns out to be unworkable, we might try to look for some other candidate for the Sheffer stroke, and eventually, if all such attempts would fail, for some wholly different set of preference data. So our two objections can be met. But it is somewhat surprising that Davidson never even considers them in his essays.

5. Problem of limited language resources

A more serious problem arises in connection with the *second* stage of interpretation – the stage at which the interpreter, after having identified the truth-functional connectives, moves on to the task of elicitation of the subject's degrees of preference and probabilities for sentences. At this stage, the interpreter is supposed to make use of the Jeffrey-Bolker methods of elicitation, as described in Jeffrey (1983). But are these methods applicable for Davidson's purposes?

Unlike Davidson, Bolker and Jeffrey start from a preference ordering on a Boolean algebra of *propositions*. Davidson seems to think that *sentences* would do just as well, provided the language contains truth-functional connectives. To be sure, truth-functional connectives are not quite Boolean operations, but there is a close correspondence between the two. For each sentence p, we can determine its equivalence class $[p]$ consisting of all the sentences that are truth-functionally equivalent to p. Then, in terms of the truth-functional connectives, it is easy to define the Boolean operations on such equivalence classes. For example, if $A = [p]$ and $B = [q]$ are such equivalence classes, then the complement of $A = [\neg p]$ and the intersection of A and $B = [p \wedge q]$. We get in this way the so-called Lindenbaum algebra, which is an example of a Boolean algebra.

However, we are not home yet. Bolker-Jeffrey representation theorem, on which Davidson relies, rests on a very strong presupposition. It presupposes that the Boolean algebra on which the preference ordering is defined is both *atomless* and *complete*.[18] To explain these notions, let us first define the notion of implication: we shall say that a state A *implies* a state B iff $A = A \cap B$.

An algebra is *atomless*, if for any non-zero state A (i.e., $A \neq 0 = A \cap -A$), A contains a non-zero state B such that B implies A but is not implied by it (we say that such a state is "strictly smaller" than A).

If states of the algebra are propositions, then the algebra is atomless if for every consistent proposition there is a stronger consistent proposition.

If X is a set of states, a *lower (upper) bound* of X is any state that implies (is implied by) every member of X. The *infimum* inf(X) of X is the greatest lower bound of X, i.e., every other lower bound of X is strictly smaller than inf(X). The *supremum* sup(X) of X is the least upper bound of X, i.e., it is strictly smaller than every other upper bound of X.

An algebra is *complete* if for any set X of states, whether finite or not, it contains the infimum and the supremum of X.

As a matter of fact, if a Boolean algebra contains the infima for all its subsets, it will of necessity contain the suprema as well. It can be shown that

$$\sup(X) = -\inf(\{-A : A \in X\}).$$

Consequently, if the states of a Boolean algebra are propositions, then the algebra is complete if for every infinite set of propositions, it contains the infimum, i.e., the proposition that corresponds to the conjunction of the propositions in that set.

Now, what about the Lindenbaum algebra? Does it satisfy the two requirements of Bolker-Jeffrey representation theorem? Roughly, this would mean (i) that for any (truth-functionally) consistent sentence p, the language contains another consistent sentence that is stronger than p (atomlessness), and (ii) that for any set of sentences, whether finite or not, the language contains a sentence that is exactly as strong as the conjunction of the sentences in the set (completeness). It is easy to see that these two requirements pull in opposite directions. The number of sentences in any language is *countable*. The same applies, therefore, to the number of equivalence classes in the corresponding Lindenbaum algebra. If that number is finite, the language will be complete but it obviously will not be atomless. And if the number of such classes is denumerable (= countably infinite), the algebra will be atomless but it will not be complete. In general, *no countable Boolean algebra can be both atomless and complete.* For a sketch of the proof, due to Sten Lindström, see Appendix.

Thus, the Bolker-Jeffrey theorem presupposes a non-denumerable algebra of propositions. The number of propositions needed for the representation theorem exceeds by far the sentential resources of any language that we might encounter. This means that Davidson's project of elicitation, with its strong dependence on Bolker-Jeffrey elicitation methods, is doomed from the start. A move from propositions to sentences is necessary if the elicitation is to build on extensional data. But it is precisely this move that makes the Bolker-Jeffrey elicitation impossible.

One might try to resist this objection. Even if full elicitation is not possible, a partial one may be all that is needed. If one could show that the Lindenbaum algebra for the language that we are trying to interpret is rich enough to allow for an *approximate* elicitation of the language user's preferences and beliefs, Davidson's project would still be viable. After all, Davidson is quite prepared to accept that interpretation cannot resolve all indeterminacies. However, to show that approximate elicitation is possible, one would have to prove that the Bolker-Jeffrey theorem would still approximately hold for a sufficiently rich algebra even if that algebra falls short of Bolker-Jeffrey requirements. Proving this result is certainly not a trivial exercise. It is surprising that Davidson has never even touched upon this problem in his long series of papers on radical interpretation.

Appendix

Theorem: No countable Boolean algebra is both atomless and complete.
Proof: Since all countable atomless Boolean algebras are isomorphic (cf. Bell and Slomson 1969, Corollary 7.7 in ch. 1, p. 30), and since Lindenbaum algebras are Boolean and countable, it is enough to prove that an infinite Lindenbaum algebra is atomless and incomplete. Then it will follow, by isomorphy, that every countable atomless Boolean algebra is incomplete.

We first prove that any infinite Lindenbaum algebra LA is *atomless*. If LA is infinite, its underlying language L must contain infinitely many atomic sentences (= sentences that are not built up from other sentences in L by means of truth-functional connectives). Therefore, for every A in LA, where $A = [p]$ for some consistent sentence p of L, we can always find an atomic sentence q that does not appear in p. Since $p \wedge q$ is stronger than p and consistent (as far as truth-functional sentential logic is concerned), $[p \wedge q]$ is a non-zero element of LA that is strictly smaller than $[p]$. Consequently, $[p]$ is not an atom of LA.

Given that LA is atomless, it must be *incomplete*. For let X be any smallest set of states in LA such that for each atomic sentence p either $[p]$ or $[\neg p]$, but not both, belongs to X. Suppose for reductio that LA is complete. Then LA contains inf(X). This, however, is impossible, since the infimum of X, if it existed, would have to be an atom of the algebra. To see this, let q be any sentence such that $[q] = \inf(X)$. By the definition of X, q must be (truth-functionally) consistent; and since for every atomic sentence p, q implies either p or $\neg p$, q cannot be extended to a stronger consistent sentence.

Notes

1. This paper is a revised version of Rabinowicz (1998). I wish to thank several people who have helped me with comments, references and suggestions: John Broome, Thorild Dahlquist, Sven Danielsson, Dagfinn Føllesdal, Sören Halldén, Bill Harper, Per-Erik Malmnäs, Paul Needham, Jan Odelstad, Rysiek Sliwinski, Howard Sobel, Fredrik Stjernberg, Fredrick Stoutland, Göran Sundholm, and Folke Tersman. I am especially grateful to Donald Davidson, who has kindly supplied historical information, and to Sten Lindström, who has proved a theorem I needed for my argument, apart from being supportive in many other ways.

2. Kanger had to impose completeness because of his choice of primitive: instead of $\succcurlyeq$, he took $\succ$ to be his only primitive and then he simply defined $\succcurlyeq$ as the complement of $\succ$. Such a definition immediately yields completeness for $\succcurlyeq$. Which shows, by the way, that $\succcurlyeq$ should *not* be defined in this way if we want to uphold a distinction between indifference and incomparability.

3. Proof: Suppose, for *reductio*, that (i) $A \succ B$, (ii) $B \succ C$, but (iii) not $A \succ C$. By the definition of $\succ$ and by the transitivity of $\succcurlyeq$, (i) and (ii) imply (iv) $A \succcurlyeq C$. Again by the definition of $\succ$, (iii) and (iv) imply (v) $C \succcurlyeq A$. Since $\succcurlyeq$ is transitive, (v) and (i) imply (vi) $C \succcurlyeq B$, which contradicts (ii).

4. Since A and B are incompatible, we could just as well replace the exclusive disjunction $A \dotplus B$ with the inclusive one: $A \cup B$. A word of warning: That (IEDI) is an intuitive condition does not mean that it is unassailable. A counter-example is provided in the next footnote.

5. In fact, a slight strengthening of (IEDI) is an axiom in the Bolker-Jeffrey theory of expected utility:

(Averaging) For all mutually incompatible states A and B,

(i) if $A \succ B$, then $A \succ (A \cup B) \succ B$; and (ii) if A $\approx$ B, then A $\approx$ (A $\cup$ B) $\approx$ B.

Since $A \cup B = A \div B$, averaging implies that an exclusive disjunction can be *strictly* interpolated between unequally valued disjuncts, if these are mutually incompatible.

The role of the averaging axiom in Bolker-Jeffrey theory is somewhat similar to the function of the axiom of *independence* in other axiomatisations of expected utility. Even though independence is a much stronger assumption than averaging, the two axioms express essentially the same idea. Consequently, the well-known Allais-type and Ellsberg-type objections to independence can be re-formulated as objections to averaging and to (IEDI). Thus, (IEDI) is not quite as innocent as it might seem to be! Here is an Allais-type counter-example: Suppose that, being risk-aversive, you prefer receiving one million dollars to a high chance of five million dollars, even though the chance is as high as, say, .9. Let A be the state to the effect that you receive one million but don't get the .9 chance for five millions, and let B be the state in which it is the other way round: instead of one million, you get the .9 chance for five millions. A and B are thus incompatible and $A \succ B$. Suppose, further, that A and B are equiprobable. Then the state $A \div B$ may be seen as a prospect of getting one million with the probability of .5, five millions with the probability of .45, and of getting nothing with the remaining probability of .05. It may well happen that you prefer B to $A \div B$ (if there is some risk of getting nothing, you may be prepared to take an additional risk in order to win a larger prize), even though you prefer A to B. Which violates (IEDI).

6. In this case, not only the exclusive but also the *inclusive* disjunction of A and B has a higher expected utility than A and B: $U(A \cup B) = U(C \cup D \cup E) = 5/3$. But unlike (IED), the Interpolation of (Inclusive) Disjunction:

(ID) For all states A, B, if $A \succ B$, then $A \succcurlyeq (A \cup B) \succcurlyeq B$,

is a principle that is consistent with (4L). To see that, consider the expected utility model just as the one we have described, in which the three equi-probable, mutually exclusive and jointly exhaustive states C, D, E are assigned U-values $1/5, 2/5, 2/5$, respectively. (4L) is still satisfied and if C, D and E are assumed to be the atoms of this state algebra, it can be checked that the model satisfies (ID).

7. Maximin and maximax are two examples of what might be called "extremal preferences". Such preferences are discussed at length in Sven Ove Hansson (1998), Ch. 7.

8. For another interpretation of "*ceteris paribus* preference" cf. Sven Ove Hansson (1998), Ch. 6. Following von Wright (1963), Hansson interprets such a preference in a very demanding way: A is *ceteris paribus* preferred to B iff for any possible realisations of these states, the A-realisation is preferred to the B-realisation, provided that other things are equal between *them*. More precisely, A is *ceteris paribus* better than B iff every complete alternative x (= Hansson's analogue of a possible world) that instantiates A but not B is preferable to a complete alternative y that instantiates B but not A, and is otherwise as similar to x as possible.

While I take *ceteris paribus* preference to be interpretable in terms of the dyadic relation of comparative similarity of alternatives (worlds) to an assumed *status quo*, Hansson interprets this kind of preference in terms of the degrees of similarity between pairs of alternatives: we are supposed to look for a pair of alternatives that are as similar to each other as possible. Thus, the underlying comparative similarity relation is four-place rather than two-place: x is more similar to y than x' to y'. It is easy to show that the preference principles validated by the two interpretations differ from each other (cf. ibid., p. 83f). (IED) is a case in point: it is valid on my interpretation but invalid on Hansson's. Both interpretations of *ceteris paribus* preference are, I think, legitimate, but the proposal I favour is considerably less demanding and therefore more common in everyday use.

9. Note that we also have it that $c(A \cup B) = x$. Thus, the present interpretation validates both (IED) and (ID).

10. For an excellent overview of this early work, see Stoutland (forthcoming).

11. These limits are delineated as follows (cf. Jeffrey 1983, sections 6.1 and 6.6): For any probability-utility pair (P, U) that provides an expected utility representation for a preference ordering $\succcurlyeq$ on the algebra of propositions (with the logically false proposition removed), (P', U') is also an expected utility representation of $\succcurlyeq$ iff for some parameters a, b, c, d such that

(i) $ad - bc > 0$,

(ii) $cU + d$ is positive for all the arguments of U,

and

(iii) if T is the logically true proposition, $cU(T) + d = 1$,

U' is a fractional linear transformation of U with respect to the parameters in question, i.e.,

$$U' = (aU + b)/(cU + d),$$

while

$$P' = P(cU + d).$$

12. For simplicity, I am here ignoring the distinction between use and mention. I hope this systematic ambiguity will not confuse the reader.

13. Even though charity considerations directly apply only at the third stage of the interpretation process, they might, when applicable, make us re-consider the interpretive hypotheses we have started with. Thus, if applying charity at the third stage to the material obtained at the earlier stages turns out to be difficult, the interpreter might at that stage come to suspect that he has made some mistakes at the outset of the process. Thus, he might decide to go back to the first stage again and try out a new interpretation of the sentential connectives. In *this* sense, it could be argued, charity is relevant to all the interpretive stages. But whether such a back-and-forth process of charity-driven interpretation would allow us to identify all pairs of probabilistically independent sentences is by no means clear.

14. For this conjecture, I am indebted to Howard Sobel. Sobel has also reminded me that Ramsey's gambles cannot be described by means of truth-functional statements.

15. Davidson (1985), fn. 5, and (1990), fn. 68: "I am indebted to Stig Kanger for showing me why an earlier attempt at a solution to this problem [= identification of the connectives] would not work." The details of their encounter have been confirmed by Davidson, in personal communication. Dagfinn Føllesdal, who was their host in Oslo, has informed me that he arranged the 1979 meeting because he felt that Kanger's logical expertise might be helpful in setting straight the formal framework for radical interpretation.

16. "[Kanger] also added some needed refinements to the present proposal." (Davidson 1985, fn. 5, and 1990, fn. 68.)

17. Strictly speaking, since it is a rationality constraint that a sentence is never held to be both true and false at the same time, condition (β) is redundant given (γ).

18. See Axiom 2 in Jeffrey (1983), Ch. 9. The preference ordering is defined on such an algebra with its zero element removed. As Jeffrey points out, the assumptions of atomlessness and completeness are used in the proof of the existence part of the representation theorem. They are not needed for the uniqueness part. For an excellent short presentation of the theorem, cf. Broome (1990). That the requirements of completeness and atomlessness might pose a problem for Davidson is suggested in Rawling (2001).

DEFINEABLE PREFERENCES: ANOTHER EXAMPLE*

Searching for a Boyfriend in a Foreign Town

Ariel Rubinstein
School of Economics, Tel Aviv University and
Department of Economics, Princeton University
rariel@post.tau.ac.il, homepage: http://www.princeton.edu/~ariel

Abstract The paper argues for the formal investigation of conditions under which that preference relations are definable in various simple languages. An example of such an investigation is given. It is shown that if a formula in the "pure language with equality" induces a preference relation on the set of T-tuples of objects in a set (independently of the number of elements in the set), then it must be that there is a sequence of preferences P_n on the set of partitions of $\{1, \ldots, T\}$ such that $(x_1, \ldots, x_T)$ is preferred to $(y_1, \ldots, y_T)$ if and only if the set of objects is of size n and $E(x_1, \ldots, x_T) P_n E(y_1, \ldots, y_T)$ where $E(x_1, \ldots, x_T)$ is the partition of $\{1, \ldots, T\}$ where i and j are in the same cell if $x_i = x_j$. Furthermore, for all large enough n the relations P_n are identical.

1. Introduction

The starting point of this paper is a view that decision makers often verbalize their considerations before making a decision. It follows that the "language" a decision maker uses to verbalize his preferences restricts his set of preferences. Thus, interesting restrictions on the richness of the decision maker's language can yield interesting restrictions on the set of an economic agent's admissible preferences.

Before proceeding to the main investigation some background is required. An economic agent in the standard economic model possesses a preference relation defined on a set of relevant consequences. These preferences provide the basis for the systematic description of his behavior and for welfare anal-

*Much of the material in this article is based on Rubinstein (1978), an early unpublished paper of mine. This paper is an "unidentical" twin to Rubinstein (1998a). The material in the two articles appears as Chapter 4 in Rubinstein (2000).

P. Gärdenfors, J. Woleński and K. Kijania-Placek (eds.), In the Scope of Logic, Methodology and Philosophy of Science, Vol. I, 235-243.

ysis. We usually assume that an economic agent is "rational" in the sense that his choice is derived from an optimization given his preferences. Since we adopt the rational man paradigm, the other constraints imposed upon an economic agent's preferences are often weak. For example, in general equilibrium theory we usually only impose conditions such as monotonicity, continuity and quasi-convexity. This restriction does not exclude a preference relation defined on the space of bundles R_2^+ which states that the bundle (x_1,x_2) is evaluated by a (utility) function $(\log(x_1+1))x_2$ whereas it does exclude the lexicographic preferences which state that the bundle (x_1,x_2) is preferred to (y_1,y_2) if $x_1 > y_1$ or ($x_1 = y_1$ and $x_2 \geqslant y_2$) and which are defined by simple logical operations.

What are the general considerations that motivate us to include or exclude preferences from the scope of analysis? One consideration is that some preferences may better explain empirical data. Although I am not an empirical economist, I am doubtful that this is a factor in the choice of restrictions imposed on preferences in the economic theory literature. Another consideration is "analytical convenience". This is a legitimate consideration but one which must be treated with the necessary caution. Another consideration relates to "bounded rationality". It may be argued that some preferences are more plausible than others since they can be derived from plausible procedures of choice. Finding such derivations is, of course, one of the main objectives of bounded rationality models (for a discussion of this point see Chapter 2 in Rubinstein (1998b)).

This analysis, however, will consider a different issue: the ability to describe the preferences in a decision maker's language. I assume that when a decision maker is involved in an intentional choice, he describes his considerations, whether to himself or to agents who operate on his behalf, using his daily language. (For a discussion of this point in the psychological literature see, for example, Shelly and Bryan (1964).) Thus, "My first priority is to obtain as many guns as possible and only after that do I worry about increasing the quantity of food" is a natural description of a preference relation which fits lexicographic preferences. "I will spend 35% of my income on food and 65% on guns" is a natural description of a rule of behavior (assuming linear budget constraints, this rule is consistent with maximizing the utility function $(\text{food})^{0.35}(\text{guns})^{0.65}$. The function $(\log(x_1+1))x_2$, on the other hand, is a "textbook" utility function that is expressed by a relatively simple mathematical formula, even though there is no rule of behavior stated in everyday language that corresponds to this utility function.

Note that when a decision maker is a collective, and decision-makers in economics are in fact often families, groups or organizations, the assumption that preferences are definable makes even more sense since in this case a decision rule must be stated in words in order to be communicated among the individuals in the collective, both during the deliberation and implementation stages.

The aim of this research project is to demonstrate that the requirement that preferences be definable can be formally analyzed. In the following, I will analyze an example which illustrates the connection between a decision maker's language and the set of definable preferences. In Rubinstein (1998a), I presented an example in which a preference is assumed to be formed from a profile of more primitive binary relations. This result was closely related to the theory of Social Choice. These two examples may serve as the first step in a much more ambitious research program to study the interaction between economic agents with "language" as a constraint on behavior, institutions, communication, etc.

2. Searching for a boyfriend in a strange town

Consider the "tale" of a girl who arrives in a strange town and wishes to find a boyfriend. She obtains information on each candidate in the form of a list of his dates during the past T days. The number T is fixed. The girl has yet to meet either the boys or the girls in town (i.e., they are considered "unknowns"). All she can glean from any two given names on the list is whether they are identical or not. The model can also be interpreted in the context of an entrepreneur comparing candidates for a job, each of whom provides a list of his past employers about whom the entrepreneur has no information.

Following are several principles which the girl might use when comparing two candidates.

(1) Prefer boy A if he dated Alice longer than B did.

(2) Prefer the boy who has dated all the girls in town.

(3) Prefer a boy if all the girls he dated, from the second date on, had previously been dated by the other boy.

(4) If the two boys dated the same girl on the first date, prefer the one who dated her longer.

(5) Prefer the boy who made the most "switches".

(6) Prefer the boy who has dated more girls.

Prior to the formal definitions, let us discuss the definability of these principles. Obviously, these principles are definable in some sense since I have just used them. However, definability depends on language and here I am interested in the decision maker who is able to use only a limited language such that:

(i) The decision maker cannot refer to a boy by his name,

(ii) She can only refer to a girl by the term "the girl he dated at time t" and

(iii) The only comparison the decision maker can make between "the girl he dated at time t" and "the girl he (or someone else) dated at time s" is whether they are the same girl or not.

Principle (1) requires the decision maker to be able to refer explicitly to the name "Alice". Principle (2) can be stated in the decision maker's limited language using only the equality relation although it requires the use of a quantifier. Principle (3, 4, 5, 6) can be stated in this simple language. However principle (3) does not provide a definition of a preference since (for the case of $T = 2$, for example) it implies that the dating history (a,b) is preferred to (b,c), which is preferred to (c,a), which is preferred to (a,b). It is easy to see that principle (4) does not define a preference either. Only principles $(5,6)$ are definable in the limited language and induce preference relations.

The assumption that the decision maker uses a *limited language* to express her preferences can have various motivations. It may be interpreted naively as a reflection of the limited language that is available to her. However, even if the language available to her is rich, it might be the case that she cannot fully use it since the available relevant information she possesses is limited (recall that she is a newcomer to town). And even if she is able to use rich language and information is readily available, she may choose, due to complexity considerations, to express her preferences using a formula which is stated in a limited language.

3. The formal model

We can now move on to the formal analysis. The mathematical tools I will be using are from standard Mathematical Logic (Boolos and Jeffrey (1989) and Crossly (1990) are good introductory books).

We characterize the definable binary relations in the "pure language with equality". This is the language of the calculus of predicates that includes symbols for equality only. In this language, an *atomic formula* is of the type $z_1 = z_2$, where z_1 and z_2 are variable names (such as $x_7 = x_3$ or $x = x_2$). A *formula* is a string of symbols constructed according to the following inductive rules: All atomic formulae are formulae; if φ and ψ are formulae, then $(\neg\varphi$, $\varphi \wedge \psi$ and $\varphi \vee \psi$, $\varphi \rightarrow \psi$ and $\varphi \leftrightarrow \psi$ are also formulae; and if x is a free variable (does not appear with a quantifier) in the formula φ, then $\exists x\varphi(x)$ and $\forall x\varphi(x)$ are formulae as well.

A *model* in this simple language is simply a set G, interpreted here as a set of girls. The validity of a formula φ with the free variables $z_1,\dots,z_K$ in a model G (when we substitute each z_k with an element $a_k \in G$) is defined by the standard truth tables and is denoted by $G \models \varphi(a_1,\dots,a_K)$.

We are interested in preferences on dating profiles of length T that are defined by a formula φ with free variables $x_1,\dots,x_T,y_1,\dots,y_T$. Given a set G, the comparison between two boys who have the dating profiles $(a_1,\dots,a_T)$ and $(b_1,\dots,b_T)$ is defined by $G \models \varphi(a_1,\dots,a_T,b_1,\dots,b_T)$. Note that Principle (2) can be expressed by the formula

$$[\forall x \bigvee_{t=1,\dots,T}(x_t = x)] \wedge [\neg\forall x \bigvee_{t=1,\dots,T}(y_t = x)]$$

while principle (3) can be expressed by the formula

$$\bigwedge_{t=2,\dots,T} \bigvee_{m=1,\dots,t-1}(x_t = y_m).$$

In the rest of this section we are interested in formulae that induce binary relations that are transitive and asymmetric under all possible circumstances, i.e. in all models.

Definition: We say that a formula φ with $2T$ free variables induces a transitive and asymmetric relation ordering if the formulae
$\forall x_1,\dots,x_T,y_1,\dots,y_T,z_1,\dots,z_T[\varphi(x_1,\dots,x_T,y_1,\dots,y_T)\wedge \varphi(y_1,\dots,y_T,z_1,\dots,z_T) \rightarrow \varphi(x_1,\dots,x_T,z_1,\dots,z_T)]$
and $\forall x_1,\dots,x_T,y_1,\dots,y_T[\varphi(x_1,\dots,x_T,y_1,\dots,y_T) \rightarrow \neg\varphi(y_1,\dots,y_T,x_1,\dots,x_T)]$
are tautologies, i.e. they are satisfied in all models.

Note that the requirement that the formula be a tautology requires that the definition be applied to all possible worlds. The definition should induce a preference even when for each possible a list of girls in length T there is a boy with this dating experience. It should also be "consistent" regardless of the number of girls in town.

4. Analysis

Thus, for any set G, the comparison of any two dating profiles of a fixed length T (i) does not depend on any comparison of the girls dated by the two boys (such as $x_3 = y_7$), (ii) depends only on the comparison between the dating stability structures of the two boys for any given size of G and (iii) is constant for sufficiently large sets.

The analysis of the definability condition, consists of three stages:

First step: We can assume that the formula φ is written *without quantifiers*.

The "pure language with equality" has the property (see, for example, Robinson (1963)) that for every model G (that is, for any set), every formula has a logically equivalent formula containing the same set of free variables with no quantifiers. To illustrate this point, consider the formula

$$\varphi(x_1,\ldots,x_T,y_1,\ldots,y_T) = [\forall x \bigvee_{t=1,\ldots,T}(x_t = x)] \wedge [\neg\forall x \bigvee_{t=1,\ldots,T}(y_t = x)],$$

which states that a boy who has dated all the girls in town is preferred to one who has not. Although the quantifier "for all" is used here, the validity of $\varphi(a_1,\ldots,a_T,b_1,\ldots,b_T)$ in a model G, depends only on the number of elements in G and the number of elements in each of the two vectors. If G includes $L < T$ elements, then, when substituting $(a_1,\ldots,a_T)$ for $(x_1,\ldots,x_T)$, the validity of the formula $[\forall x \bigvee_{t=1,\ldots,T}(x_t = x)]$ is the same as that of a formula stating that in the vector $(x_1,\ldots,x_T)$ there are L different elements, which can be written without quantifiers. (Take the disjunction of conjunctions, each of which corresponds to a partition of $\{1,\ldots,T\}$ into L non-empty sets $\{e_1,\ldots,e_L\}$ and state that $x_i = x_j$ for any i and j that are in the same e_k and $\neg x_i = x_j$, if i and j are in distinct cells of the partition.)

The fact that for any model G, φ has an equivalent formula with no quantifiers implies that for any given model G, φ is equivalent to a disjunction of configurations of the variables $\{x_1,\ldots,x_T,y_1,\ldots,y_T\}$, where a *configuration* of $\{x_1,\ldots,x_T,y_1,\ldots,y_T\}$ is the "description of which variables are equal." (In other words, it is a conjunction of formulae where, for each pair of variables z_1 and z_2 in $\{x_1,\ldots,x_T,y_1,\ldots,y_T\}$, either $z_1 = z_2$ or $\neg z_1 = z_2$ appears in the conjunction with the following constraint: If $z_1 = z_2$ and $z_2 = z_3$ appear in the conjunction, $z_1 = z_3$ is a conjunct as well.)

Second step: Characterization of the preference relations given a set G

For any given cardinality of G, the only preferences that are definable (in this simple language) are those that are induced by preferences on the "dating stability profile" of each candidate. The term "dating stability profile" refers to a partition of $\{1,\ldots,T\}$, in which i and j are in the same element of the partition if and only if the boy has dated the same girl at time i and time j. In other words, given a vector $(a_1,\ldots,a_T)$, define $E(a_1,\ldots,a_T)$ to be the partition of $\{1,\ldots,T\}$ in which i and j are in the same partition if and only if $a_i = a_j$.

For example, if a boy is fickle and dates a girl for only one period, the corresponding dating structure is the partition $\{\{1\},\ldots,\{T\}\}$; a "faithful" boy is characterized by the dating structure $\{\{1,2,\ldots,T\}\}$.

> *Claim*: If φ defines an ordering in a model G, then for any two dating profiles with $E(a_1,\ldots,a_T)=E(b_1,..,b_T)$, it is not true that $G\models\varphi(a_1,\ldots,a_T,b_1,\ldots,b_T)$; in other words, the decision maker is indifferent between the two profiles $(a_1,\ldots,a_T)$ and $(b_1,\ldots,b_T)$.

Proof: We will show that for any dating structure, if two dating profiles have this structure, the defined preference cannot prefer one to the other. For notational simplicity, we consider the case in which the boy dates T distinct girls.

For any configuration $\psi(x_1,\ldots,x_T,y_1,\ldots,y_T)$ in the disjunctive normal form of φ, where $x_i\neq x_j$ and $y_i\neq y_j$ for all $i\neq j$, denote

$$N(\psi)=\{t\mid \text{there is an } s \text{ such that } x_t=y_s \text{ appears in } \psi\}.$$

Let ψ^* be such a configuration with the lowest number of elements in $N(\psi)$. It is impossible that $N(\psi^*)=0$, since ψ^* would then be the conjunction of all formulae of the type $\neg z_1=z_2$ for all $z_1\neq z_2\in\{x_1,\ldots,x_K,y_1,\ldots,y_K\}$ and the asymmetry tautology would not hold.

It follows that

$$\forall x_1,\ldots,x_T,y_1,\ldots,y_T[\psi^*(x_1,\ldots,x_T,y_1,\ldots,y_T)\wedge\psi^*(y_1,\ldots,y_T,z_1,\ldots,z_T)\rightarrow \varphi(x_1,\ldots,x_T,z_1,\ldots,z_T)]$$

is a tautology. Let ψ be a configuration which satisfies that for any t and r either $x_t=y_r$ or $\neg x_t=y_r$ appears within the configuration and $x_t=y_r$ appears in the conjunction if and only if there is an s such that both $x_t=y_s$ and $x_s=y_r$ appear in ψ^*. Then ψ must be a configuration in the disjunctive normal form of $\varphi(x_1,\ldots,x_T,y_1,\ldots,y_T)$. $N(\psi)$ is a subset of $N(\psi^*)$ and thus, by the minimality of $N(\psi^*)$ it must be the case that $N(\psi^*)=N(\psi)$. Denote by σ the permutation on the set $N(\psi^*)$, defined by $\sigma(i)=j$ if $x_i=y_j$ appears in ψ^*. There must be an integer n such that σ^n is the identical permutation. The formula φ is the disjunction of conjunctions and one of these must be the conjunction of $x_i=y_i$ for $i\in N(\psi^*)$ and the negation of all other equalities, contradicting the asymmetry tautology.

Third Step: For large enough sets "the same rule applies"

We are still left with the possible dependency of the defined preference on the cardinality of the set G: for any integer n there is a formula

$$\varphi_n(x_1,\ldots,x_T,y_1,\ldots,y_T)$$

with no quantifiers such that

$$G \models \forall x_1,\ldots,x_T,y_1,\ldots,y_T[\varphi_n(x_1,\ldots,x_T,y_1,\ldots,y_T) \leftrightarrow \varphi(y_1,\ldots,y_T,x_1,\ldots,x_T)]$$

for all models G with cardinality n. In other words, the decision maker's preference relation on dating profiles may depend on the number of elements in G. This brings me to the **third** and last point: By the compactness argument there is some n^* such that if G's cardinality is at least n^*, $\varphi(x_1,\ldots,x_T,y_1,\ldots,y_T)$ is equivalent to $\varphi_{n^*}(x_1,\ldots,x_T,y_1,\ldots,y_T)$.

Summarizing:

Proposition: If $\varphi(x_1,\ldots,x_T,y_1,\ldots,y_T)$ is a formula which induces a transitive and asymmetric binary relation, then for any n, there is an ordering $>_n$ on the set of "dating structures" of $\{1,\ldots,T\}$ such that if the size of G is n,
$(a_1,\ldots,a_T)$ relates to $(b_1,\ldots,b_T)$ iff $E(a_1,\ldots,a_T) >_n E(b_1,..,b_T)$
and there exists a number n^* such that all $>_n$ are identical for n greater than some n^*.

5. Conclusion

In this paper I have presented an example of an application of a definability constraint on the set of admissible *preference relations*. One could think of similar investigations of other objects such as the choice problems facing the decision maker, or the set of available strategies in a game. Imposing constraints on the class of situations, the individuals' preference relations and set of strategies may yield interesting results especially in interactive situations such as games. This would be a very challenging project. The aim of this paper, however, was only to persuade the reader that the subject can be discussed within formal models and that the definability assumption on preferences is not only natural but also analytically tractable.

References

Boolos, G.S., Jeffrey, R.C.: (1989), *Computability and Logic*, Cambridge University Press, Boston, NY and London.

Crossley, J.N., Stillwell, J.C., Brickhill, C.J., Williams, N.H., Ash, C.J.: (1990), *What Is Mathematical Logic?*, Dover Publications.

Robinson, A.: (1963), *Introduction to Model Theory and to the Meta Mathematics of Algebra*, North Holland, Dordrecht.

Rubinstein, A.: (1978), 'Definable Preference Relations – Three Examples', Center of Research in Mathematical Economics and Game Theory, R.M. 31, the Hebrew University, Jerusalem.

Rubinstein, A.: (1998a), 'Definable Preferences: An Example', *European Economic Review* **42**, pp. 553–560.

Rubinstein, A.: (1998b), *Modeling Bounded Rationality*, MIT, Cambridge.

Rubinstein, A.: (2000), *Economics and Language*, Cambridge University Press, Cambridge.

Shelly, M.Y., Bryan, G.L.: (1964), 'Judgments and the Language of Decisions', in Shelly, M.Y., Bryan, G.L. (eds.), *Human Judgment and Optimality*, John Wiley, New York.

IX

PHILOSOPHY OF LOGIC, MATHEMATICS, AND COMPUTER SCIENCE

CALCULATIONS BY MAN & MACHINE: MATHEMATICAL PRESENTATION*

Wilfried Sieg
Carnegie Mellon University
Pittsburgh, PA 15213

0. Introduction

These investigations aim to provide a characterization of computations by machines that is as general and convincing as that of computations by human computors given by Turing. The groundwork was laid by Gandy in his thought-provoking 1980 paper *Church's Thesis and Principles for Mechanisms* – a rich and difficult, but sometimes unnecessarily (and maddeningly) complex paper. The structure of Turing's argument actually guided Gandy's analysis. If Gandy had been concerned with just sequential computations, his analysis would have joined Turing's perfectly with *exactly one* difference: physical restrictions for a machine, instead of sensory limitations of a human computor, would have motivated crucial boundedness and locality conditions. However, Gandy realized through conversations with J.C. Shepherdson that the analysis "must take parallel working into account".[1]

In a comprehensive survey article, published ten years after Gandy's paper, Leslie Lamport and Nancy Lynch argued that the theory of sequential computing "rests on fundamental concepts of computability that are independent of any particular computational model". They emphasized that, in contrast, the "fundamental formal concepts underlying distributed computing", if there were any, had not yet been developed. "Nevertheless", they wrote, "one can make some informal observations that seem to be important":

> Underlying almost all models of concurrent systems is the assumption that an execution consists of a set of discrete events, each affecting only part of the system's state. Events are grouped into processes, each process being a more or less completely sequenced set of events sharing some common locality in terms

*During the International Congress I had fruitful and engaging discussions with Yuri Gurevich, mostly, on the conceptual issues underlying this paper. – A comparative study of Gurevich's ASMs, but also of standard models of parallel computations (as in Brassard and Bratley's book) is the topic of a third paper on "Calculations by man & machine" with the subtitle "Comparisons and applications"

P. Gärdenfors, J. Woleński and K. Kijania-Placek (eds.), In the Scope of Logic, Methodology and Philosophy of Science, Vol. I, 247-262.

> of what part of the state they affect. For a collection of autonomous processes to act as a coherent system, the processes must be synchronized. (p. 1166)

Gandy's analysis of parallel computations is conceptually convincing and provides a sharp mathematical form of the informal assumption(s) "underlying almost all models of concurrent systems". Gandy takes as the paradigmatic parallel computation the evolution of the Game of Life or of other cellular automata.

The definitional preliminaries in Gandy's paper are rather lengthy; Shepherdson wrote in 1988: "Although Gandy's principles were obtained by a very natural analysis of Turing's argument they turned out to be rather complicated, involving many subsidiary definitions in their statement. In following Gandy's argument, however, one is led to the conclusion that that is in the nature of the situation." The "nature of the situation" is actually not as complex. The presentation can be simplified by choosing definitions appropriately and focusing sharply on the central informal ideas. Steps in clarifying and streamlining Gandy's mathematical development were taken in earlier work by John Byrnes and me.[2] In early 1999, I succeeded in simplifying further the mathematical apparatus, but my investigations took also a methodological turn that is articulated in Sieg (2001). The conceptual analysis of calculations by man and machine given there, is complemented here by a systematic mathematical presentation.

Having learned Turing's lesson, that we have to take into account the computing agent or device (when analyzing a notion of computability), we have to ask "Parallel computations by whom or what?". Gandy focused on *discrete mechanical devices* possibly excluding other physical devices, for example, analogue machines. The only physical presuppositions about such devices are a lower bound on the size of atomic components and an upper bound on the speed of signal propagation.[3] Calculations proceed in discrete steps; thus such mechanical devices are "in a loose sense" digital computers. Finally, the behavior of these devices is uniquely determined, once a full description of their initial state is given, i.e., they are deterministic. After setting up in section 1 the basic mathematical frame and defining *Turing Computors*, I give in section 2 a straightforward, but re-oriented definition of *Gandy Machines*.

I view the definition of a Gandy machine as an abstract definition of the same character as that of a group or topological space, with many interestingly different models. The *only* axiomatic principle characterizing these machines is a restricted version of Gandy's principle of local causation. The proof of the central theorem,[4] needed for the reduction of Gandy to Turing machines, is corrected and presented in section 3. It is the crucial step to obtain a suitable representation theorem: any process carried out by a model of the axioms, i.e., by a particular Gandy machine, can be simulated by a Turing machine.

My concluding remarks in section 4 contain that formulation and some further observations.

1. Turing's computor

The crucial step, taken by Post and Turing in their analysis of calculability, was this: they moved away from directly characterizing calculable number theoretic functions (as proposed, for example, by Church using Gödel's equational calculus) and turned attention to the underlying mechanical operations on finite syntactic configurations; the story of this development is told in my (1994) and also, briefly, in Sieg (2001). The computation models proposed by Turing and Post in 1936 were almost identical, but only Turing argued for the adequacy of his model; for Post it was a "working hypothesis". The real systematic confluence of Turing's and Post's work occurred in 1947, when Post gave a most elegant way of describing Turing machines via his production systems (on the way to solving, negatively, the word-problem for semi-groups).[5] The configurations of a Turing machine are given by *instantaneous descriptions* (or ids) of the form $\alpha q_i s_k \beta$, where α and β are possibly empty strings of symbols in the machines's alphabet; more precisely, an id contains exactly one state symbol q_i, and to the right of it there must be at least one symbol s_k. Such ids express that the current tape content is $\alpha s_k \beta$, the machine is in state q_i, and it scans a square with symbol s_k. Quadruples $q_i s_k c_1 q_m$ of the machine's program express that the machine, when observing symbol s_k in internal state q_i, is to carry out operation c_1 and change its internal state to q_m. Such quadruples are incorporated into rules operating on ids; for example, if the operation c_1 is the machine's command *print* 0, the corresponding rule is:

$$\alpha q_i s_k \beta \Rightarrow \alpha q_m 0 \beta.$$

Such rules can be formulated, obviously, for all the different operations of a Turing machine. One just has to append 0 to α (β) in case c_1 is the operation *move to the left (right)* and α (β) is the empty string; this reflects the expansion of the finite, but potentially infinite tape by a blank square.[6]

Coming back to Turing's conceptual analysis: it attempts to arrive at the most basic, local operations on a bounded number of configurations. If one combines this with the Post presentation of Turing machines, restrictive conditions can be stated in great generality as follows:

(B) (Boundedness) *There is a fixed bound on the number of configurations a computor can immediately recognize.*

(L) (Locality) *A computor can change only immediately recognizable (sub-) configurations.*

Computors proceed deterministically; consequently, the computing process has to satisfy:
(D) (Determinacy) *The immediately recognizable (sub-)configuration determines uniquely the next computation step (and id).*
Discrete dynamical systems provide an elegant way of formulating these considerations mathematically. Let D be a class of states (ids or syntactic configurations), and let F be a function from D to D transforming a given state into the next one, i.e., describing the evolution of the system (the computation of the machine). Using already Gandy's framework, states are presented as non-empty hereditarily finite sets over an infinite set U of atoms. Such sets reflect states of physical devices just as mathematical structures represent states of nature. Thus, any isomorphic structure will do as well; one should notice that this reflection is done somewhat indirectly, as only the $\in$-relation is available.

A class S of states is called *structural*, if S is closed under $\in$-isomorphisms; if D is a class of states, I denote by S_D the corresponding structural class. The lawlike connections between states are given by *structural operations* G from S to S. Structural operations satisfy the following condition:[7] for all permutations π on U and all $x \in S$, $G(x^\pi)$ is $\in$-isomorphic over x^π to $G(x)^\pi$ for any state z, z^π stands for $\pi^*(z)$, where π^* is the uniquely determined $\in$-isomorphism on the hereditarily finite sets extending π. "x is $\in$-isomorphic over z to y" means that the $\in$-isomorphism between x and y is the identity on the atoms in the transitive closure of z, i.e., in the support of z (briefly: $\mathrm{Sup}(z)$). I will use the abbreviation "$x \cong_z y$"; if a special underlying permutation σ is to be pointed out, I write "$x \cong_z y$ via σ".

If x is a given state, regions of the next state are *locally* determined. Thus it is important to describe suitable substructures of x on which the computor can operate. Proper subtrees y of the $\in$-tree for x are called *parts for* x, briefly $y <^* x$, if they are specified as follows:[8] $y \neq x$ and y is a non-empty subset of

$$\{v \mid (\exists z)(v <^* z \;\&\; z \in x)\} \cup \{r \mid r \in x\}.$$

If the non-empty subset in this $\in$-recursive definition consists at each stage of exactly one element, y is a *path through* x. Paths through x are of the form $\{r\}^n$, for a natural number n and some atom r. A collection C of parts for x is a *cover for* x just in case for every path y through x there is a $z \in C$, such that y is a path through z. These last two definitions are not needed for the definition of a Turing computor; also the following definitions will be given in greater generality than needed for that purpose. (I just want to avoid duplicating matters for Gandy machines in the next section.)

The local operations are given by a structural operation G that works on certain parts y for x. Each y lies in one of a finite number of isomorphism classes (or stereotypes). So let T be a fixed, finite class of stereotypes: a part for x that is a member of a stereotype of T is called, naturally enough, a *T-part for x*.

A T-part y for x is a *causal neighborhood for* x given by T, briefly $y \in \mathrm{Cn}(x)$, if there is no T-part y^* for x such that y is $\in$-embeddable into y^*. G operates on such causal neighborhoods. The values of G, however, are in general not exactly what is needed for assembling the next state. For that purpose, we introduce *determined regions* of a state z obtained from causal neighborhoods for $x: v \in \mathrm{Dr}(z,x)$ if and only if $v <^* z$ and there is a $y \in \mathrm{Cn}(x)$, such that $G(y) \cong_y v$ and $\mathrm{Sup}(v) \cap \mathrm{Sup}(x) \subseteq \mathrm{Sup}(y)$. The last condition for Dr guarantees that new atoms in $G(y)$ correspond to new atoms in v, and that the new atoms in v are new for x. If one requires G to satisfy similarly $\mathrm{Sup}(G(y)) \cap \mathrm{Sup}(x) \subseteq \mathrm{Sup}(y)$, then the condition "$G(y) \cong_y v$" can be strengthened to "$G(y) \cong_x v$". The new atoms are thus always taken from $U \setminus \mathrm{Sup}(x)$.[9] One final definition: for given states z and x let $A(z,x)$ stand for $\mathrm{Sup}(z) \setminus \mathrm{Sup}(x)$. Note that the number of new atoms introduced by G is bounded, i.e., $| A(G(y), \mathrm{Sup}(x)) | < n$ for some natural number n (any $x \in S$ and any causal neighborhood y for x). Now we define, finally, a Turing computor, keeping the Post-description and the general boundedness and locality conditions firmly in mind.

Definition $M = \langle S; T, G \rangle$ is a *Turing computor on* S, where S is a structural class, T a finite set of stereotypes, and G a structural operation on T, if and only if, for every $x \in S$ there is a $z \in S$, such that

(LC.0) $(\exists! y)\, y \in \mathrm{Cn}(x)$

(LC.1) $(\exists! v <^* z)\, v \cong_x G(\mathrm{cn}(x))$;

(GA.1) $z = (x \setminus \mathrm{cn}(x)) \cup \mathrm{dr}(z,x)$.

$\mathrm{cn}(x)$ and $\mathrm{dr}(z,x)$ denote the sole causal neighborhood of x, respectively the determined region of z. – It is easily seen that Turing machines are Turing computors under an appropriate set theoretic representation; I indicate one such representation in the Appendix. Notice that the next state z is determined uniquely up to $\in$-isomorphism. I denote it by $M(x)$ and define what it means for a dynamical system to be (Turing) computable. (Recall that S_D is the structural class corresponding to D.)

Definition Let $\langle D, F \rangle$ be a discrete dynamical system; F is called (Turing) *computable* if and only if there is a Turing computor M on S_D, such that for each $x \in D$: $F(x) \cong_x M(x)$.

2. Parallel computing

Can we generalize, in a suitable way, these considerations from sequential to parallel computations? Keeping in mind the considerations reviewed in the Introduction, we want to have a computational device that determines the evolution of dynamical systems in parallel. In analogy to the definition of Turing computability above, assuming that the next state z of a Gandy machine is also determined uniquely up to $\in$-isomorphism over x, we set tentatively:

Definition Let $\langle D, F\rangle$ be a discrete dynamical system; F is called *computable in parallel* if and only if there is a Gandy machine M on S_D, such that for each $x \in D$: $F(x) \cong_x M(x)$.

Thus, in contrast to Gandy (1980) and Sieg & Byrnes (1999B), Gandy machines are no longer viewed as discrete dynamical systems satisfying certain restrictive principles, but rather as restricted devices to *compute in parallel* (the evolution of) such systems up to $\in$-isomorphism over x. In addition to the stereotypes and structural operation working on causal neighborhoods, we have here a second set of stereotypes and a second structural operation: the latter allow us to "assemble" the determined regions.

Definition $M = \langle S; T_1, G_1, T_2, G_2\rangle$ is a *Gandy Machine on S*, where S is a structural class, T_i a finite set of stereotypes, G_i a structural operation on T_i, if and only if, for every $x \in S$ there is a $z \in S$, such that

(LC.1) $(\forall y \in \mathrm{Cn}_1(x))(\exists! v \in \mathrm{Dr}_1(z,x))\, v \cong_x G_1(y)$;

(LC.2) $(\forall y \in \mathrm{Cn}_2(x))(\exists v \in \mathrm{Dr}_2(z,x))\, v \cong_x G_2(y)$;

(GA.1) $(\forall C)[C \subseteq \mathrm{Dr}_1(z,x))\&\bigcap\{\mathrm{Sup}(v)\cap A(z,x) \mid v \in C\} \neq \emptyset \rightarrow$
$(\exists w \in \mathrm{Dr}_2(z,x))(\forall v \in C) v <^* w]$;[10]

(GA.2) $z = \bigcup \mathrm{Dr}_1(z,x)$.

LC stands for *Local Causation*, whereas GA abbreviates *Global Assembly*. – The central fact for Gandy machines (z is determined uniquely up to $\in$-isomorphism over x) follows easily from the next theorem:[11]

Theorem Let M be $\langle S; T_1, G_1, T_2, G_2\rangle$ as above and $x \in S$; if there are z and z' in S satisfying principles (LC.1–2), (GA.1), and if $\mathrm{Dr}_1(z,x)$ and $\mathrm{Dr}_1(z',x)$ cover z and z', then $\mathrm{Dr}_1(z,x) \cong_x \mathrm{Dr}_1(z',x)$.

In the following $\mathrm{Dr}_1, \mathrm{Dr}'_1, A$, and A' will abbreviate $\mathrm{Dr}_1(z,x), \mathrm{Dr}_1(z',x), A(z,x)$, and $A(z',x)$ respectively. Note that Dr_1 and Dr'_1 are finite. Using (LC.1) and (LC.2) one can observe that there is a natural number m and there are sequences v_i and $v'_i, i < m$, such that $\mathrm{Dr}_1 = \{v_i \mid i < m\}, \mathrm{Dr}'_1 = \{v'_i \mid i < m\}$, and v'_i is the unique part of z' with $v_i \cong_x v'_i$ via permutations π_i (for all $i < m$). Here is a picture of the situation:

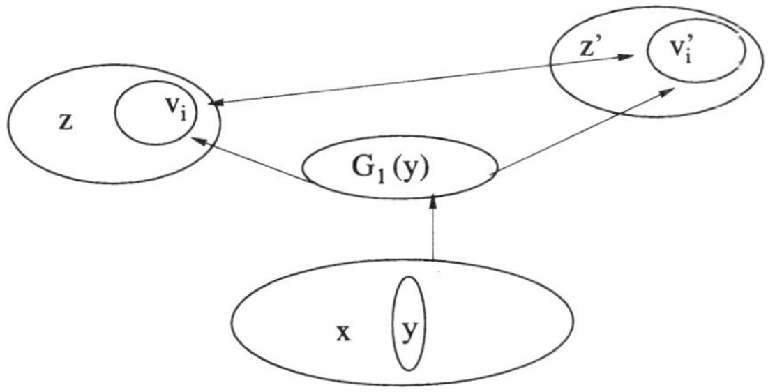

To establish the Theorem, we have to find a *single* permutation π that extends to an $\in$-isomorphism over x for all v_i and v'_i simultaneously. Such a π must obviously satisfy for all $i < m$:

(i) $v_i \cong_x v'_i$ via π

and, consequently,

(ii) $\pi[\mathrm{Sup}(v_i)] = \mathrm{Sup}(v'_i)$;

As π is an $\in$-isomorphism over x, we have:

(iii) $\pi[A] = A'$.

Condition (ii) implies for all $i < m$ and all $r \in A$ the equivalence between $r \in \mathrm{Sup}(v_i)$ and $r^{\pi} \in \mathrm{Sup}(v'_i)$. This can be also expressed by

(ii*) $\mu(r) = \mu'(r^{\pi})$, for all $r \in A$,

where $\mu(r) = \{i \mid r \in \mathrm{Sup}(v_i)\}$ and $\mu'(r) = \{i \mid r \in \mathrm{Sup}(v'_i)\}$; these are the *signatures* of r with respect to z and z'.

To obtain such a permutation, the considerations are roughly as follows: (i) if the v_i do not overlap, then the π_i will do; (ii) if there is overlap, then an equivalence relation $\cong$ ($\cong'$) on A (A') is defined by $r_1 \cong r_2$ iff $\mu(r_1) = \mu(r_2)$, and analogously for $\cong'$; (iii) then we prove that the "corresponding" equivalence classes $[r]_{\cong}$ and $[s]_{\cong'}$ (the signatures of their elements are identical) have the same cardinality. $[r]_{\cong}$ can be characterized as $\bigcap\{\mathrm{Sup}(v_i) \cap A \mid i \in \mu(r)\}$; similar for $[s]_{\cong'}$. This characterization is clearly independent of the choice of representative by the very definition of the equivalence relation(s). With this in place, a global $\in$-isomorphism can be defined. These considerations are made precise through the proofs of the combinatorial lemma and two corollaries in the next section.

3. Global assembly

All considerations in this section are carried out under the assumptions of the Theorem: $M = \langle S; T_1, G_1, T_2, G_2 \rangle$ is an arbitrary Gandy machine and $x \in S$ an arbitrary state; we assume furthermore that z and z' are in S, the principles (LC.1–2) and (GA.1) are satisfied, and Dr_1 and Dr'_1 cover z and z'. We want to show that $\mathrm{Dr}_1 \cong_x \mathrm{Dr}'_1$, knowing already that there are sequences v_i and v'_i of length m, such that $\mathrm{Dr}_1 = \{v_i \mid i < m\}, \mathrm{Dr}'_1 = \{v'_i \mid i < m\}$, and v'_i is the unique part of z' with $v_i \cong_x v'_i$ via permutations π_i (for all $i < m$). I start out with the formulation of a key lemma concerning overlaps.

Overlap Lemma Let $r_0 \in A$ and $\mu(r_0) \neq \emptyset$; then there is a permutation ρ on U with $v_i \cong_x v'_i$ via ρ for all $i \in \mu(r_0)$.

Proof. We have $\{v_i \mid i \in \mu(r_0)\} \subseteq \mathrm{Dr}_1$; as r_0 is in A and in $\mathrm{Sup}(v_i)$ for each $i \in \mu(r_0) \neq \emptyset$, we have also that $\bigcap\{\mathrm{Sup}(v_i) \cap A \mid i \in \mu(r_0)\} \neq \emptyset$. The antecedent of (GA.1) is satisfied, and we conclude that there is a $w \in \mathrm{Dr}_2$ such that $v_i <^* w <^* z$, for all $i \in \mu(r_0)$. Using (LC.2) we obtain a $w' \in \mathrm{Dr}'_2$ with $w \cong_x w'$. This $\in$-isomorphism over x is induced by a permutation ρ and yields for all $i \in \mu(r_0)$

$$v_i^\rho <^* w^\rho = w' <^* z'.$$

So we have, $v_i \cong_x v_i^\rho$ and $v_i^\rho <^* z'$, thus – using (LC.2) – $v_i^\rho = v'_i$; that holds for all $i \in \mu(r_0)$. Q.E.D.

Note that the condition $\mu(r) \neq \emptyset$ is satisfied in our considerations for any $r \in A$, as Dr_1 is a cover of z; so we have for any such r an appropriate *overlap permutation* ρ^r for r. The crucial combinatorial lemma we have to establish is this:

Combinatorial Lemma For $r_0 \in A : |\{r \in A \mid \mu(r_0) \subseteq \mu(r)\}| = |\{s \in A' \mid \mu(r_0) \subseteq \mu'(s)\}|$.

Proof. Consider $r_0 \in A$. I establish first the claim

$$\rho[\{r \in A \mid \mu(r_0) \subseteq \mu(r)\}] \subseteq \{s \in A' \mid \mu(r_0) \subseteq \mu'(s)\},$$

where ρ is an overlap permutation for r_0. The claim follows easily from

$$r \in A \;\&\; \mu(r_0) \subseteq \mu(r) \rightarrow \mu(r_0) \subseteq \mu'(r^\rho),$$

by observing that r^ρ is in A'. Assume, to establish this conditional indirectly, for arbitrary $r \in A$ that $\mu(r_0) \subseteq \mu(r)$ and $\neg(\mu(r_0) \subseteq \mu'(r^\rho))$. The first assumption implies that $r \in \mathrm{Sup}(v_i)$ for all $i \in \mu(r_0)$, and the construction of ρ yields then:

$$r^\rho \in \mathrm{Sup}(v'_i) \text{ for all } i \in \mu(r_0). \qquad (\heartsuit)$$

The second assumption implies that there is a k in $\mu(r_0) \setminus \mu'(r^\rho)$. Obviously, $k \in \mu(r_0)$ and $k \notin \mu'(r^\rho)$. The first conjunct $k \in \mu(r_0)$ and ($\heartsuit$) imply that $r^\rho \in \mathrm{Sup}(v'_k)$; as the second conjunct $k \notin \mu'(r^\rho)$ means that $r^\rho \notin \mathrm{Sup}(v'_k)$, we have obtained a contradiction.

Now I'll show that $\rho[\{r \in A \mid \mu(r_0) \subseteq \mu(r)\}]$ cannot be a *proper* subset of $\{s \in A' \mid \mu(r_0) \subseteq \mu'(s)\}$. Assume, to obtain a contradiction, that it is; then there is $s^* \in A'$ that satisfies $\mu(r_0) \subseteq \mu'(s^*)$ and is not a member of $\rho[\{r \in A \mid \mu(r_0) \subseteq \mu(r)\}]$. As $\mu(r_0) \subseteq \mu'(s^*)$, s^* is in $\mathrm{Sup}(v'_i)$ for all $i \in \mu(r_0)$; the analogous fact holds for all $r \in A$ satisfying $\mu(r_0) \subseteq \mu(r)$, i.e., all such r must be in $\mathrm{Sup}(v_i)$ for all $i \in \mu(r_0)$. As $v_i \cong_x v'_i$ via ρ for all $i \in \mu(r_0)$, s^* must be obtained as a ρ-image of some r^* in $\mathrm{Sup}(x)$ or in A (and, in the latter case, violating $\mu(r_0) \subseteq \mu(r^*)$). However, in either case we have a contradiction. The assertion of the Lemma is now immediate. Q.E.D.

Next I establish two consequences of the Combinatorial Lemma, the second of which is basic for the definition of the global isomorphism π.

Corollary 1 For any $I \subseteq \{0,1,\dots,m-1\}$ with $I \subseteq \mu(r_0)$ for some r_0 in A,

$$\mid \{r \in A \mid I \subseteq \mu(r)\} \mid = \mid \{s \in A' \mid I \subseteq \mu'(s)\} \mid .$$

Proof. Consider an arbitrary $I \subseteq \mu(r_0)$ for some r_0 in A. If $I = \mu(r_0)$, then the claim follows directly from the Combinatorial Lemma. If $I \subset \mu(r_0)$, let $r^0,\dots,r^{k-1}$ be elements r of A with $I \subset \mu(r)$ and require that $\mu(r^j) \neq \mu(r^{j'})$, for all $j, j' < k$ and $j \neq j'$, and for every $r \in A$ with $I \subset \mu(r)$ there is a unique $j < k$ with $\mu(r) = \mu(r^j)$. The Combinatorial Lemma implies, for all $j < k$,

$$\mid \{r \in A \mid \mu(r^j) \subseteq \mu(r)\} \mid = \mid \{s \in A' \mid \mu(r^j) \subseteq \mu'(s)\} \mid .$$

Now it is easy to verify the claim of Corollary 1:

$$\begin{aligned} &\mid \{r \in A \mid I \subseteq \mu(r)\} \mid = \\ &\mid \{r \in A \mid (\exists j < k)\, \mu(r^j) \subseteq \mu(r)\} \mid = \\ &\mid \{s \in A' \mid (\exists j < k)\, \mu(r^j) \subseteq \mu'(s)\} \mid = \\ &\mid \{s \in A' \mid I \subseteq \mu'(s)\} \mid . \end{aligned}$$

This completes the proof of Corollary 1. Q.E.D.

The second important consequence of the Combinatorial Lemma can be obtained now by an inductive argument.

Corollary 2 For any $I \subseteq \{0,1,\dots,m-1\}$ with $I \subseteq \mu(r_0)$ for some r_0 in A

$$\mid \{r \in A \mid I = \mu(r)\} \mid = \mid \{s \in A' \mid I = \mu'(s)\} \mid .$$

Proof. (by downward induction on $\mid I \mid$). Abbreviating $\mid \{r \in A \mid I = \mu(r)\} \mid$ by v_I and $\mid \{s \in A' \mid I = \mu'(s)\} \mid$ by v'_I, the argument is as follows:

Base case ($|I|=m$): In this case there are no proper extensions I^* of I, and we have

$$\begin{aligned}
\nu_I &= |\{r \in A \mid I = \mu(r)\}| \\
&= |\{r \in A \mid I \subseteq \mu(r)\}|, \text{ as there is no proper extension of } I, \\
&= |\{s \in A' \mid I \subseteq \mu'(s)\}|, \text{ by Corollary 1}, \\
&= |\{s \in A' \mid I = \mu'(s)\}|, \text{ again, as there is no proper extension}, \\
&= \nu'_I
\end{aligned}$$

Induction step ($|I|) < m$): Assume that the claim holds for all I^* with $n+1 \leqslant |I^*| \leqslant m$ and show that it holds for I with $|I| = n$. Using the induction hypothesis we have, summing up over all proper extensions I^* of I:

$$\sum_{I^*} \nu_I^* = \sum_{I^*} \nu'_{I^*} \qquad (\clubsuit)$$

Now we argue as before:

$$\begin{aligned}
\nu_I &= |\{r \in A \mid I = \mu(r)\}| \\
&= |\{r \in A \mid I \subseteq \mu(r)\}| - \sum_{I^*} \nu_{I^*} \\
&= |\{s \in A' \mid \subseteq \mu'(s)\}| - \sum_{I^*} \nu'_{I^*} \quad \text{by Corollary 1 and } (\clubsuit), \\
&= |\{s \in A' \mid = \mu'(s)\}| \\
&= \nu'_I
\end{aligned}$$

This completes the proof of Corollary 2. Q.E.D.

Finally, we can define an appropriate global permutation π. Given an atom $r \in A$, there are only finitely many different overlap permutations, i.e., different on $\bigcup\{\mathrm{Sup}(v_i) \mid i \in \mu(r)\}$. Let us select one of them and call it ρ^r. That permutation can be restricted to $[r]_{\approx} = \bigcap\{\mathrm{Sup}(v_i) \cap A \mid i \in \mu(r)\}$; let ρ^* denote this restriction. Because of Corollary 2, ρ^* is a bijection between $[r]_{\approx}$ and $[\rho^*(r)]_{\approx'}$. – The desired global permutation is now defined as follows for any atom $r \in \bigcup\{\mathrm{Sup}(v_i) \mid i < m\}$:

$$\pi(r) = \begin{cases} \rho^*(r) & \text{if } r \in \bigcap\{\mathrm{Sup}(v_i) \cap A \mid i \in \mu(r)\}, \\ r & \text{otherwise} \end{cases}$$

π is a well-defined bijection with $\pi[A] = A'$ and $\mu(r) = \mu'(r^{\pi})$. It remains to establish the

Claim: For all $i < m, v_i \cong_x v'_i$ via π.

For the *Proof* consider an arbitrary $i < m$. By the basic set-up of our considerations, we have $\pi_i(v_i) = v_i'$. If v_i does not contain in its support an element of A, then π and π_i coincide; if v_I's support contains an element of A that is possibly even in an overlap, the argument proceeds as follows. Notice that *all* elements of $[r]_\approx$ are in $\mathrm{Sup}(v_i)$ as soon as one $r \in A$ is in $\mathrm{Sup}(v_i)$. Taking this fact into account and using the pruning operation[12] $\uparrow$ we have, by the definition of π and of $v_i \uparrow [r]_\approx$: $\pi(v_i \uparrow [r]_\approx) = \rho^*(v_i \uparrow [r]_\approx)$. The definition of ρ^* and the fact that $\rho^r(v_i) = v_i'$ allow us to infer that $\rho^*(v_i \uparrow [r]_\approx) = v_i' \uparrow [\rho^*(r)]_{\approx'}$. As $\mu'(\rho^*(r)) = \mu'(\pi_i(r))[= \mu(r)]$ we can extend this sequence of identities by $v_i' \uparrow [\rho^*(r)]_{\approx'} = v_i' \uparrow [\pi_i(r)]_{\approx'}$. Consequently, as $\pi_i(v_i) = v_i'$, we have $v_i' \uparrow [\pi_i(r)]_{\approx'} = \pi_i(v_i \uparrow [r]_\approx)$.

These considerations hold for all $r \in \mathrm{Sup}(v_i) \cap A$; we can conclude $\pi(v_i) = \pi_i(v_i)$ and, with $\pi_i(v_i) = v_i'$, we have $\pi(v_i) = v_i'$. Q.E.D. (of claim)

This concludes, finally, the argument for the Theorem that was formulated already in section 2.

4. Concluding remarks

If the operation F of a dynamical system is computable in parallel by a Gandy machine M, then it is also Turing computable. It follows from the Theorem and (GA.2) that $M(x)$ is uniquely determined, up to isomorphism, for any M and any M-state x: we are dealing only with (finitely many) finite objects, and the axiomatic conditions for a Gandy machine are decidable. Thus, a search will allow us to find $M(x)$. This fact is best understood, it seems to me, as a representation theorem in the given axiomatic setting.

Let me reiterate, what I emphasized in Sieg (2001). It may be that Gandy machines can be simplified further, for example, by a graph theoretic presentation or a category theoretic description.[13] But what is needed most, in my view, is their further mathematical investigation, e.g., for issues of complexity and speed-up, and their use in significant applications, e.g., for the analysis of DNA computations or of parallel distributed processes. The latter was done by De Pisapia in his (2000) for many important kinds of artificial neural nets.[14] The consequence is that artificial neural nets of these varieties can be simulated by Turing machines. Analogous results are obtained, using quite different techniques, by Siegelmann in (1999).

However, the most difficult and subtle aspect of Gandy machines, namely the addition of new atoms, is not used at all for these neural nets, as they have a fixed number of nodes. So there are two obvious questions: "Is there a natural subclass of Turing computable functions in which these neural nets lie?", and "Are there mental processes for whose representation this aspect of Gandy machines might be crucial?". These are important and most appealing questions.

Appendix

In this appendix I sketch a set theoretic presentation of a Turing machine as a Turing computor, but also – even more briefly – that of the Game of Life as a Gandy machine. Consider a Turing machine with symbols $s_0, \ldots, s_k$ and internal states $q_0, \ldots, q_l$; its program is given as a finite list of quadruples of the form $q_i s_j c_k q_l$, expressing that the machine is going to perform action c_k and change into internal state q_l, when scanning symbol s_j in state q_i. The tape is identified with a set of overlapping pairs

$$\mathrm{Tp} := \{\langle b,b\rangle, \langle b,c\rangle, \ldots, \langle d,e\rangle, \langle e,e\rangle\}$$

where b,c,d,e are distinct atoms; c is the leftmost square of the tape (with a possibly non-blank symbol on it), d its rightmost one. The symbols and states are represented by $\underline{s}_j := \{r\}^{(j+1)}, 0 \leqslant j \leqslant k$; the internal states are given by $\underline{q}_j := \{r\}^{(k+1)+(j+1)}, 0 \leqslant j \leqslant l$. The tape content is given by

$$\mathrm{Ct} := \{\langle \underline{s}_{j0}, c\rangle, \ldots, \langle \underline{s}_{jr}, d\rangle\}$$

and, finally, the id is represented as the union of Tp, Ct, and $\{\langle \underline{q}_i, c\rangle\}$. So the structural set S of states is obtained as the set of all ids closed under $\in$-isomorphisms. Stereotypes (for each program line given by $q_i s_j$) consist of parts like

$$\{\langle \underline{q}_i, r\rangle, \langle \underline{s}_j, r\rangle, \langle t,r\rangle, \langle r,u\rangle\};$$

these are the causal neighborhoods on which G is to operate. Consider the program line $q_i s_j s_k q_l$ (print s_k); applied to the above causal neighborhood G is to yield

$$\{\langle \underline{q}_l, r\rangle, \langle \underline{s}_k, r\rangle, \langle t,r\rangle, \langle r,u\rangle\}.$$

For the program line $q_i s_j R q_l$ (move Right) two cases have to be distinguished. In the first case, when r is not the rightmost square, G yields

$$\{\langle \underline{q}_l, u\rangle, \langle \underline{s}_j, r\rangle, \langle t,r\rangle, \langle r,u\rangle\};$$

in the second case, when r is the rightmost square, G yields

$$\{\langle \underline{q}_l, *\rangle, \langle \underline{s}_j, r\rangle, \langle \underline{s}_0, *\rangle, \langle t,r\rangle, \langle r,*\rangle, \langle *,u\rangle\};$$

where $*$ is a new atom. The program line $q_i s_j L q_l$ (move Left) is treated similarly. – It is easy to verify that a Turing machine presented in this way is a Turing Computor.

Gandy considered playing Conway's Game of Life as a paradigmatic case of parallel computing. It is being played on subsets of the plane, more precisely, subsets that are constituted by finitely many connected squares. For reasons that will be obvious in a moment, the squares are also called *internal cells*; they can be in two states, *dead* or *alive*. In my presentation the internal cells are surrounded by one layer of *border cells*, the latter in turn by an additional layer of *virtual cells*. Border and virtual cells are dead by convention. Internal cells and border cells are jointly called *real*. The layering ensures that each real cell is surrounded by a full set of eight neighboring cells. For real cells the game is played according to the rules:

(1) living cells with 0 or 1 (living) neighbor die (from isolation);

(2) living cells with 4 or more (living) neighbors die (from overcrowding);

(3) dead cells with exactly 3 (living) neighbors become alive.

In all other cases the cell's state is unchanged.
A real cell a with neighbors $a_1, \ldots, a_8$ and state $s(a)$ is given by

$$\{a, s(a), \langle a_1, \ldots, a_8 \rangle\}.$$

(The neighbors are given in "canonical" order starting with the square in the leftmost top corner and proceeding clockwise; $s(a)$ is $\{a\}$ in case a is alive, otherwise $\{\{a\}\}$.) The T_1-causal neighborhoods of real cells are of the form

$$\{\{a, s(a), \langle a_1, \ldots, a_8 \rangle\}, \{a_1, s(a_1)\}, \ldots, \{a_8, s(a_8)\}\}.$$

It is obvious, how to define the structural operation G_1 on the causal neighborhoods of internal cells; the case of border cells requires attention. There is a big number of stereotypes that have to be treated, so I will discuss only one simple case that should, nevertheless, bring out the principled considerations. In the following diagram we start out with the cells that have letters assigned to them; the diagram should be thought of extending at the left and at the bottom. The v's indicate virtual cells, the b's border cells, the $\{a\}$'s internal cells that are alive, and the $*$'s new atoms that are added in the next step of the computation. Let's see how that comes about.

$*_0$	$*_1$	$*_2$	$*_3$	$*_4$	$*_5$	$*_6$	$*_7$		
v_0	v_1	v_2	v_3	v_4	v_5	v_6	v_7	v_8	v_9
b_0	b_1	b_2	b_3	b_4	b_5	b_6	b_7	b_8	v_{10}
$\{a_0\}$	$\{a_1\}$	$\{a_2\}$	$\{a_3\}$	$\{a_4\}$	$\{a_5\}$	$\{a_6\}$	$\{a_7\}$	b_9	v_{11}

Consider the darkly shaded square b_3 with its neighbors, i.e., its presentation

$$\{b_3, \{\{b_3\}\}, \langle v_2, \ldots, b_2\rangle\};$$

applying G_1 to its causal neighborhood yields

$$\{\{b_3, \{b_3\}, \langle v_2, \ldots, b_2\rangle\}, \{v_3, \{\{v_3\}\}, \langle *_2, *_3, *_4, v_4, b_4, b_3, b_2, v_2\rangle\}\},$$

where $*_2$, $*_3$, and $*_4$ are new atoms (and v_3 has been turned from a virtual cell into a real one, namely a border cell). Here the second set of stereotypes and the second structural operation come in to ensure that the new squares introduced by applying G_1 to "adjacent" border cells (whose neighborhoods overlap with the neighborhood of b_3) are properly identified in the next state. Consider as the appropriate T_2-causal neighborhood the set consisting of the T_1-causal neighborhoods of b_2, b_3, and b_4; G_2 applied to it yields the set with presentations of the cells v_2, v_3, and v_4.

Notes

1. Gandy (1980), p. 125.

2. Sieg (1990), Byrnes (1993), and Sieg & Byrnes (1999B).

3. That is emphasized on p. 126, but also on pp. 135–136 of Gandy (1980). – How the purely physical considerations lead to boundedness and locality conditions is discussed at great length in Mundici & Sieg.

4. That is Lemma 5.5 on p. 144 of Gandy (1980).

5. Post's way of looking at Turing machines underlies also the presentation of Davis (1958); for a more detailed discussion the reader is referred to that classical text.

6. Turing recognized the significance of Post's presentation for achieving mathematical results, but also for the conceptual analysis of calculability: as to the former, Turing extended in his (1950) Post's and Markov's result concerning the unsolvability of the word problem for semi-groups to semi-groups with cancellation; as to the latter, the reader should consult the semi-popular and wonderfully informative presentation of 'Solvable and Unsolvable Problems' (1953).

7. For motivation of this particular condition see p. 154 of Sieg & Byrnes (1999B) and p. 128 of Gandy (1980).

8. I am deviating quite consciously from Gandy's terminology; my "part" (is more general than, but) corresponds roughly to "located subassembly" in Gandy (1980) and to "subassembly" in Sieg & Byrnes (1999B). The reader should also compare it to Gandy's $\subseteq^*$, p. 136. Gandy remarks: "If one considers y as a tree of its $\in$-chains, then $u \subseteq^* y$ implies that u is a subtree with the same vertex as y." The relation is defined by the condition $(\exists s \subseteq Tc(y))u = y \upharpoonright s$.

9. This operation is an operation on x and y, as it introduces, possibly, new atoms – new for x. It has in this very weak sense a "global" aspect; however, as it is a structural operation, the precise choice of the atoms does not matter at all.

10. In Gandy's set-up the finiteness of the C has to be explicitly required; here it is a trivial consequence of the finiteness of Dr_1. Furthermore, principle (GA.1) implies a fixed upper bound on the number of determined regions that have new atoms in common, as there is clearly an upper bound on the number of distinct parts of elements $\mathrm{Dr}_2(z,x)$.

11. In Gandy (1980) this uniqueness up to $\in$-isomorphism over x is achieved in a much more complex way, mainly, because parts of a state are proper subtrees, in general non-located. Given an appropriate definition of cover, a collection C is called an *assembly for x*, if C is a cover for x and the elements of C are maximal. The fact that C is an assembly for exactly one x, if indeed it is, is expressed by saying that *C uniquely assembles to x*; for details see Sieg & Byrnes (1999B), p. 157. In my setting, axiom (GA.2) is equivalent to the claim that $\mathrm{Dr}_1(z,x)$ uniquely assembles to z.

12. The pruning operation applies to an element x of HF and a subset Y of its support: $x \uparrow Y$ is the subtree of x that is built up exclusively from atoms in Y. The $\in$-recursive definition is: $(x \cap Y) \cup [\{y \uparrow (Y \cap Tc(y)) \mid y \in x\} \backslash \{\emptyset\}]$. Cf. Sieg and Byrnes (1999B), pp. 155–156.

13. A graph theoretic presentation was proposed in Byrnes & Sieg (1996). On the topic of a category theoretic definition Gandy wrote in his (1980), p. 147: "The heavy use made of restrictions [...] suggests that a treatment using concepts analogous to those of sheaf theory or topos theory might be worth developing. However, it seems to me that the concepts from category theory which would be necessary would be too abstract to allow one to use them (as we have used the more concrete notions of set theory) as a justification for the main thesis of this paper." Perhaps this issue should be revisited twenty years after its original formulation. A starting-point can be found in Herron (1995).

14. Here is a possibility for rich interaction with classical and contemporary work in the foundations of mathematics, namely, the formalization of analysis in very weak formal frameworks.

References

Brassard, G., Bratley, P.: (1996), *Fundamentals of Algorithmics*, Prentice Hall, Englewood Cliffs, New Jersey.

Byrnes, J.: (1993), *Notions of Computation*, M.S. Thesis, Carnegie Mellon University, Pittsburgh.

Byrnes, J., Sieg, W.: (1996), 'A graphical presentation of Gandy's parallel machines' (Abstract), *Bulletin of Symbolic Logic* **2**, pp. 452–453.

Davis, M.: (1958), *Computability and Unsolvability*, McGraw-Hill, New York.

De Pisapia, N.: (2000), *Gandy Machines: an abstract model of parallel computation for Turing Machines, the Game of Life, and Artificial Neural Networks*, M.S. Thesis, Carnegie Mellon University, Pittsburgh.

Gandy, R.: (1980), 'Church's Thesis and principles for mechanisms', in J. Barwise, H.J. Keisler, K. Kunen (eds.), *The Kleene Symposium*, North-Holland, pp. 123–148.

Gurevich, Y.: (1999), 'Sequential ASM Thesis', MSR-TR-99-09, Microsoft Research, Redmond.

Herron, T.: (1995), 'An alternative definition of pushout diagrams and their use in characterizing K-graph machines', Carnegie Mellon University, Pittsburgh.

Kolmogorov, A., Uspensky, V.: (1963), 'On the definition of an algorithm', AMS Translations **21**, 2, pp. 217–245.

Lamport, L., Lynch, N.: (1990), 'Distributed Computing: Models and Methods', in J. van Leeuwen (ed.), *Handbook of Theoretical Computer Science*, Elsevier Science Publisher B.V., Amsterdam, pp. 1157–1199.

Mundici, D., Sieg, W.: (1995), 'Paper machines', *Philosophia Mathematica* **3**, pp. 5–30.

Post, E.: (1936), 'Finite combinatory processes. Formulation 1', *Journal of Symbolic Logic* **1**, pp. 103-105.

Post, E.: (1947), 'Recursive unsolvability of a problem of Thue', *Journal of Symbolic Logic* **12**, pp. 1-11.

Shepherdson, J.C.: (1988), 'Mechanisms for computing over arbitrary structures', in R. Herken (ed.), *The Universal Turing Machine – a Half-Century Survey*, Oxford University Press, Oxford, pp. 581-601.

Sieg, W.: (2001), 'Calculations by man & machines: conceptual analysis', Technical Report 104, Philosophy, Methodology, Logic, September 2000; to appear in W. Sieg, R. Sommer, C. Talott (eds.), *Reflections on the Foundations of Mathematics* (Essays in honor of Solomon Feferman), volume 15 of Lecture Notes in Logic, Association for Symbolic Logic.

Sieg, W., Byrnes, J.: (1999A), 'Gödel, Turing, & K-graph machines', in A. Cantini, E. Casari, P. Minari (eds.), *Logic and Foundations of Mathematics*, Synthese Library 280, Kluwer, pp. 57-66.

Sieg, W., Byrnes, J.: (1999B), 'An abstract model for parallel computations: Gandy's Thesis', *The Monist* **82** (1), pp. 150-164.

Siegelmann, H.T.: (1999), *Neural Networks and Analog Computation – Beyond the Turing Limit*, Birkhäuser, Boston, Basel, Berlin.

Tamburrini, G.: (1987), *Reflections on Mechanism*, Ph.D. Thesis, Department of Philosophy, Columbia University, New York.

Turing, A.: (1936), 'On computable numbers, with an application to the Entscheidungsproblem', *Proc. London Math. Soc.*, series 2, 42, pp. 230–265.

Turing, A.: (1950), 'The word problem in semi-groups with cancellation', *Ann. of Math.* **52**, pp. 491–505.

Turing, A.: (1953), 'Solvable and unsolvable problems', *Science News* **31**, pp. 7–23.

STEPS TOWARD A DYNAMIC CONSTRUCTIVISM

Giovanni Sambin
Dipartimento di Matematica Pura ed Applicata, Università di Padova
via Belzoni 7 – 35131 Padova (Italy)
sambin@math.unipd.it

Prologue. The philosophy of mathematics and logic has been an object of my reflections ever since I first undertook serious study of these subjects. Over many years I have slowly developed a global way of thinking about the foundations, not only of mathematics, but of abstract concepts in general (see Sambin (1987), Sambin (1991), Sambin (1998) and the introduction of Sambin and Valentini (1998)). Recently, the fertility of such a global approach, at which I have arrived independently of the usual schools, has been confirmed by two specific and novel technical developments in logic and in mathematics. This fact has finally convinced me to address a wider audience, and, moreover, in English.[1]

Since I cannot convey everything in a few pages, this paper is necessarily an overview. I hope in the future to find the energy to write up a complete version, which will include all the details, arguments, connections and references omitted here. Also, if my exposition is somewhat ideological and my language nontechnical, it is because my goal is to take the initial steps towards a new conception.

1. A general claim

Right from the start, the philosophy of mathematics of Brouwer – the founder of intuitionism – was rejected by most because of its links with mysticism. This seems correct. Certainly it is not a sound practice to base scientific knowledge on any form of religious belief.

Now the real problem is that, if by religion we mean any form of unprovable belief in something beyond our perceptions, then apparently also all other existing philosophies of mathematics are based no less on religion. Logicism, which claims that mathematics can be reduced to logic, gives no explanation of why logic should stand by itself, given a priori. Formalism purposely deprives mathematics of all intentionality and thus accepts it, without any discussion, as

P. Gärdenfors, J. Woleński and K. Kijania-Placek (eds.), In the Scope of Logic, Methodology and Philosophy of Science, Vol. I, 263-286.

given in its totality, since its aim is to justify all of it by reducing it to its 'objective' external appearance, that is, to signs. In general, all of the present day approaches seem still to be permeated by some form of platonism: something, be it called 'truth', 'existence', 'objectivity', or the like, is assumed to exist independently of our perceptions and of our minds. Assumptions of this sort may be understandable from a psychological point of view, since they relieve the anxieties of emptiness and uncertainty as well as furnish an aim for our trials and errors, but it must be emphasized that they have no scientific basis.

Of course, we all remain free to believe whatever we wish. This does not mean, however, that all is relative: the important things in the end are the consequences one draws from such beliefs. In my view the problem with religion is that it has often been accepted quite uncritically, in the form of 'blind faith', and this can lead to the making of very bad, even catastrophic mistakes. Among these, a very bad example – but one providing an excellent illustration of my claim – is the massacres by Cortes and Pizzarro of several million people in Central and South America on the supposition that they did not possess souls. Similarly, the supposition, dogma even, that an absolute truth must exist and that classical axiomatic set theories such as ZF must form a part of this truth, has resulted in a number of cultural errors – intellectual atrocities even – of which the Banach-Tarski paradox is just another very bad – or excellent – example. More than just being philosophically shaky, ZF has severely impaired our intuition by splitting it from what ZF would have us believe. (To see this, just try to teach certain statements of ZF to any nonmathematician equipped with common sense; cf. also Boolos (1998)).

This is a pathological state which requires healing. It has been present for at least a century, but little has been done about it. I believe the reason for this is the lack of a good general philosophy, capable of providing sound guiding principles. But now such a thing seems possible. We must go beyond our founding fathers Frege, Peano, Russell, Hilbert, Brouwer, Gödel. They did their task; now it is our turn.

What do we know today that was not known sixty years ago, in 1939? We must recognize that the last sixty years 1939–1999 have been much less productive of the philosophy of mathematics and logic than have the previous sixty, 1879–1939. This being the case, where should we now seek new ideas? On what can we base our hope for a freer, healthier attitude? Many novel ideas have come from *biology*, chiefly since the discovery of DNA in the 50s. Of the greatest interest here is the fact that, after fifty years, it is now generally believed by biologists and by neuroscientists that nothing beyond biology and evolution is theoretically necessary to explain human beings, that is, their bodies *and* their minds (it must be stressed that this is to be meant in a theoretical sense, independently of present inability to give a detailed account of phenomena like consciousness and creativity). No metaphysics, no ghosts, no homunculi. This is a big intellectual change, which has led to a new global cultural attitude.[2] My general claim is simply that the same holds for logic and mathematics. That is, a naturalistic, evolutionary attitude is fully sufficient, and, as I hope to show, actually convenient for explaining not only the human body and mind, but also all human intellectual products, including logic and mathematics, which are just the most exact of these. That is, mathematics is a product of our minds and so to explain it we require no more than what is needed by biologists to explain the mind.

I hasten to point out that this does not mean replacing old religions and previous principles by new ones, the principle of evolution, say. In fact, we shall deduce nothing from the assumption of evolutionary theory. Evolution is only a useful guide for finding good explanations and new results. When these are convincing by themselves, they can stand alone, under no assumptions, and evolution can be discarded. We shall see two examples of this in sections 5 and 6.

The change of attitude I propose towards mathematics is similar to the recent change of attitude towards mind among biologists. It is a matter of fact that each of us possesses both a brain, and a mind, whatever *that* is; it has not been proved, and I believe it cannot be proved, that we have a spirit, distinct from mind in the sense that it can live independently of the body. Biology helps us to accept this as a matter of fact. Similarly, it is a matter of fact that human beings have produced some kind of mathematics; it is by no means proved, and I claim it is in fact unprovable, that it exists in itself, independently of our minds, in some metaphysical mathematical universe.

Of course, while accepting this, the debate remains open and many other points should be made. Let me here recall just one experimental fact about our brains, since it may help to remove a common source of resistance to a biological explanation of mind. It has been argued that accepting a biological, i.e., materialistic, explanation of mind means asserting that the human brain is a machine, that is, in the end, a Turing machine, and hence that there is no room for human creativity. Basing their claims on experimental evidence, certain neuroscientists[3] strongly deny that the brain works like a Turing machine. I see no solid grounds for doubting this. However, even assuming that the brain is a Turing machine, its mere quantitative complexity is so high that it totally eludes our own control and certainly contains room for anything, including that which we term human creativity and the elusiveness of mathematics. For consider: according to recent estimates, the human cerebral cortex contains 10^{11} neurons, each possessing perhaps 10^4 synaptical buttons. If 10 pages are needed to describe one neuron with all its synapses, then the whole cortex would require 10^{12} pages, that is $10,000$ libraries, each with $100,000$ books of $1,000$ pages each. And this is still probably a tiny number when compared with that of potential brain states ...

2. The familiar questions

I would like to make clear that the evolutionary-naturalistic frame of mind should not be imported passively from biology, as a *deus ex machina* for solving problems in the foundations of logic and mathematics. In actuality, this frame of mind was built from below, step-by-step as it were, over a very long reflective journey, starting from questions familiar to anybody interested in foundations. It was only after reading several books on evolution that I became aware, a posteriori, of the perfect compatibility with my own conclusions. Some of the familiar questions are:

> What is the nature of mathematical entities? Where do they come from? Are they discovered or invented?
>
> What is mathematical and logical truth? Is it absolute? If so, how can we attain it? Why is a proof convincing? What is objectivity? And if objectivity is really there and accessible to everybody, why has all discussion not simply ceased?
>
> What is the meaning of existence in mathematics? Which entities exist? Where, when, and how do they exist? Is it possible to delineate the difference in the modes of existence of Nero Wolfe and Robin Hood? Of black holes and ether? Of 17 and an inaccessible cardinal?

Also intriguing are certain specific questions:

> Which set theory should we accept? Why should we believe in ZF? and in the axiom of choice? and in paradoxes such as Banach-Tarski?
>
> Which logic is 'correct', classical or intuitionistic? Who was right, Brouwer or Hilbert?
>
> etc.

This list has been made precisely to show that they are just familiar questions which anyone can put to himself. Perhaps what characterizes my approach is an aversion to answers 'by authority', an incorrigible aversion to concede anything as true unless I see it with my own eyes. Yes, I admit, I am exactly like a child of 4 or 5 who goes on asking why? why? why? and doesn't stop until he is satisfied. And even when his father, or any father, asserts that one should just believe, and no further questions are tolerated, that child may fall silent, but still doesn't stop thinking. No silence or act of faith has ever convinced me; I have always been seeking something more reliable, and actually so reliable that it can withstand any possible question. Taken in this way, any of the above questions has the potential to lead us to new domains of thought. It takes no more than a couple of defenses of common sense against widespread opinion, apparently, for one to be considered a revolutionary!

It is only relatively recently that I have, after many years, been able to reach a satisfactory understanding, that is, answers to these questions which in my opinion satisfy even the demands of my incorrigible aversion. It is important to emphasize that I mean *all* such and similar questions, because I believe that this is required of any good philosophy of mathematics (otherwise one can easily cheat, for example, by assuming the existence of God, or the like). But, of course, I do not expect to be able to furnish every detail (the mere mention of which seems to me a bit megalomaniac). My answers are of a philosophical, nontechnical nature, and yet genuine answers, since the feelings of anxiety sometimes engendered by the questions have been appreciably lowered! Gradually, I have acquired a whole new way of thinking, one which works seven days a week, and I have also been freed of all traces of guilt for not believing in something I simply don't see, such as classical logic and set theory. I thank life for having given me all this.

The answers which I have found to these questions now seem to me so satisfying, so simple and natural, that I can hardly see the distance I have moved from more widespread opinions. Among the many things I should say, it is now difficult to make the right choices, and in the right order.

It seems to me that the principal change has been the move to seeing everything from a dynamic, rather than a static, viewpoint. This applies in particular to foundations – so maybe the word 'foundation' itself should be changed, since it refers to something reliable insofar as it is static. I mean what I say: the price that must be paid is that traditional concepts such as truth, existence, objectivity, usually taken as pillars, must also be regarded as the results of a process. Nothing in mathematics exists in itself, or by divine *fiat*.

This may look at first as too high a price to pay, as a lunatic attack on the sources of any certainties we may happen to possess, an attack explicable only as the product of a self-destructive attitude. But, thank heavens!, just the opposite is the case. It is a false impression, similar to that produced when one contemplates giving up smoking, or freeing oneself of any other form of addiction. At first, the idea of quitting looks like a suicidal jump into emptiness, but after a while you see that you don't need the addictive substance, and that actually the world (in this case, the world of mathematics) is much nicer, more complex, more interesting and – what matters most – more alive than it appeared to be before. So, in the end, the gain far outweighs the pain of withdrawal! One should not, therefore, expect from me relief from that pain – much less in the form of another, stronger addiction. The only help I can give is to offer myself as an example: it should show that it is possible to free the self of addictions or pathologies, and yet remain a mathematician or logician.

3. The dynamic process

Any abstract concept, as expressed by a word, is the result of a dynamic process of abstraction. Actually, it is not the result, but is, rather, the dynamic process itself, going on in each of our minds and hence in the community. It would take quite long to justify this claim properly, and I hope I have done it already in my previous paper Sambin (1991). So let me here give just some examples with comments.

Even understanding such common words as 'apple' requires a lot of mental work. Each of us somehow reconstructs, in his or her mind, the process by which the concept of apple has been formed. So we must abstract not only from time, position, ... but also from defects, colour, taste and so on. The concept of apple, however precise, still contains some element of idealization, that is, some property which does not correspond to reality, but only serves the purpose of convenience. An example might help. A drawing like Fig. 1 shows the axis of symmetry of our concept of apple. In reality, there is no such

Figure 1.

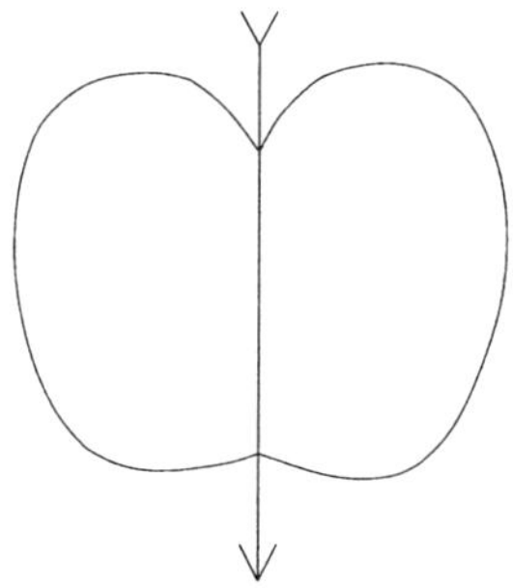

symmetry, and all real apples are slightly asymmetric. To test this, put an apple on a table, and place a piece of cardboard on top of it. You will see that the plane of the cardboard is not parallel to that of the table's surface, which shows that the apple is not perfectly symmetric. So why is it symmetric in our mind? Is it we who are 'wrong'? Certainly not; it is just a matter of saving mental energy. Keeping in mind a concept of apple which includes being symmetric-but-not-perfectly-so would be costly to the brain. So we add symmetry even if in reality it is not perfect, revealing it to be just an idealization we impose on the concept to make it simpler.

This shows that while of course the concept of apple has a lot to do with apples, in nature only apples but no concepts of apple are available. So there is no identification, and not even an a priori fixed link, between our concept and what we call the 'object in itself'. The concept is constructed by us, socially and individually, and it changes continually in a dynamic way, in response to changes in our needs or our knowledge. This last statement can be checked experimentally. I bet that I would be able to change, for example, the concept of pipe in the reader's mind, if I were given just a few minutes of attention. My confidence in being able to achieve this, of course, derives from the fact that I know enough about pipes to be able to increase the reader's knowledge (by talking about shapes, names, materials, etc. or by explaining why each smoker has so many pipes, etc.), even if he is a pipe smoker (by talking about my pipes and how I took care of them, etc., when I used to smoke). Even such a small experience would suffice to attach new data to the word 'pipe' and thus change for ever the concept of pipe in the reader's mind (and for me to win the bet). This shows just how dynamic concepts are.

The third example shows that an excess of idealization may be not convenient. It comes from my recent experience. Fig. 2 gives the shape of the

Figure 2.

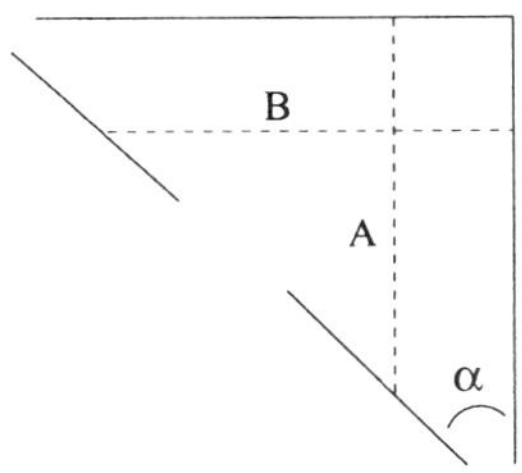

kitchen in the old flat I am renovating. It was impossible to measure the length of the walls, since the angles were still under reconstruction. So I measured A, and it came out at 219 cms. Measuring was not an easy matter, because neither plaster nor floor was yet in place, and one had to make complicated estimates. Then I assumed B was the same length as A. But if that were the case, then fridge and dishwasher could not both be put in B, which was the only way I could have both. I was so disappointed that I measured B directly, just to check. This turned out to be 228 cms, which was surely enough! I measured it three times to make sure. What was wrong with my first idealization? Simply that the angle α was not, as I had implicitly assumed, 45 degrees. I think the whole story is typical. I needed to measure my kitchen, that is to make an idealization. The first was too simple, and it did not meet my needs. Further data allowed me to make it more precise, until it was adequate. Note that there is no such thing as the perfect idealization, the platonic idea of my kitchen! I stopped when I had the idealization which just sufficed for my purposes.

The conclusion I suggest is that it is always thus: all concepts, including the most abstract ones like those of logic and of mathematics, are the result of a dynamic process of abstraction. Indeed, why should they be essentially different from the familiar, everyday concepts in the first place?

4. A dynamic constructivism

I know very well that the attitude I am describing requires a radical change of perspective, a turning upside-down of the customary sources of certainty. It is usually thought that the certainty and reliability of mathematics is guaranteed by its possession of some special ingredients – objectivity, absolute truth, etc. I propose to reverse this, and say that it is *we* who call mathematics whatever is certain and reliable to a fully satisfactory degree (which we usually express by saying that it is objectively true or the like).

Such a change of perspective is a liberating experience which is usually undergone on first learning about the evolutionary explanation of the natural world, or, more generally, in passing from a static to a dynamic view of the world. I can only give an example here. It is usually taken for granted that we all have a strong will to live, and that this is a primitive instinct (whatever that means) which cannot be explained. The change of perspective in this case consists just in realizing that no special assumption is necessary, and that the explanation is quite simple: all those who had a little weaker will to live by natural selection are all already dead! As R. Dawkins reminds us, there are many more ways to be dead than to be alive. Those who found the way to be still alive today, now possess what we call a strong will to live. But there is no need to assume it at the beginning, as an a priori.

The situation is quite similar also in the mathematical world. Consider for instance the natural numbers. To be able to rely on certainty and objectivity of facts pertaining them, there is no need to think of them as given a priori (however indirectly, through sets, for example). It is enough to recall that they are the result of an extremely long evolution, right up to Dedekind's and Peano's insights on induction. All those conceptions which were even just a tiny bit less precise or less efficient (such as Roman notation, lack of zero, improper induction principles, and so on) have already been selected out by history and are now defunct.

I have spoken of the natural numbers, but it should be clear that I mean to claim that the same applies to all of mathematics. All of it is produced through a dynamic process, and the feeling of certainty, stability, objectivity, etc. however strong it may be, is only something that we attain a posteriori. It is neither a matter of pure convention nor of existence a priori, but of survival of the fittest: what remains stable is only what works best (also in respect of our own demands on conceptualization), all the rest is sooner or later wiped out.

At this point the next question is naturally: if all is constructed, what meaning should one attach to words already mentioned such as *objectivity*, *truth*, *existence* in mathematics? Let us begin with *objectivity*.

What does objectivity become in this new perspective? If it includes being absolutely sure of what we say, how can we attain it? From our new standpoint, there is no such absolute notion of objectivity and so it is the concept of objectivity itself which must be changed and newly defined. The word 'objectivity' clearly comes from 'object', and the reason also seems to me clear: objectivity is reached when we get the same feeling of certitude and the same independence of the subject that we get with what we call concrete, or external objects. So let us look closely at the source of our certitude when we speak about objects.

We must admit that even what we call a concrete object does not exist in itself, independently of our mind. What exists in itself is only a certain distribution of matter in a portion of space at a certain moment, which causes certain sensory inputs to enter our brains. When we speak of an object as existing in itself, what we are actually referring to is a mental construction, obtained by a process of abstraction from what it is convenient to count as contingent, such as place, time, illumination, our mood, etc. We attribute existence in itself simply for reasons of convenience: it is much quicker, more efficient, and reliable, exactly as in the case of all those mental functions which have become almost automatic, like pieces of 'hardware'. While the strong agreement we can reach concerning objects is ultimately based on the fact that we have the same sensory organs, still it must include much mutual adaptation. What we consider to be 'objectively true', that is part of the external world and hence necessary, is often just one possible way of organizing our perceptions, and hence our knowledge of the world. Witness, for example, the fact that different people may see different colours (typical is the pair blue-green) and that different languages use different concepts for the same reality (Italian has three words – vetro, bicchiere, occhiali – for glass and its plural, while Chinese has no word for wrist).

Similarly, objectivity in mathematics is reached when some entity is considered an object, that is, acted upon by automatic mental functions similar to those used for the (internal) image of an (external) object. As with concrete objects, to reach this stage the process of abstraction must include the interaction with other individuals, and with the outcomes of *their* mental processes. This is the principal, and deepest difference from Brouwer. It is through a democratic, though occasionally turbulent, dynamic process of achieving consensus that we arrive at what a posteriori we call objective or objectively true.

This is my understanding of the word 'objectivity'. Since no objects are assumed to exist by themselves, the meaning of objectivity becomes very close to that of inter-subjectivity and also of complete certainty. Of course, one may disagree, but note that at least I am giving an explanation. In the absence of dynamics, it is very difficult to talk about objectivity without assuming it to be universally clear, that is without falling exactly into what one tries to avoid, the – very subjective – claim that objective is what is considered as true in a certain period and by a certain community, usually ours! Here is how unjustified assumptions, such as set theory or logic taken as a priori, creep in.

The explanation of what is to be meant by truth and by existence follows the same general idea, namely that none of them is absolute, that our feeling that they are independent of us is due to reasons of convenience and that a new, positive meaning must be found.

The traditional definition of truth as *adaequatio rei et intellectus*, equality of reality and intellect, is essentially correct, once it is understood in a dynamic way, as something to be achieved. So it is not a case of two terms being equal, but rather of two processes, the results of which are to be made to converge (this is confirmed by the fact that the prefix *ad* in *adaequatio* gives the idea of movement) (see Fig. 3). Reality is first taken in (which is the original meaning

Figure 3.

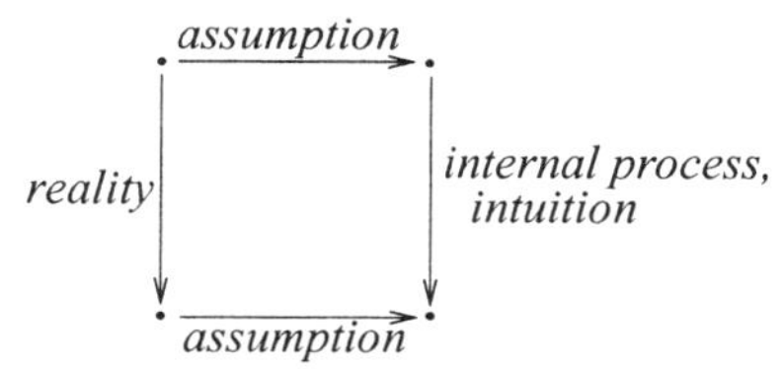

of 'assuming') and elaborated by pure thought, or intuition, or internal mental processes; if the outcome agrees with a second test on reality, then truth is confirmed. Otherwise, something must be changed.

This point of view allows one to see also how agreement on truth is built up. Since reality also includes other individuals – possibly figures of authority such as teachers, authors, etc. – when we are presented with something which is put forward as true, we have an alternative: either it agrees with our intuition and we also recognize it as true, thus confirming it as a piece of truth and contributing to its becoming universal, or we reject it and look for something different. This is how the dynamic process, and hence truth, is kept alive.

Such dynamics involving many different subjects is also needed, in my view, to explain how new concepts and new theories are created, and how the meaning of words can change. Moreover, the dynamics involved is immediately visible in some cases, both individually, as in learning and in doing research, and socially, in historical development. In these cases a static view is clearly not sufficient. But unfortunately dynamics is not an ingredient like salt, which can be added at the last minute, after tasting; it is rather like yeast, which must be present right from the start.

And finally, what is existence? It is we who give existence to mathematical entities, just as it is we who create abstract concepts and who attain objectivity. This is more easily grasped in certain cases such as large cardinals, ultrafilters, or infinitesimals: their form of existence is not so different from that of Santa Claus or of Nero Wolfe, in that they have been created by human beings and exist only in the sense that certain human beings (mathematicians, children, readers of novels) keep them alive, both in their minds and in their discussions.

But, if one looks carefully, one sees that all of mathematics is like that. Consider for instance the number $10^{10^{10}}$; this probably exceeds the number of electrons in the known physical universe, and thus I feel safe in saying that it does not exist in nature, in the sense that it cannot be used to count something existing concretely (of course, it is a relatively small number when our abstract constructions, such as permutations, possible states, subsets, etc. are to be counted). So one can see that the natural numbers too are kept alive by our minds; the strong feeling of external existence is due to their very long history, their universal utility and their conceptual simplicity (they are just the simplest never-ending procedure). In other words, the natural numbers are now completely reliable, and almost everybody believes that they are perfectly acceptable as they stand. But if this were considered as a sufficiently good reason to declare them to exist 'in themselves', a priori, should we believe that also, say, large inaccessible cardinals exist a priori? and if not, where should we put the border between the natural numbers and large cardinals? Any choice would be artificial and easily criticized.

However psychologically difficult at first, the correct explanation is that existence is never an absolute attribute, independent of our minds; it follows also that different forms of existence are possible. Thus in a certain sense the natural numbers 'exist more strongly' than do large cardinals, in something like the same way that Pinocchio, Lewis Carroll's Alice and Donald Duck 'exist more strongly' than other less well-known characters. One could – paradoxically – even say that, from this point of view, Nero Wolfe 'exists more strongly' than Robin Hood does, since we know him better![4]

It should by now be clear that the familiar dilemma 'discovered or invented?' has in my opinion a definite answer: *invented*. But, as with abstract concepts and with objectivity, it is often much more efficient to look for something in the belief that it has to be discovered rather than invented. (This is quite similar to the well known fact that, given a problem admitting a yes-no answer, it is much easier to find a full proof if the pure answer yes or no is known.)

The feeling that one has truly discovered something can on occasion be quite strong and fully justified. This is particularly the case when that thing is just a logical consequence of previous knowledge. Here 'discovery' is just the bringing to awareness of a detail which was implicit in a whole invented picture. That is, it is again only a local psychological phenomenon, which does not change the global, simple substance: the whole mathematical universe exists only in the minds of human beings. Does the self-deception of believing that it exists in itself make it nobler, or more interesting, or more reliable?

One can begin to see that reflecting on familiar questions has actually produced more than a collection of isolated answers. The answers can now be seen as arising from a unitary and global philosophical attitude towards logic and mathematics. This attitude I call *dynamic constructivism*. *Constructivism*, because at the root of everything I see, as did Brouwer, the mental constructions of an individual; and *dynamic* to emphasize that such constructions involve the interplay between intuition and reality (which includes other individuals) and not just pure introspection, as in Brouwer's intuitionism.

I hope that what I have said so far is sufficient at least to suggest that what I call dynamic constructivism has some kind of internal consistency. Assuming this, another way to clarify what I mean is to characterize it in terms of the changes that should be made to be able to reach it from any one of the common views in the philosophy of mathematics. I do this very briefly, in terms of rough but, I hope, suggestive images.

Brouwer's main claim, namely that no mathematical object or truth is given and that all must be constructed, is, in my view, correct. Actually, I believe that most of Brouwer's insights are correct. In the last analysis, my claim is that to obtain dynamical constructivism it is enough to drop Brouwer's justification for the intersubjectivity of mathematics, which is essentially the direct mystical inspiration of the creative subject, and simply replace it with the existence of other individuals. Contrary to Brouwer, I am happy to recognize the existence of my fellow human beings (I must confess, however, that this is not always achieved without effort, and that there are certain human beings whose existence I am happier to accept than that of some others). Each of us is a creative subject, or could be. But then, instead of hoping the One to give the same inspiration to every subject, it is simpler to explain intersubjectivity as a dialectical process, the communication – debate, agreement, fight … – between many creative subjects. And this may take place also less directly, in the form of adjustment, within the same creative subject, between intuition (content, meaning, what has to be expressed) and its expression in an intersubjective language. These are the reasons why

I have added the adjective dynamic; but perhaps one should speak just of present-day constructivism, liberated from useless and troublesome assumptions.

The aim of formalism is apparently to reach the certitude of objectivity by cutting off whatever involves a subject, and so all intentions, intuitions, meanings, etc.; that is why mathematics is reduced to signs and formal systems. This is not only unsatisfactory, since it fails to explain the meaning of mathematics and how it is done, but also illusory: formalism seems to ignore the fact that as soon as a language (or formal system) is considered, a subject reading that language (which involves, at least, recognizing the signs) is automatically on the scene. So the change here would be to put metalanguage back in its proper place, which automatically means restoring the role of subject, with its intentions, intuitions and all such nuisances, unavoidable when human beings are involved. The one safe path to objectivity is to *transcend* the subject; pretending that there is no subject is a false shortcut, which in the end is similar to Brouwer's pretence that there is only one.

In a certain sense, it is easier to obtain dynamic constructivism from platonism, at least in verbal terms, since it is only a matter of subtracting: drop the static view, the fact that concepts, or ideas, are statically given somewhere. Maybe it is easier to grasp this in the converse direction: think of platonism as obtained from dynamic constructivism by postulating that the process of construction of a concept actually has an aim, a télos, so that all our trials and errors are just approximations converging to something which exists already. My suggestion is then to do without the useless, troublesome assumption of the convergence point, and accept that there is *only* the process of trial and error, and that this *is* the concept. In other words, I share the platonists' conviction that we need concepts, but, less optimistically, I expect that we must work hard to get them, and that they are never to be found (*where* would that be?) already prepared by someone (*who* could have prepared them?) for us (*why* would they be so obliging to have done so?).

So far the philosophical setting has been presented mainly in terms of cultural attitude; we will see that it also has a technical continuation into logic and mathematics. Here by 'continuation' I mean not just the drawing of consequences, but also a process of confirmation, since the new cultural attitude has allowed us to find new structures, and these in turn prove that it is possible to develop logic and mathematics in consonance with the general philosophy, thereby confirming it.

A vital difference between dynamic constructivism and other approaches is the stress placed on a *multiplicity* of individuals who must perforce communicate with one another. As a consequence, crucial importance is here attached to the interplay between the self and the others, between what the self has in mind and what he/she is able to communicate, or, more abstractly, between meaning and expression, content and form, metalanguage and language (see Fig. 4). Usually, the discrepancy between meaning and its expression is taken

Figure 4.

I		*others*
what I have in mind	→	*what I communicate*
meaning		*expression*
content	←	*form*
metalanguage		*language*

as a misfortune and simply ignored – intuitionism and formalism each escaping this misfortune simply by denying one half of the picture, to wit, language and, respectively, metalanguage. By contrast, in the present framework this discrepancy is accepted as an essential fact of life, thus revealing it as the very source of the dynamics and the underlying engine for the creation of concepts. It has been a deeply moving experience to see how well this philosophical expectation is confirmed by rigorous mathematical development in the case of logic: forcing the convergence of language and metalanguage turns out to be precisely the way to generate all logical connectives, in a fully precise sense which is outlined in section 5.

It is often assumed that adopting a 'weaker' logic and foundation gives a 'smaller' body of truths. This makes sense only if truth is considered to be independent of us and immutable; then of course one can see no real reason for any form of self-limitation, and it must be said that some constructivists have contributed to nurturing this false impression. Qualifying constructivism as dynamic means in particular rejecting the idea that a fixed foundation should be given once and for all (since this would make it static!). So, it is not a matter of 'rescuing' mathematics in terms of such a foundation: there are no magic wands, named ZFC, category theory or even constructive type theory, which are able automatically to transform whatever is expressed in their language into meaningful mathematics. In fact, the complexity of mathematics cannot be reduced to fixed ingredients (logic, formulae, intuition, sets, categories, ...). Rather, one by one, each piece of mathematics, at the moment it is developed, should be provided with the simplest and most convincing underlying conceptual structure, by means of which the peculiar nature of that fragment is faithfully explained. In a dynamic approach, a 'weaker' foundation is simply a means of maintaining finer distinctions and achieving better control of the idealization, that is, of those elements of information which have been retained as well as those which have been discarded (see also Sambin and Valentini (1998)). A bonus is that sometimes a deep structure can emerge, whose presence was previously hidden by foundational assumptions. A typical example of this is what I call the basic picture, that is a structure underlying topology. This is sketched in section 6 in just enough detail to allow some other general comments.

5. Basic Logic

Since Brouwer's revolution, which has introduced plurality into the space of logics, any foundation claiming to adequacy is faced with an alternative: *either* a good argument has to be found showing that there can be only one logic, which is 'truer' than all others – and by 'good' here I mean that the argument should not presuppose that same logic in some equivalent form – *or* the existence of a variety of logics must be explicitly acknowledged and explained.

It is difficult to accept a plurality of logics and still believe that logic deals with a static, a priori, and hence univocal notion of truth. If instead truth is understood as constructed, and hence also relative to a degree of abstraction, then it is possible to conceive of a logic as a stock of rules allowing one to pass, at that level of abstraction, from a given state of (grasped) knowledge to a new piece of knowledge and making no use of further data. In this more open view, what remains to be done is to find some structure in the space of possible logics, in other words, to explicate just how a logic is created. This we now do.

First comes the metalanguage, then the language, just as first comes the content we want to express and only then the means of its expression (the word 'meaning' in Latin is *significatum*, that is 'what has been made into a sign'). So we assume we have propositions A, B, C, ... ; it is not necessary here to know exactly what a proposition is, so long as it is sharply distinguished from a judgement, or assertion, based on it, which I write as *A is*, *B is* (the common denominator of assertions like *A is true*, *A is available*, *A is measurable*, ...). The proposition A is at the object level, the assertion *A is* at the metalevel. In addition, we need just two metalinguistic links, *and* and *yields*, for building compound assertions.

This setting is sufficient for introducing and explaining all the notation of a calculus of sequents, in the style of Gentzen (1935). In fact, a sequent $\Gamma \vdash \Delta$ usually abbreviates $C_1, \ldots, C_m \vdash D_1, \ldots, D_n$, which in turn here is an abbreviation of (C_1 *is and*... *and* C_m *is*) *yields* (D_1 *is and*... *and* D_n *is*). Note that $\Gamma \vdash A$ and $C \vdash A$ abbreviate (C_1 *is and*... *and* C_m *is*) *yields A is* and *C is yields A is*, respectively. A rule like

$$\frac{\Gamma \vdash \Delta}{\Gamma' \vdash \Delta'}$$

is also an abbreviation, of $(\Gamma \vdash \Delta)$ *yields* $(\Gamma' \vdash \Delta')$, and similarly in the case of two premises, where the empty space abbreviates *and*.

All the assumptions we need are:
identity $A \vdash A$

composition: on the left $$\frac{\Gamma \vdash A \quad A, \Gamma' \vdash \Delta}{\Gamma, \Gamma' \vdash \Delta}$$

on the right $$\frac{\Gamma \vdash \Delta, A \quad A \vdash \Delta'}{\Gamma \vdash \Delta, \Delta'}$$

for any A, Γ, Γ', Δ, Δ'. Composition is just the formal expression of a very intuitive idea: if there is a product (or conclusion) Δ which results from some ingredients (assumptions) A, Γ', then we can substitute the ingredient A by the ingredients Γ of a previous production of A, and have the same product Δ. The same is assumed on the other side of *yields*, on succedents.

Now I can show, under these assumptions alone, that all connectives are created in accordance with a single, general scheme. First we have what I call a definitional equation, for example

$$\Gamma \vdash A\&B \quad \textit{if and only if} \quad \Gamma \vdash A \textit{ and } \Gamma \vdash B$$

The connective we want to create (& in the example) appears on the left side, which I then call *definiendum*, while the *definiens* is on the right. I say that the connective (& in the example) is the reflection of a metalinguistic link (*and* in the example) in a certain configuration. Thus the definitional equation is still a desideratum: it just expresses the wish for a connective producing a new proposition ($A\&B$ in the example) which behaves, when asserted in a given situation (at the right of $\vdash$, in the example), just as the compound assertion ($\Gamma \vdash A$ *and* $\Gamma \vdash B$, in the example) with that link. I say that & reflects *and* at the right of $\vdash$ and that & obeys the principle of reflection, since, as we shall now see, its inference rules are obtained by solving the definitional equation.

The direction from definiens to definiendum gives what I call the formation rule (in the example, it is

$$\frac{\Gamma \vdash A \quad \Gamma \vdash B}{\Gamma \vdash A\&B}$$

recalling that *and* is expressed here by the empty space). The other direction, from definiendum to definiens (in the example, $\Gamma \vdash A\&B / \Gamma \vdash A$ and $\Gamma \vdash A\&B / \Gamma \vdash B$), does not yield a good rule, since the connective to be defined appears in the premises. I call it *implicit reflection* since it expresses the required information only implicitly. We can however solve the equation by finding an acceptable rule which is equivalent to implicit reflection. First we trivialize the premise of implicit reflection to obtain an axiom (in the example, replacing Γ with $A\&B$ we obtain the &-axioms $A\&B \vdash A$ and $A\&B \vdash B$). Then by composition we reach the result that the connective to be defined appears in the conclusion, but on the other side (in the example, if $A \vdash \Delta$ then from the &-axiom by composition also $A\&B \vdash \Delta$). This yields the rule I call *explicit reflection* (in the example, $A \vdash \Delta / A\&B \vdash \Delta$), which is perfectly good. Now we can see that we can also proceed in the converse direction, from explicit reflection to the axiom by trivializing the premise (in the example, putting $\Delta = A$ one obtains $A\&B \vdash A$) and from the axiom to implicit reflection by composition (in the example, from $A\&B \vdash A$ to $\Gamma \vdash A\&B / \Gamma \vdash A$, and similarly when the succedent is B). So the three rules are all equivalent. Therefore we have found inference rules, those of formation and of explicit reflection, which taken together are equivalent to the definitional equation. So, finally, the corresponding wish is satisfied, and the new connective fully defined.

The principal discovery is that all connectives in the most familiar logics are created according to the principle of reflection and, moreover, the solution of the definitional equation always adheres to the same pattern as above.[5] Looking again at the above example for &, one sees immediately that all arguments continue to apply if we interchange the left-hand and right side of each sequent; in fact, all our assumptions are fully symmetric. So the equation $A \vee B \vdash \Delta$ *if and only if* $A \vdash \Delta$ *and* $B \vdash \Delta$ is also solved; it gives the usual inference rules for disjunction $\vee$ (symmetric to those for &). So, interestingly enough, no metalinguistic link for disjunction is required – fortunately, as it happens, since such a link would not be easy to explain – once we are ready to pay the price of accepting a formation rule for $\vee$ acting on the left.

A less common connective is the solution of a simpler definitional equation, that is $A \otimes B \vdash \Delta$ *if and only if* $A, B \vdash \Delta$; here it is perhaps easier to grasp that $\otimes$ reflects *and* on the left of $\vdash$, since in that position the assertion $A \otimes B$ *is* can equivalently replace *A is and B is*. The solution here yields the connective $\otimes$ of Girard's linear logic. Actually this is not exactly the case, since in linear logic the inference rules have contexts on both sides, that is they are obtained by solving the definitional equation $\Gamma, A \otimes B \vdash \Delta$ *if and only if* $\Gamma, A, B \vdash \Delta$, with two free parameters Γ and Δ, one on each side. Definitional equations with only one free parameter (Δ in the example of $\otimes$) give rise to a new logic, strictly below linear logic. I have called it *basic logic*, since it is the least logic satisfying the principle of reflection. This means it is as weak as possible but still possess a clear and simple conceptual structure. The sequent calculus **B** for basic logic is similar to a two-sided sequent calculus for linear logic (without exponentials), except for the absence of all contexts of active formulae. The standard reference for basic logic, containing more motivation and of course all technical details, is Sambin, Battilotti, and Faggian (2000).

Basic logic furnishes a simple structure in which all familiar logics find a place. In fact, by performing on the sequent calculus **B** all the combinations of actions **L**, of restoring contexts on the left, **R**, of restoring contexts on the right, and **S**, of adding the structural rules of weakening and contraction, one obtains a cube of logics in which the equivalences shown in Fig. 5 hold.

Figure 5.

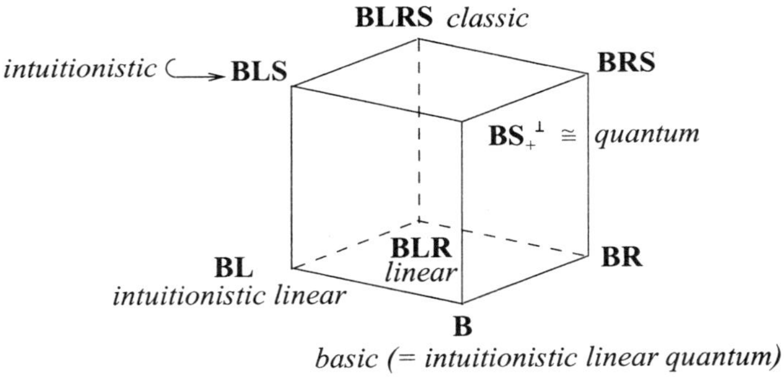

(N.B. **BLS** is intuitionistic logic equipped with an extra connective which is symmetric of implication; equipped with a primitive negation, **BS** gives basic orthologic, one of the weak quantum logics).

The fertility of basic logic from a technical standpoint is confirmed by two results, both of which had been sought for a long time. One is an easy theorem of normalization of proofs (cut-elimination) for orthologic, one of the best-known quantum logics, see Faggian and Sambin (1998). The other is a single cut-elimination procedure which works simultaneously for all logics in the cube, once they are formulated as subsystems of the same sequent calculus, see Faggian and Sambin (1999).

But in my view it is within the conceptual domain that basic logic has consequences of the greatest interest. The rules governing connectives are sharply distinguished from structural rules, including those for handling contexts. Basic logic is the minimal core, with no structural rules, and it brings fully to view the idea that any logic is characterized by a choice of structural rules,[6] that is, rules operative at the metalevel, dealing just with assertions. In this way the important problem of clarifying (and classifying) the diverse notions of truth associated with the variety of logics is reduced simply to the analysis of structural rules.

It seems to me that the theory of meaning for the connectives implicit in the principle of reflection is an improvement on that of Dummett and Prawitz. For here the meaning of each connective is given, not by an introduction rule, but by a definitional equation, which provides a clear description of the role of the connective and specifies exactly how it is created. As a consequence, one can derive not only the elimination rules but also the introduction rules (note that 'introduction' is sometimes what I have called 'formation', sometimes what

I have called 'reflection', and this is why the terminology had to be changed). So one sees that all rules of inference are derivable from meaning; moreover, this has been achieved by means of tools which have been introduced explicitly beforehand, and which include just two metalinguistic links.

In every logic of the cube, each connective is defined by a definitional equation, whose solution is obtained by always employing the same method. Thus all logics are born in the same way, which is independent of the choice of structural rules. So we can conclude that the dynamic relationship between metalanguage and language provides a structure which is common to all logics – the search for which is an important philosophical goal – and which is deeper than any specific notion of truth.

These brief remarks should at least suggest how basic logic can be of help in 'putting some order' into the space of logics. One should however be careful not to expect any new absolute to appear. For example, noncommutative basic logic is quite possible, it is strictly below basic logic, and some linguists have already shown interest in it.

6. The basic picture

A fully constructive (intuitionistic and predicative) theory of subsets is obtained by conceiving of a subset of a given set S as a predicate, or propositional function, over S, that is a proposition $U(a)$ depending on a variable a ranging over S (see Sambin and Valentini (1998)). For any element $a \in S$, we write $a \,\varepsilon\, U$ to mean that $U(a)$ holds. The next step is to consider two sets, let's call them X and S, and the weakest link between them, namely a binary relation $\Vdash$, that is a propositional function $x \Vdash a$ with two arguments, $x \in X$ and $a \in S$. I call this a *basic pair*.

For any subset $U \subseteq S$ we define the existential anti-image of U along the relation $\Vdash$ to be the subset

$$\mathsf{ext}\, U \equiv \{x \in X : (\exists a)(x \Vdash a \,\&\, a \,\varepsilon\, U)\}.$$

Similarly,

$$\mathsf{rest}\, U \equiv \{x \in X : (\forall a)(x \Vdash a \rightarrow a \,\varepsilon\, U)\}$$

is called the universal anti-image of U along $\Vdash$. Notice the strict logical duality here: $x \,\varepsilon\, \mathsf{rest}\, U$ is expressed by a formula which is obtained from that expressing $x \,\varepsilon\, \mathsf{ext}\, U$ by formally replacing $\exists$ and $\&$ with their duals $\forall$ and $\rightarrow$, respectively.

In a fully symmetric way, the existential and the universal image along $\Vdash$ of a subset $Z \subseteq X$ are defined by

$$\Diamond Z \equiv \{a \in S : (\exists x)(x \Vdash a \,\&\, x \,\varepsilon\, Z)\}$$

and

$$\Box Z \equiv \{a \in S : (\forall x)(x \Vdash a \to x \,\varepsilon\, Z)\},$$

respectively. This suffices to express the first discovery. Think of X as a set of points, S as a set of names for formal neighborhoods, and $x \Vdash a$ as x lies in a, or x belongs to the extension of a. Then the usual definition of the interior of a subset $Z \subseteq X$ is

$$\mathsf{int}\, Z \equiv \{x \in X : (\exists a)(x \Vdash a \,\&\, (\forall z)(z \Vdash a \to z \,\varepsilon\, Z))\};$$

we now see immediately that this is just the composition of ext after $\Box$, that is

$$\mathsf{int}\, Z = \mathsf{ext}\,(\Box Z).$$

One then quickly sees also that the closure of Z, defined as usual by

$$\mathsf{cl} Z \equiv \{x \in X : (\forall a)(x \Vdash a \to (\exists z)(z \Vdash a \,\&\, z \,\varepsilon\, Z))\},$$

is exactly the logical dual of the interior, so that we have also

$$\mathsf{cl} Z = \mathsf{rest}\,(\Diamond Z).$$

By symmetry, one can define another pair of operators, defined on subsets of S, by putting

$$\mathcal{A}U \equiv \Box(\,\mathsf{ext}\, U) \text{ and } \mathcal{J}U \equiv \Diamond(\,\mathsf{rest}\, U)$$

for any $U \subseteq S$. The operator $\mathcal{A}$ also has a topological content, since it is none other than the principal operator of formal topology, yielding the formal open subsets of S. The novelty here is that the operator $\mathcal{J}$, to which we have been led solely by considerations of symmetry and logical duality, also has a topological content, namely that of furnishing the definition of formal closed subsets of S. We say that Z is *concrete open* if $Z = \mathsf{int} Z$ and *concrete closed* if $Z = \mathsf{cl} Z$. Symmetrically, we say that U is *formal open* if $U = \mathcal{A}U$ and *formal closed* if $U = \mathcal{J}U$. Like int, $\mathcal{J}$ is, by symmetry, an interior operator and, like cl, $\mathcal{A}$ is, *ipso facto*, a closure operator. Thus concrete and formal, open and closed subsets form complete lattices. The whole structure is depicted in Fig. 6. The isomorphism between the lattice of concrete open subsets and that of formal open subsets, and similarly for closed subsets, confirms that the definitions are correct.

To obtain a topology in the usual sense on the set X, it is enough to add the condition that the neighborhood system at each point be convergent; this is related to distributivity of the lattice of formal open subsets. So the structure illustrated above underlies both usual, or pointwise, and formal, or pointfree, topology. I have called it the *basic picture*. It also includes a similar symmetric structure which underlies the notion of continuity. It shows, among other

Figure 6.

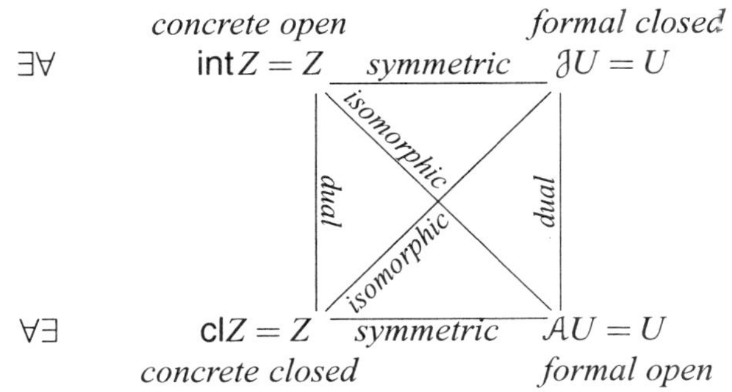

things, that one can develop the field of nondistributive topology, in which the symmetry between concrete and formal, and that between open and closed, is still unbroken. A first paper on the basic picture is Sambin and Gebellato (1999); a full exposition, together with developments, is in preparation.

I hope this brief summary at least makes the reader curious. It should be enough, in any case, to show that topology arises simply in passing from one to two sets (linked by a relation), and that it is deeply connected with logic. One could even maintain that topological notions are just a quick, convenient way of treating combinations of quantifiers.

It is also important to note that a new foundational attitude like the present one, engenders not only new understanding, but also new mathematics. An example is the operator $\mathfrak{J}$, which yields the new notion of binary positivity predicate. This is the dual of a formal cover, also in the sense that it is generated by co-induction, rather than by induction.

7. Comments and open problems

Basic logic and the basic picture have emerged as two hitherto unnoticed natural portions of the landscape of logic and of mathematics. Their principal ingredients are basic and universal concepts: symmetry, logical duality, convergence between language and metalanguage. The logic needed is just intuitionistic predicate logic and any set theory is sufficient, as long as a set can be the domain of quantifications. Moreover, they help to explain many facts about logic and topology by inserting them in a general, but quite elementary structure. Actually, these structures – in particular the fact that $\mathsf{int} = \mathsf{ext}\,\Box$ and $\mathsf{cl} = \mathsf{rest}\,\Diamond$ – are so elementary that they could have been discerned at any time since the 20s or 30s. But, incredible at it seems, as far as I know they have remained unnoticed up to now. So it is very natural to ask: why now? On reflection, this fact is less surprising when the rigidity of ideological attitudes is taken into account. Paradoxically, it is precisely the fact that no specific foundation is involved which made their discovery harder rather than easier. Consider the basic picture. Could an orthodox intuitionist have seen it? Certainly not, because of his idiosyncrasy against formal language: it is virtually impossible to see symmetries and logical dualities without ever writing, for instance, signs like $\forall$ and $\exists$ for quantifiers. A formalist believes wholeheartedly in classical logic, which makes the whole structure collapse and become invisible, because, if one assumes the principle of double negation, the operator cl becomes definable as $-\mathsf{int}-$, and similarly for $\Box$, just as $\forall$ is definable as $\neg\exists\neg$. And finally, if one believes in the powerset, platonists certainly do, the discovery would even never be conceived, since there would be no need to present the base of a topological space as indexed by a second set, like S above. Similar remarks apply to basic logic; just note that technically nothing else is required besides a metalanguage, a fact which has to be accepted since the time of Gödel's theorems, and sequent calculus, which has been known since Gentzen.

So it would seem that ideological blocks have been chiefly responsible for the previous failure to discover these structures. The conclusion is, I believe, that a freer, less dogmatic frame of mind, or foundation, keeps open the possibility of new discoveries. A 'weaker' foundation, with less rigid assumptions concerning the nature of mathematics, can be more consonant with our mental processes, and so may help us to see deep structures which are rendered invisible, and hence destroyed, by a 'stronger' foundation. In this precise sense, it is the 'stronger' foundation which becomes an unreasonable self-limitation, something like making love while wearing boxing gloves.

The two discoveries I have briefly described are just the beginning of the development of logic and of mathematics free of metaphysical constraints. Still, they show that such a development is possible. Thus, in connection with dynamic constructivism, there is a lot of further technical work to be done and many open problems to be solved. This shows how vital the new philosophy is. Among the many open problems, I conclude with three which seem to me particularly meaningful.

We have seen that logic can be justified simply in terms of the principle of reflection, that is of the dynamical interplay between language and metalanguage. I expect that some form of set theory can be explicated in a similar way. So the first open problem is to find the mathematical development of this philosophical expectation. A solution of course would be extremely interesting and relevant for the development of mathematics with no metaphysical assumptions. My impression is that one will arrive at a fragment of constructive type theory.

The second problem is to find a mathematical, complete semantics for basic logic. Since basic logic yields all known logics in a modular way, one would expect to be able to formulate a mathematical semantics which, in a similarly modular way, yields the well-known semantics for the well-known logics. It is natural to expect the basic picture to furnish the right framework for the correct definition of validity, since, like basic logic, it is nondistributive and based on symmetry.

A theorem of Gödel and Kreisel says that there is no possible complete naive semantics for intuitionistic logic. But the extension of basic logic BLS is not exactly the same as intuitionistic logic, since it has an extra connective, symmetric of implication. So by adding this connective, and imposing a new notion of naive semantics (possibly in which refutability and not only validity is defined for formulae), one might hope to obtain a naive complete semantics. This is the third problem, a positive solution to which would be an important extra benefit conferred by dynamic constructivism.

Epilogue. So we reach our conclusion. As you can see, I have chosen a new name, but at the same time it has not been my purpose to come up with a new recipe to replace those already on the market, a new 'final foundation', for the simple reason that there is no such thing. I have sometimes spoken very subjectively, placing myself and my experience (my memories of my father, my present problems with my kitchen, ...) in center stage. I am aware of the fact that this is unorthodox in a scientific communication, or perhaps simply inelegant. But it coheres fully with the viewpoint I have tried to put forward. It is only through a subject, with her or his intuitions and subjectivity, that new truths can be attained. Truth is born as subjective truth; only if later accepted by other subjects, can it become universal and finally objectified.

In more specific terms, everything I said in this paper comes from what I have been able to see, things that have caused me suffering, but which I now firmly believe to be true and for which finally I have tried to argue. Now it is your turn: you should evaluate it, choose what to reject and what to make yours, find better answers, and so on. Feedback from you will also modify my views. And precisely in this way the dynamic process has no end.

Notes

1. I am sincerely grateful to John L. Bell for his invaluable role in tailoring my text into an English adequate for expressing what I had in mind and to Juliette Kennedy for an intense correspondence leading to several improvements of exposition.

2. Certain philosophers now share this opinion: see e.g. Dennett (1995).

3. E.g. Edelman (1992).

4. I am not suggesting that existence has degrees measured by numbers. Still, it might be an instructive exercise to develop a formal theory with values of existence ranging over for instance the real interval $[0, 1]$, so as to facilitate our understanding some key aspects of what we call existence, in the same way as the various many-valued logics may help to understand the concept of truth.

5. For lack of space here, I cannot reproduce a visual explanation which is possible in an actual talk and is obtained by first showing a slide with the common, bare structure, and then superposing on it a slide for each of the connectives.

6. This is also formally precise if the sequent calculus **UB** of Faggian and Sambin (1999) is adopted.

References

Boolos, G.: (1998), 'Must We Believe in Set Theory?', in G. Boolos, *Logic, Logic, and Logic*, Harvard UP, Cambridge, MA, pp. 120–132.

Dennett, D.C.: (1995), *Darwin's Dangerous Idea: Evolution and the Meanings of Life*, Simon & Schuster, New York.

Edelman, G.M.: (1992), *Bright Air, Brilliant Fire: On the Matter of the Mind*, Basic Books, New York.

Faggian, C., Sambin, G.: (1998), 'From basic logic to quantum logics with cut-elimination', *International Journal of Theoretical Physics* **37**(1), pp. 31–37.

Faggian, C., Sambin, G.: (1999), 'A unified approach to logical calculi and normalization of proofs', Lecture Notes of the 11th European Summer School in Logic, Language and Information (ESSLLI), Utrecht University.

Gentzen, G.: (1935), 'Untersuchungen über das logische Schliessen', *Mathematische Zeitschrift* **39**, pp. 176–210. Eng. trans. in M.E. Szabo, (ed.), *The collected papers of Gerhard Gentzen*, North-Holland (1969), pp. 68–131.

Sambin, G.: (1987), 'Alla ricerca della certezza perduta (Forma-contenuto nei fondamenti della matematica)', in Longo, O. (ed.), *Forma Rappresentazione Struttura*, Laboratorio servizio tecnologia, Napoli, pp. 17–35.

Sambin, G.: (1991), 'Per una dinamica nei fondamenti', in G. Corsi, G. Sambin (eds.), *Atti del Congresso 'Nuovi problemi della logica e della filosofia della scienza'*, Bologna, CLUEB, pp. 163–210.

Sambin, G.: (1998), 'Matematica, logica e verità, verso un risanamento', in M. Emmer (ed.), *Matematica e Cultura*, Springer, Italia, pp. 35–41.

Sambin, G., Battilotti, G., Faggian, C.: (2000), 'Basic logic: reflection, symmetry, visibility', *Journal of Symbolic Logic* **65(3)**, pp. 979–1013.

Sambin, G., Gebellato, S.: (1999), 'A preview of the basic picture: a new perspective on formal topology', in T. Altenkirch, W. Naraschewski, Reus, B. (eds.), *International Workshop, TYPES '98, Lecture Notes in Computer Science 1657*, Springer, pp. 194–207.

Sambin, G., Valentini, S.: (1998), 'Building up a toolbox for Martin-Löf's type theory: subset theory', in G. Sambin, J. Smith (eds.), *Twenty-five years of constructive type theory*, Oxford UP, pp. 221–244.

ON THE DISTINCTION PROOF-TRUTH IN MATHEMATICS

Roman Murawski
Adam Mickiewicz University
Faculty of Mathematics and Comp. Sci.
Matejki 48/49, 60–769 Poznań, Poland
rmur@math.amu.edu.pl

Abstract The paper is devoted to some historical, philosophical and logical considerations connected with the distinction between proof and truth in mathematics. The crucial rôle of Gödel's incompleteness theorems as well as of the undefinability of truth vs. definability of provability and the rôle of finitary vs. infinitary methods are stressed. The problem of the need of extending the available methods by new rules of inference and new axioms is also considered.

Concepts of proof and truth are (even in mathematics) ambiguous. It is commonly accepted that proof is the ultimate warrent for a mathematical proposition, that proof is a source of truth in mathematics. One can say that a proposition A is true if it holds in a considered structure or if we can prove it. But what is a proof? And what is truth?

The axiomatic method was considered (since Plato, Aristotle and Euclid) to be the best method to justify and to organize mathematical knowledge. The first mature and most representative example of its usage in mathematics were *Elements* of Euclid. They established a pattern of a scientific theory and a paradigm in mathematics. Since Euclid till the end of the nineteenth century mathematics was developed as an axiomatic (in fact rather a quasi-axiomatic) theory based on axioms and postulates. Proofs of theorems contained several gaps – in fact the lists of axioms and postulates were not complete, one freely used in proofs various "obvious" truths or referred to the intuition. Proofs were informal and intuitive, they were rather demonstrations and the very concept of a proof was of a psychological (and not of a logical) nature. Note that almost no attention was paid to the precization and specification of the language of theories – in fact the language of theories was simply the unprecise colloquial language. One should also note here that in fact till the end of the nineteenth century mathematicians were convinced that axioms and postulates

P. Gärdenfors, J. Woleński and K. Kijania-Placek (eds.), In the Scope of Logic, Methodology and Philosophy of Science, Vol. I, 287-303.

should be true statements. It seems to be connected with Aristotle's view that a proposition is demonstrated (proved to be true) by showing that it is a logical consequence of propositions already known to be true. Demonstration was conceived here of as a deduction whose premises are known to be true and a deduction was conceived of as a chaining of immediate inferences.

Basic concepts underlying the Euclidean paradigm have been clarified on the turn of the nineteenth century. In particular the intuitive (and rather psychological in nature) concept of an informal proof (demonstration) was replaced by a precise notion of a formal proof and of a consequence. This was the result of the development of mathematical logic and of a crisis of the foundations of mathematics on the turn of the nineteenth century which stimulated foundational investigations.

One of the directions of those foundational investigations was the program of David Hilbert and his *Beweistheorie*. Note at the very beginning that "this program was never intended as a comprehensive philosophy of mathematics; its purpose was instead to legitimate the entire corpus of mathematical knowledge" (cf. Rowe 1989, p. 200). Note also that Hilbert's views were changing over the years, but always took a formalist direction.

Hilbert sought to justify mathematical theories by means of formal systems, i.e., using the axiomatic method. He viewed the latter as holding the key to a systematic organization of any sufficiently developed subject. In "Axiomatisches Denken" (1918, p. 405) Hilbert wrote:

> When we put together the facts of a given more or less comprehensive field of our knowledge, then we notice soon that those facts can be ordered. This ordering is always introduced with the help of a certain *network of concepts* (*Fachwerk von Begriffen*) in such a way that to every object of the given field corresponds a concept of this network and to every fact within this field corresponds a logical relation between concepts. The network of concepts is nothing else than the *theory* of the field of knowledge.

By Hilbert the formal frames were contentually motivated. First-order theories were viewed by him together with suitable non-empty domains, *Bereiche*, which indicated the range of the individual variables of the theory and the interpretations of the nonlogical vocabulary. Hilbert, as a mathematician, was not interested in establishing precisely the ontological status of mathematical objects. Moreover, one can say that his program was calling on people to turn their mathematical and philosophical attention away from the problem of the object of mathematical theories and turn it toward a critical examination of the methods and assertions of theories. On the other hand he was aware that once a formal theory has been constructed, it can admit various interpretations. Recall here his famous sentence from a letter to G. Frege of 29th December 1899 (cf. Frege 1976, p. 67):

> Yes, it is evident that one can treat any such theory only as a network or schema of concepts besides their necessary interrelations, and to think of basic elements as being any objects. If I think of my points as being any system of objects, for example the system: love, law, chimney-sweep [...], and I treat my axioms as [expressing] interconnections between those objects, then my theorems, e.g. the theorem of Pythagoras, hold also for those things. In other words: any such theory can always be applied to infinitely many systems of basic elements.

The essence of the axiomatic study of mathematical truths consisted for him in the clarification of the position of a given theorem (truth) within the given axiomatic system and of the logical interconnections between propositions.

Hilbert sought to secure the validity of mathematical knowledge by syntactical considerations without appeal to semantic ones. The basis of his approach was the distinction between the unproblematic 'finitistic' part of mathematics and the 'infinitistic' part that needed justification. As is well known Hilbert proposed to base mathematics on finitistic mathematics via proof theory (*Beweistheorie*). The latter was planned as a new mathematical discipline in which mathematical proofs are studied by mathematical methods. Its main goal was to show that proofs which use ideal elements (in particular actual infinity) in order to prove results in the real part of mathematics always yield correct results. One can distinguish here two aspects: consistency problem and conservation problem. The consistency problem consists in showing (by finitistic methods, of course) that the infinitistic mathematics is consistent; the conservation problem consists in showing by finitistic methods that any real sentence which can be proved in the infinitistic part of mathematics can be proved also in the finitistic part. One should stress here the emphasis on consistency (instead of correctness).

To realize this program one should formalize mathematical theories (even the whole of mathematics) and then study them as systems of symbols governed by specified and fixed combinatorial rules.

The formal axiomatic system should satisfy three conditions: it should be complete, consistent and based on independent axioms. The consistency of a given system was the criterion for mathematical truth and for the very existence of mathematical objects. It was also presumed that any consistent theory would be categorical, that is, would (up to isomorphism) characterize a unique domain of objects. This demand was connected with the completeness.

The meaning and understanding of completeness by Hilbert plays a crucial rôle from the point of view of our subject. Note at the beginning that in the *Grundlagen der Geometrie* completeness was postulated as one of the axioms (the axiom was not present in the first edition, but was included first in the French translation and then in the second edition of 1903). In fact the axiom V(2) stated that it is not possible to extend the system of points, lines and planes by adding new entities so that the other axioms are still satisfied. In

Hilbert's lecture at the Congress at Heidelberg in 1904 (cf. 1905a) one finds such an axiom system for the real numbers. Later there appears completeness as a property of a system. In lectures "Logische Principien des mathematischen Denkens" (1905, p. 13) Hilbert explains the demand of the completeness as the demand that the axioms suffice to prove all "facts" of the theory in question. He says: "We will have to demand that all other facts (Thatsachen) of the given field are consequences of the axioms". On the other hand one can say that Hilbert's early conviction as to the solvability of every mathematical problem – expressed for example in his 1900 Paris lecture (cf. Hilbert 1901) and repeated in his opening address "Naturerkennen und Logik" (cf. Hilbert 1930a) before the Society of German Scientists and Physicians in Königsberg in September 1930 – can be treated as informal reflection of a belief in completeness.

In his 1900 Paris lecture Hilbert spoke about completeness in the following words (see the second problem): "When we are engaged in investigating the foundations of a science, we must set up a system of axioms which contains an exact and complete description of the relations subsisting between the elementary ideas of that science."

One can take the "exact and complete description" to be complete enough to decide the truth or falsity of every statement. Semantically such completeness follows from categoricity, i.e., from the fact that any two models of a given axiomatic system are isomorphic; syntactically it means that every sentence or its negation is derivable from the given axioms. Hilbert's own axiomatizations were complete in the sense of being categorical. But notice that they were not first-order, indeed his axiomatization of geometry from *Grundlagen* as well as his axiomatization of arithmetic published in 1900 were second-order.

The demand discussed here would imply that a complete (in this sense) system of axioms is possible only for sufficiently advanced theories. On the other hand Hilbert called for complete systems of axioms also for theories being developed. One should also add here that Hilbert admitted the possibility that a mathematical problem may have a negative solution, i.e., that one can show the impossibility of a positive solution on the basis of a considered axiom system (cf. Hilbert 1901).

In Hilbert's lectures from 1917–18 (cf. Hilbert 1917–18) one finds completeness in the sense of maximal consistency, i.e., a system is complete if and only if for any non-derivable sentence, if it is added to the system then the system becomes inconsistent. In his lecture at the International Congress of Mathematicians in Bologna in 1928 Hilbert stated two problems of completeness: one for the first-order predicate calculus (completeness with respect to validity in all interpretations, hence the semantic completeness) and the second for a system of elementary number theory (formal completeness, in the sense of maximal consistency, i.e. Post-completeness, hence the syntactical completeness) (cf. Hilbert 1930).

Hilbert's emphasis on the finitary and syntactical methods together with the demand of (and belief in) the completeness of formal systems seem to be the source and reason of the fact that, as Gödel put it (cf. Wang 1974, p. 9), "[...] formalists considered formal demonstrability to be an *analysis* of the concept of mathematical truth and, therefore were of course not in a position to *distinguish* the two". Indeed, the informal concept of truth was not commonly accepted as a definite mathematical notion at that time. As Gödel wrote in a crossed-out passage of a draft of his reply to a letter of the student Yossef Balas: "[...] a concept of objective mathematical truth as opposed to demonstrability was viewed with greatest suspicion and widely rejected as meaningless" (cf. Wang 1987, 84–85). Therefore Hilbert preferred to deal in his metamathematics solely with the forms of the formulas, using only finitary reasonings which were considered to be safe – contrary to semantical reasonings which were non-finitary and consequently not safe. Non-finitary reasonings in mathematics were considered to be meaningful only to the extent to which they could be interpreted or justified in terms of finitary metamathematics.[1]

On the other hand there was no clear distinction between syntax and semantics at that time. Recall for example that, as indicated earlier, the axiom systems came by Hilbert often with a built-in interpretation. Add also that the very notions necessary to formulate properly the difference syntax–semantics were not available to Hilbert.

The problem of the completeness of the first-order logic, i.e., the fourth problem of Hilbert's Bologna lecture, was also posed as a question in the book by Hilbert and Ackermann *Grundzüge der theoretischen Logik* (1928). It was solved by Kurt Gödel in his doctoral dissertation (1929, cf. also 1930) where he showed that the first-order logic is complete, i.e., every true statement can be derived from the axioms. Moreover he proved that, in the first-order logic, every consistent axiom system has a model. More exactly Gödel wrote that by completeness he meant that "every valid formula expressible in the restricted functional calculus [...] can be derived from the axioms by means of a finite sequence of formal inferences". And added that this is equivalent to the assertion that "Every consistent axiom system [formalized within that restricted calculus] [...] has a realization" and to the statement that "Every logical expression is either satisfiable or refutable" (this is the form in which he actually proved the result). The importance of this result is, according to Gödel, that it justifies the "usual method of proving consistency". One should notice here that the notion of truth in a structure, central to the very definition of satisfiability or validity, was nowhere analyzed in either Gödel's dissertation or his published revision of it. There was in fact a long tradition of use of the informal notion of satisfiability (compare the work of Löwenheim, Skolem and others).

Some months later, in 1930, Gödel solved three other problems posed by Hilbert in Bologna by showing that arithmetic of natural numbers and all richer theories are essentially incomplete (provided they are consistent) (cf. Gödel 1931). It is interesting to see how Gödel arrived at this result.

Gödel himself wrote on his discovery in a draft reply to a letter dated 27th May 1970 from Yossef Balas, then a student at the University of Northern Iowa (cf. Wang 1987, pp. 84–85). Gödel indicated there that it was precisely his recognition of the contrast between the formal definability of provability and the formal undefinability of truth that led him to his discovery of incompleteness. One finds also there the following statement:

> [...] long before, I had found the *correct* solution of the semantic paradoxes in the fact that truth in a language cannot be defined in itself.

On the base of this quotation one can argue that Gödel obtained the result on the undefinability of truth independently of A. Tarski (cf. Tarski 1933).[2]

Note also that Gödel was convinced of the objectivity of the concept of mathematical truth. In a latter to Hao Wang (cf. Wang 1974, p. 9) he wrote:

> [...] it should be noted that the heuristic principle of my construction of undecidable number theoretical propositions in the formal systems of mathematics is the highly transfinite concept of 'objective mathematical truth' as *opposed* to that of 'demonstrability'.

In this situation one should ask why Gödel did not mention the undefinability of truth in his writings. In fact, Gödel even avoided the terms "true" and "truth" as well as the very concept of being true (he used the term "richtige Formel" and not the term "wahre Formel"). In the paper 'Über formal unentscheidbare Sätze [...]' (1931) the concept of a true formula occurs only at the end of Section 1 where Gödel explains the main idea of the proof of the first incompleteness theorem (but again the term "inhaltlich richtige Formel" and not the term "wahre Formel" appears here). Indeed, talking about the construction of a formula which should express its own unprovability invokes the interpretation of the formal system.

On the other hand the term "truth" occurred in Gödel's lectures on the incompleteness theorems at the Institute for Advanced Study in Princeton in the spring of 1934. He discussed there, among other things, the relation between the existence of undecidable propositions and the possibility of defining the concept "true (false) sentence" of a given language in the language itself. Considering the relation of his arguments to the paradoxes, in particular to the paradox of "The Liar", Gödel indicates that the paradox disappears when one notes that the notion "false statement in a language B" cannot be expressed in B. Even more, 'the paradox can be considered as a proof that "false statement in B" cannot be expressed in B.'

What were the reasons of avoiding the concept of truth by Gödel? An answer can be found in a crossed-out passage of a draft of Gödel's reply to the letter of the student Yossef Balas (mentioned already above). Gödel wrote there:

> However in consequence of the philosophical prejudices of our times 1. nobody was looking for a relative consistency proof because [it] was considered axiomatic that a consistency proof must be finitary in order to make sense, 2. a concept of objective mathematical truth as opposed to demonstrability was viewed with greatest suspicion and widely rejected as meaningless.

Hence it leads us to the conclusion formulated by S. Feferman in (1984) in the following way:

> [...] Gödel feared that work assuming such a concept [i.e., the concept of mathematical truth – R.M.] would be rejected by the foundational establishment, dominated as it was by Hilbert's ideas. Thus he sought to extract results from it which would make perfectly good sense even to those who eschewed all non-finitary methods in mathematics.

Though Gödel tried to avoid concepts not accepted by the foundational establishment, his own philosophy of mathematics was in fact platonist. He was convinced that (cf. Wang 1996, p. 83):

> It was the anti-Platonic prejudice which prevented people from getting my results. This fact is a clear proof that the prejudice is a mistake.

Gödel's theorem on the completeness of first-order logic and his discovery of the incompleteness phenomenon together with the undefinability of truth vs. definability of formal demonstrability showed that formal provability cannot be treated as an analysis of truth, that the former is in fact weaker than the latter. It was also shown in this way that Hilbert's dreams to justify classical mathematics by means of finitistic methods cannot be fully realized. Those results together with Tarski's definition of truth (in the structure) and Carnap's work on the syntax of a language led also to the establishing of syntax and semantics in the 1930s.

On the other hand it should be added that Gödel shared Hilbert's "rationalistic optimism" (to use Hao Wang's term) insofar as informal proofs were concerned. In fact Gödel retained the idea of mathematics as a system of truth, which is complete in the sense that "every precisely formulated yes-or-no question in mathematics must have a clear-cut answer" (cf. Gödel 1970). He rejected however – in the light of his incompleteness theorem – the idea that the basis of these truths is their derivability from axioms. In his Gibbs lecture of 1951 Gödel distinguishes between the system of all true mathematical propositions from that of all demonstrable mathematical propositions, calling them, respectively, mathematics in the objective and subjective sense. He claimed also that it is objective mathematics that no axiom system can fully comprise.

Gödel's incompleteness theorems and in particular his recognition of the undefinability of the concept of truth indicated a certain gap in Hilbert's programme and showed in particular, roughly speaking, that (full) truth cannot be comprised by provability and, generally, by syntactic means. The former can be only approximated by the latter. Hence there arose a problem: how should Hilbert's finitistic point of view be extended?

Hilbert in his lecture in Hamburg in December 1930 (cf. Hilbert 1931) proposed to admit a new rule of inference. This rule was similar to the ω-rule, but it had rather informal character (a system obtain by admitting it would be semi-formal). In fact Hilbert proposed that whenever $A(z)$ is a quantifier-free formula for which it can be shown (finitarily) that $A(z)$ is a correct (richtig) numerical formula for each particular numerical instance z, then its universal generalization $\forall x A(x)$ may be taken as a new premise (Ausgangsformel) in all further proofs.

Gödel pointed in many places that new axioms are needed to settle both undecidable arithmetical and set-theoretic propositions. In (1931?, p. 35) he stated that "[...] there are number-theoretic problems that cannot be solved with number-theoretic, but only with analytic or, respectively, set-theoretic methods". And in (1933, p. 48) he wrote: "there are arithmetic propositions which cannot be proved even by analysis but only by methods involving extremely large infinite cardinals and similar things". In (1970) Gödel proposed "cultivating (deepening) knowledge of the abstract concepts themselves which lead to the setting up of these mechanical systems". In (1972) (this paper was a revised and expanded English version of (1958)) Gödel claimed that concrete finitary methods are insufficient to prove the consistency of elementary number theory and some abstract concepts must be used in addition. In the paper (1946) Gödel explicitely called for an effort to use progressively more powerful transfinite theories to derive new arithmetical theorems.

Also Zermelo proposed to allow infinitary methods to overcome restrictions revealed by Gödel. According to Zermelo the existence of undecidable propositions was a consequence of the restriction of the notion of proof to finitistic methods (he said here about "finitistic prejudice"). This situation could be changed if one used a more general "scheme" of proof. Zermelo had here in mind an infinitary logic, in which there were infinitely long sentences and rules of inference with infinitely many premises. In such a logic, he insisted, "*all* propositions are decidable!". He thought of quantifiers as infinitary conjunctions or disjunctions of unrestricted cardinality and conceived of proofs not as formal deductions from given axioms but as metamathematical determinations of the truth or falsity of a proposition. Thus syntactic considerations played no role in his thinking.

To give a rough account of how those suggestions and proposals to extend the finitistic point of view do in fact work let us quote some technical results. We restrict ourselves to the case of the arithmetic of natural numbers, more exactly to Peano arithmetic PA.

Generally speaking one can obtain completions of PA by:

- admitting the ω-rule,
- adding new axioms (in particular reflection principles),
- adding (partial) notion(s) of truth.

Let us start by considering the case of the ω-rule, i.e., of the following rule:

$$\frac{\varphi(0), \varphi(\overline{1}), \varphi(\overline{2}), \ldots, \varphi(\overline{n}), \ldots \quad (n \in \mathbf{N})}{\forall x \varphi(x)}$$

Denote by $(\mathrm{PA})_\omega$ Peano arithmetic PA with the ω-rule. One can easily see that $(\mathrm{PA})_\omega$ is complete – it follows from the fact that its unique model up to isomorphism is the standard model $\mathcal{N}_0 = \langle \mathbf{N}, S, 0, +, \cdot \rangle$. Hence $(\mathrm{PA})_\omega = \mathrm{Th}(\mathcal{N}_0)$.

One can ask: how many times must the ω-rule be applied to obtain a complete extension of PA? To give an answer let us define the following hierarchy of theories where T is any first-order theory in the language L(PA) of Peano arithmetic:

$$\mathrm{T}^0 = \mathrm{T},$$

$$\mathrm{T}^{\alpha+\frac{1}{2}} = \mathrm{T}^\alpha \cup \{\varphi : \varphi \text{ is of the form } \forall x \psi(x) \text{ and } \psi(\overline{n}) \in \mathrm{T}^\alpha \text{ for every } n \in \mathbf{N}\},$$

$\mathrm{T}^{\alpha+1}$ = the smallest set of formulas containing $\mathrm{T}^{\alpha+\frac{1}{2}}$ and closed under the rules of inference of PA,

$$\mathrm{T}^\lambda = \bigcup_{\alpha<\lambda} \mathrm{T}^\alpha \text{ for } \lambda \text{ limit.}$$

One can now prove that

Theorem 1 $\mathrm{Th}(\mathcal{N}_0) = (\mathrm{PA})_\omega = \mathrm{PA}^\omega$.

Recall the hierarchy of formulas of the language L(PA). Let $\Sigma_0^0 = \Pi_0^0 = \Delta_0^0$ be the set of all quantifier free formulas and all formulas with bounded quantifiers. Define Σ_{n+1}^0 to be the set of all formulas of the form $\exists x \psi$ for $\psi \in \Pi_n^0$ and Π_{n+1}^0 to be the set of all formulas of the form $\forall x \psi$ for $\psi \in \Sigma_n^0$. We also define Δ_n^0 as the set of all formulas equivalent (in PA) to a Σ_n^0 formula and to a Π_n^0 formula. One can prove that

Theorem 2 *For every $n \in \mathbf{N}$ the theory PA^n is complete with respect to Σ_{2n+1}^0 sentences.*

In PA one can define partial notions of truth, i.e., one can define satisfaction and truth for formulas of a given class of the arithmetical hierarchy. Denote by $Sat_{\Sigma_n^0}$ the definition of satisfaction for Σ_n^0 formulas; similarly let $Sat_{\Pi_n^0}$ denote the definition of satisfaction for Π_n^0 formulas. Note that $Sat_{\Sigma_0^0}$ and $Sat_{\Pi_0^0}$ are Σ_1^0 formulas and that $Sat_{\Sigma_n^0}$ and $Sat_{\Pi_n^0}$ (for $n \geqslant 1$) are Σ_n^0 and Π_n^0, respectively. Let further $Tr_{\Sigma_n^0}$ and $Tr_{\Pi_n^0}$ denote truth predicates for Σ_n^0 and Π_n^0 sentences.[3] In the sequel we shall identify formulas defining satisfaction and truth and their extensions in the standard model $\mathcal{N}_0$.

The previous theorem can now be formulated as:

$$\mathrm{PA}^n \supseteq \mathrm{PA} + Tr_{\Sigma_{2n+1}^0}.$$

In the definition of the hierarchy T^α no restriction was put on formulas to which the ω-rule was applied. Consider now a hierarchy in which such a restriction is put. So let T be any theory in the language L(PA). Define the following hierarchy of theories (cf. Niebergall 1996):

$$\begin{aligned}
\mathrm{T}^{(0)} &= \mathrm{T}, \\
\mathrm{T}^{(\alpha+\frac{1}{2})} &= \mathrm{T}^{(\alpha)} \cup \{\varphi : \varphi \text{ is of the form } \forall x \psi(x) \text{ and } \psi(x) \in \Sigma_{2\alpha+1}^0 \text{ and } \psi(\overline{n}) \in \\
&\qquad \mathrm{T}^{(\alpha)} \text{ for every } n \in \mathbf{N}\}, \\
\mathrm{T}^{(\alpha+1)} &= \text{the smallest set of formulas containing } \mathrm{T}^{(\alpha+\frac{1}{2})} \text{ and closed under} \\
&\qquad \text{the rules of inference of PA}, \\
\mathrm{T}^{(\lambda)} &= \bigcup_{\alpha<\lambda} \mathrm{T}^{(\alpha)} \text{ for } \lambda \text{ limit.}
\end{aligned}$$

Hence the ω-rule is now applied at stage n to Σ_{2n+1}^0 formulas only.

One has the following

Theorem 3 (Niebergall 1996) *For any $n \in \mathbf{N}$,*

$$\mathrm{PA}^{(n)} = \mathrm{PA} + Tr_{\Sigma_{2n+1}^0}.$$

The above theorems[4] indicate interconnections between Peano arithmetic augmented with the ω-rule and the partial truths. Other connections between them can be formulated in the language of interpretability. So let $\mathrm{S} \preccurlyeq \mathrm{T}$ denote that a theory S is relatively interpretable in the theory T (in the sense of Tarski (1953)). We have now the following facts (cf. Niebergall 1996):

Theorem 4 *Let* Con_S *(for an appropriate theory S) denote a statement of* L(PA) *stating that S is consistent. Then*

$$\mathrm{PA}^n \preccurlyeq \mathrm{PA} + Tr_{\Sigma_{2n+1}^0} + \mathrm{Con}_{\mathrm{PA}^n},$$

$$\mathrm{PA}^n \preccurlyeq \mathrm{PA} + Tr_{\Sigma_{2n+2}^0},$$

$$\mathrm{PA} + \mathrm{Con}_{\mathrm{PA}+Tr_{\Sigma^0_n}} \preccurlyeq \mathrm{PA} + Tr_{\Sigma^0_2},$$
$$\mathrm{PA} + \mathrm{Con}_{\mathrm{PA}^n} \preccurlyeq \mathrm{PA} + Tr_{\Sigma^0_2}.$$

The above theorems indicate that the arithmetical truth, i.e., the set $\mathrm{Th}(\mathcal{N}_0)$ of all arithmetical sentences true in the standard model $\mathcal{N}_0$, can be approximated by syntactical methods, i.e., by demonstrability – though not by finitary means (one uses here the ω-rule).

So far we have considered Peano arithmetic and partial truths. Ask now: what about PA and the full truth? Gödel's and Tarski's theorem shows that the truth predicate for L(PA) cannot be defined in PA. But one can extend the language L(PA) by adding a new binary predicate S called satisfaction class and characterizing it axiomatically by adding to Peano arithmetic PA axioms being an appropriate modification of Tarski's definition of satisfaction (cf. Krajewski (1976), where this notion was introduced, or Murawski (1997)). Note that since those axioms form a finite set of axioms one can write them as a single formula of the language $\mathrm{L(PA)} \cup S$. Denote this theory as PA + "S is a satisfaction class". One can extend this theory by adding new axioms stating special properties of S. In particular one can demand that S is full, i.e., S decides any formula of L(PA) on any valuation or that S is Γ-inductive for Γ being a given class of formulas of the language $\mathrm{L(PA)} \cup S$, i.e., that the induction axiom holds for all formulas of the class Γ (if Γ is the class of all formulas of $\mathrm{L(PA)} \cup S$ then one says that the satisfaction class S is inductive).

Since theories T of the indicated type are extensions of PA one can ask what about natural numbers can be proved in T, i.e., one can consider theories of the type

$$\mathrm{PA}^{\mathrm{T}} = \{\varphi \in \mathrm{L(PA)} : \mathrm{T} \vdash \varphi\}.$$

Theorems of PA^{T} are those sentences of the language L(PA) of Peano arithmetic (hence sentences about natural numbers) which can be proved in the stronger theory T. A natural problem of finding an axiomatics of the theory PA^{T} arises.

One can easily see that the following theories are conservative extensions of PA:

(a) PA + "S is a satisfaction class",

(b) PA + "S is a full satisfaction class",

(c) PA + "S is an inductive satisfaction class".

This means that one can prove in those theories exactly the same theorems about natural numbers (i.e., formulas of the language L(PA)) as in Peano arithmetic PA. Hence the addition of a new notion, i.e., of a notion of a satisfaction class (and consequently a notion of truth), with properties indicated in (a)–(c) does not increase the prooftheoretical power of a theory with respect to sentences of the language L(PA). On the other hand the assumption that a satisfaction class is full and Δ^0_0-inductive gives a nonconservative extension of PA. In fact one can prove in this theory the consistency of PA.

The theories PA^{T} for T being PA + "S is a full (Σ^0_m-)inductive satisfaction class" can be characterized by transfinite induction or the consistency of appropriate ω-logics. Denote by $\Gamma - \mathrm{PA}(S)$ the theory PA + "S is a full Γ-inductive satisfaction class" and by PA(S) the theory PA + "S is a full inductive satisfaction class".

Consider the following sequence of formulas of the language L(PA) (one uses here arithmetization):

$$\Gamma_0(\varphi) = \text{“PA} \vdash \varphi\text{“},$$

$$\Gamma_{n+\frac{1}{2}}(\varphi) = \text{“}\varphi \textit{ is of the form } \eta \vee \forall z \psi(z) \textit{ and } \forall z \Gamma_n(\eta \vee \psi(S^z 0))\text{“},$$

$$\Gamma_{n+1}(\varphi) = \text{“}\textit{there exists a proof of the formula } \varphi \textit{ based on } \mathrm{PA} \cup \{\psi : \Gamma_{n+\frac{1}{2}}(\psi)\}\text{“}.$$

Observe that in this system of ω-logic only the application of the ω-rule increases the degree of complexity of a proof.

Theorem 5 (Kotlarski 1986) $\mathrm{PA}^{\Delta^0_0 - \mathrm{PA}(S)} = \mathrm{PA} \cup \{\neg \Gamma_n(0 = 1) : n \in \mathbf{N}\}$.

It can also be proved (cf. Kotlarski 1986) that the theory $\Delta^0_0 - \mathrm{PA}(S)$ is equal to the theory

$$\mathrm{PA} + S \text{ is a full satisfaction class} + \forall \varphi[(\mathrm{PA} \vdash \varphi) \rightarrow S(\varphi)].$$

The last sentence can be read as: "S makes all theorems of PA true". It is equivalent to the Δ^0_0-inductiveness of the satisfaction class S.

The system of ω-logic described above can be iterated in the transfinite and one can axiomatise theories $\mathrm{PA}^{\Sigma^0_m - \mathrm{PA}(S)}$ ($m \in \mathbf{N}$) and $\mathrm{PA}^{\mathrm{PA}(S)}$ by consistency statements of appropriate systems of this logic (cf. Kotlarski and Ratajczyk 1990).

Define for an ordinal α a sequence $\omega_m(\alpha)$ in the following way: $\omega_0(\alpha) = \alpha$, $\omega_{m+1}(\alpha) = \omega^{\omega_m(\alpha)}$ and put $\omega_n = \omega_n(\omega)$. Let now $TI(\rho)$, where ρ is an ordinal, denote the scheme of transfinite induction up to ρ. Then the following theorem holds.

Theorem 6 (Kotlarski and Ratajczyk 1990a) *Let m be a natural number. Then*

$$PA^{\Sigma^0_m - PA(S)} = PA \cup \{TI(\varepsilon_{\omega_m(k)}) : k \in \mathbf{N}\}$$

$$PA^{PA(S)} = PA \cup \{TI(\varepsilon_{\omega_k}) : k \in \mathbf{N}\}.$$

The above theorems show how strong is Peano arithmetic augmented with an appropriate notion of satisfaction (and truth). One can see that only by assuming that the added notion of satisfaction (truth) is full and at least Δ^0_0-inductive one obtains a proper extension of PA. It is interesting that such extensions are equivalent to PA extended by appropriate forms of transfinite induction or by the statements of the consistency of appropriate systems of ω-logic. In other words, the above theorems show in particular that what can be proved about natural numbers using Peano axioms and the notion of satisfaction (truth) that is assumed to be full and Σ^0_m-inductive is exactly the same as what can be proved in PA plus transfinite induction for ordinals $\varepsilon_{\omega_m(k)}$ (for all $k \in \mathbf{N}$) or in PA plus appropriate consistency statements. Similarly for PA plus full inductive satisfaction (truth) on the one hand and PA plus transfinite induction for ordinals ε_{ω_k} (for all $k \in \mathbf{N}$) or PA plus appropriate consistency statements on the other. They show also that by adding to PA the notion of satisfaction (truth) and assuming that it is full and makes all theorems of PA true one obtains a theory with exactly the same theorems about natural numbers as by taking PA augmented with a concept of a full and Δ^0_0-inductive satisfaction (truth) or PA plus appropriate consistency statements.

In the above considerations we restricted ourselves to formal proofs and to the semantical notion of truth *in* mathematics. We tried to show how was developing the awareness of differences between them (from the hopes that formal proofs provide sufficient means to exhaust the mathematical truth to the discovery of various limitations of them). Let us finish with some general remarks.

Concepts of proof and truth are (even in mathematics) ambiguous. One should distinguish between working proofs of everyday mathematics and idealized formal proofs used by logicians. On the other hand a proof in mathematics has various aspects and can be studied from various points of view. One can distinguish psychological, social, cultural and logical aspects of proofs. A proof can be studied as a mathematical or as an epistemological object. The former is precisely defined on the basis of mathematical logic, the latter is a vague concept. The former is an idealization of proofs occurring in a research practice of mathematicians, is a reconstruction of them. Recently one can observe in the philosophy of mathematics a tendency to concentrate on the actual research practice of mathematicians rather than on idealized foundational reconstructions of it and consequently to study the methods actually used by mathematicians.

Similar distinctions can be made with respect to the concept of truth. The semantical concept of truth precisely defined by Tarski is in fact a mathematical notion. It provides a definition of truth *in* mathematics,[5] it is the concept of truth for a model in a formal language (its essential feature is to define truth in terms of reference or satisfaction on the basis of a particular kind of syntactico-semantical analysis of the language). But one can also speak about epistemic truth – cf. for example Isaacson (1987) and (1992) where it is argued that Peano arithmetic is complete with respect to an epistemic notion of arithmetical truth.

The distinction between proof and truth in mathematics presupposes of course some philosophical assumptions. In fact for pure formalists and for intuitionists there exists no truth/proof problem. For them a mathematical statement is true just in case it is provable, and proofs are syntactic or mental constructions of our own making. In the case of a platonist (realist) philosophy of mathematics the situation is different. One can say that platonist approach to mathematics enabled Gödel to state the problem and to be able to distinguish between proof and truth, between syntax and semantics.[6]

Notes

1. Cf. Gödel's letter to Hao Wang dated 7th December 1967 – see Wang (1974), p. 8.
2. For the problem of the priority of proving the undefinability of truth see Woleński (1991) and Murawski (1998).
3. Construction of $Sat_{\Sigma^0_n}$ and $Sat_{\Pi^0_n}$ can be found in Kaye (1991) and Murawski (1999).
4. Note that many of those theorems hold not only for Peano arithmetic PA but for a broad class of theories – cf. Niebergall (1996).
5. One should distinguish the truth *in* mathematics and the truth *of* mathematics.
6. Note that, as indicated above, Hilbert was not interested in philosophical questions and did not consider them.

References

Feferman, S.: (1984), 'Kurt Gödel: conviction and causation', in *Philosophia Naturalis* **1**, a special issue: Weingartner, P. et al. (eds.), *Philosophy of Science – History of Science. A Selection of Contributed Papers of the 7th International Congress of Logic, Methodology and Philosophy of Science, Salzburg, 1983*, Verlag Anton Hain, Meisenheim/Glan, pp. 546–562. Reprinted in Shanker, S.G. (ed.), *Gödel's Theorem in Focus*, Croom Helm, London 1988, pp. 96–114.

Frege, G.: (1976), *Wissenschaftlicher Briefwechsel*, transl. by G. Gabriel, H. Hermes, F. Kambartel, Ch. Thiel, A. Veraart, Felix Meiner Verlag, Hamburg.

Gödel, K.: (1929), 'Über die Vollständigkeit des Logikkalküls', doctoral dissertation, submitted in 1929; published and translated in K. Gödel, *Collected Works*, vol. I, ed. by S. Feferman, et al., Oxford University Press, New York and Clarendon Press, Oxford, 1986, pp. 60–101.

Gödel, K.: (1930), 'Die Vollständigkeit der Axiome des logischen Funktionenkalküls', *Monatshefte für Mathematik und Physik* **37**, pp. 349–360.

Gödel, K.: (1931), 'Über formal unentscheidbare Sätze der 'Principia Mathematica' und verwandter Systeme. I', *Monatshefte für Mathematik und Physik* **38**, pp. 173–198.

Gödel, K.: (1931?), 'Über unentscheidbare Sätze', first published (German text and English translation 'On undecidable sentences') in K. Gödel, *Collected Works*, vol. III, ed. by S. Feferman, et al., Oxford University Press, New York and Oxford, 1995, pp. 30–35.

Gödel, K.: (1933), 'The present situation in the foundations of mathematics'; first published in K. Gödel, *Collected Works*, vol. III, ed. by S. Feferman, et al., Oxford University Press, New York and Oxford, 1995, pp. 45–53.

Gödel, K.: (1934), *On undecidable propositions of formal mathematical systems* (mimeographed lecture notes, taken by S.C. Kleene and J.B. Rosser), Princeton; reprinted with revisions in M. Davis (ed.), *The Undecidable: Basic Papers on Undecidable Propositions, Unsolvable Problems, and Computable Functions*, Raven Press, Hewlett, N.Y., 1965, pp. 39–74.

Gödel, K.: (1946), 'Remarks before the Princeton Bicentennial Conference on Problems in Mathematics', 1–4; first published in Davis, M. (ed.), *The Undecidable: Basic Papers on Undecidable Propositions, Unsolvable Problems, and Computable Functions*, Raven Press, Hewlett, N.Y., 1965, pp. 84–88.

Gödel, K.: (1951), 'Some basic theorems on the foundations of mathematics and their implications'; first published in K. Gödel, *Collected Works*, vol. III, ed. by S. Feferman, et al., Oxford University Press, New York and Oxford, 1995, pp. 304–323.

Gödel, K.: (1958), 'Über eine bisher noch nicht benützte Erweiterung des finiten Standpunktes', *Dialectica* **12**, pp. 280–287.

Gödel, K.: (1970), 'The modern development of the foundations of mathematics in the light of philosophy'; first published (German text and English translation) in K. Gödel, *Collected Works*, vol. III, ed. by S. Feferman, et al., Oxford University Press, New York and Oxford, 1995, pp. 374–387.

Gödel, K.: (1972), 'On an extension of finitary mathematics which has not yet been used', revised and expanded English version of Gödel (1958), to have appeared in *Dialectica*, first published in K. Gödel *Collected Works*, vol. II, ed. by S. Feferman, et al., Oxford University Press, New York-Oxford, 1990, pp. 271–280.

Hilbert, D.: (1899), *Grundlagen der Geometrie. Festschrift zur Feier der Enthüllung des Gauss-Weber-Denkmals*, B.G. Teubner, Leipzig, pp. 3–92.

Hilbert, D.: (1900), 'Über den Zahlbegriff', *Jahresbericht der Deutschen Mathematikervereinigung* **8**, pp. 180–184.

Hilbert, D.: (1901), 'Mathematische Probleme', *Archiv der Mathematik und Physik* **1**, pp. 44–63 and 213–237.

Hilbert, D.: (1902/03), 'Über den Satz von der Gleichheit der Basiswinkel im gleichschenkligen Dreieck', *Proceedings of the London Mathematical Society* **35**, pp. 50–68.

Hilbert, D.: (1903), *Grundlagen der Geometrie*, 2nd edition, Teubner, Leipzig.

Hilbert, D.: (1905), 'Logische Principien des mathematischen Denkens', Lecture notes by Ernst Hellinger, Mathematisches Institut, Georg-August-Universität Göttingen, Sommer-Semester 1905, unpublished manuscript.

Hilbert, D.: (1905a), 'Über die Grundlagen der Logik und der Arithmetik', in A. Krazer (ed.), *Verhandlungen des dritten Internationalen Mathematiker-Kongresses in Heidelberg vom 8. bis 13. August 1904*, Teubner, Leipzig, pp. 174–185.

Hilbert, D.: (1917–18), *Prinzipien der Mathematik, Lecture notes by Paul Bernays*, Mathematisches Institut, Georg-August-Universität Göttingen, Wintersemester 1917–18, unpublished typescript.

Hilbert, D.: (1918), 'Axiomatisches Denken', *Mathematische Annalen* **78**, pp. 405–415.

Hilbert, D.: (1930), 'Probleme der Grundlegung der Mathematik', *Mathematische Annalen* **102**, pp. 1–9.

Hilbert, D.: 1930a, 'Naturerkennen und Logik', *Naturwissenschaften* **18**, pp. 959–963.

Hilbert, D.: (1931), 'Die Grundlegung der elementaren Zahlentheorie', *Mathematische Annalen* **104**, pp. 485–494.

Hilbert, D., Ackermann, W.: (1928), *Grundzüge der theoretischen Logik*, Verlag von Julius Springer, Berlin.

Isaacson, D.: (1987), 'Arithmetical truth and hidden higher-order concepts', in The Paris Logic Group (eds.), *Logic Colloquium'85*, Elsevier Science Publishers B.V., (North-Holland), Amsterdam, pp. 147–169.

Isaacson, D.: (1992), 'Some considerations on arithmetical truth and the ω-rule', in M. Detlefsen (ed.), *Proof, Logic and Formalization*, Routledge, London and New York, pp. 94–138.

Kaye, R.: (1991), *Models of Peano Arithmetic*, Clarendon Press, Oxford.

Kotlarski, H.: (1986), 'Bounded induction and satisfaction classes', *Zeitschrift für Mathematische Logik und Grundlagen der Mathematik* **32**, pp. 531–544.

Kotlarski, H., Ratajczyk, Z.: (1990), 'Inductive full satisfaction classes', *Annals of Pure and Applied Logic* **47**, 199–223.

Kotlarski, H., Ratajczyk, Z.: (1990a), 'More on induction in the language with a full satisfaction class', *Zeitschrift für Mathematische Logik und Grundlagen der Mathematik* **36**, pp. 441–454.

Krajewski, S.: (1976), 'Non-standard satisfaction classes', in W. Marek, M. Srebrny, A. Zarach (eds.), *Set Theory and Hierarchy Theory*, Proc. Bierutowice Conf. 1975, Lecture Notes in Mathematics 537, Springer Verlag, Berlin–Heidelberg–New York, pp. 121–144.

Murawski, R.: (1997), 'Satisfaction classes – a survey', in R. Murawski, J. Pogonowski (eds.), *Euphony and Logos*, Editions Rodopi, Amsterdam–Atlanta, GA, pp. 259–281.

Murawski, R.: (1998), 'Undefinability of truth. The problem of the priority: Tarski vs Gödel', *History and Philosophy of Logic* **19**, pp. 153–160.

Murawski, R.: (1999), *Recursive Functions and Metamathematics*, Kluwer Academic Publishers, Dordrecht/Boston/London.

Niebergall, K.G.: (1996), *Zur Metamathematik nichtaxiomatisierbarer Theorien*, Centrum für Informations- und Sprachverarbeitung Ludwig-Maximilians-Universität München, Bericht 96–87.

Rowe, D.E.: (1989), 'Klein, Hilbert, and the Göttingen mathematical tradition', *Osiris* **(2) 5**, pp. 186–213.

Tarski, A.: (1933), *Pojęcie prawdy w językach nauk dedukcyjnych* (The Notion of Truth in Languages of Deductive Sciences), Nakładem Towarzystwa Naukowego Warszawskiego, Warszawa. Eng. trans. as 'The Concept of Truth in Formalized Languages', in *Logic,*

Semantics, Metamathematics: Papers from 1923 to 1938 by Alfred Tarski, Oxford University Press, Oxford 1956; second edition with corrections and emendations, ed. by J. Corcoran, Hackett, Indianapolis, IN, 1983, pp. 152–278.

Tarski, A.: (1953), 'A general method in proofs of undecidability', in A. Tarski, A. Mostowski, R.M. Robinson, *Undecidable Theories*, North-Holland Publ. Comp., Amsterdam 1953, pp. 1–35.

Wang, Hao: (1974), *From Mathematics to Philosophy*, Routledge and Kegan Paul, London.

Wang, Hao: (1987), *Reflections on Kurt Gödel*, MIT Press, Cambridge, Mass.

Wang, Hao: (1996), *A Logical Journey. From Gödel to Philosophy*, MIT Press, Cambridge, Mass. and London, England.

Woleński, J.: (1991), 'Gödel, Tarski and the undefinability of truth', in *Yearbook 1991 of the Kurt Gödel Society*, Wien, pp. 97–108.

X

PHILOSOPHY OF THE PHYSICAL SCIENCES

TOPOS THEORY AS A FRAMEWORK FOR PARTIAL TRUTH

Jeremy Butterfield
All Souls College
Oxford OX1 4AL
jb56@cus.cam.ac.uk

Abstract This paper develops some ideas from previous work (coauthored, mostly with C.J. Isham). In that work, the main proposal is to assign as the value of a physical quantity in quantum theory (or classical physics), not a real number, but a certain kind of set (a sieve) of quantities that are functions of the given quantity. The motivation was in part physical – such a valuation illuminates the Kochen-Specker theorem; in part mathematical – the valuations arise naturally in the theory of presheaves; and in part conceptual – the valuations arise from applying to propositions about the values of physical quantities some general axioms governing partial truth for any kind of proposition.

In this paper, I give another conceptual motivation for the proposal. I develop (in Sections 2 and 3) the notion of a topos (of which presheaves give just one kind of example); and explain how this notion gives a satisfactory general framework for making sense of the idea of partial truth. Then I review (in Section 4) how our proposal applies this framework to the case of physical theories.

1. Introduction

This paper develops some ideas from previous work (coauthored, mostly with C.J. Isham).[1] In that work, the main proposal is to assign as the value of a physical quantity in quantum theory (or classical physics), not a real number, but a certain kind of set (a sieve) of quantities that are functions of the given quantity.

In Isham and Butterfield (1998), we motivated this proposal in terms of (i) its relevance to foundations of quantum theory – such a valuation illuminates the Kochen-Specker theorem; and (ii) its mathematical naturalness – the valuations arise naturally in the theory of presheaves, which are a natural generalization of the notion of a set. Roughly speaking, a presheaf is a set whose membership varies from one context to another; and in that paper, a context was given by a quantum-theoretic quantity (a bounded self-adjoint operator). In Butter-

P. Gärdenfors, J. Woleński and K. Kijania-Placek (eds.), In the Scope of Logic, Methodology and Philosophy of Science, Vol. I, 307-329.

field and Isham (1999), we developed some conceptual motivations for the proposal. We explained how the proposed valuations arise much more generally than just for quantum physics. In particular, they arise just as naturally in classical physics; and more generally, they arise from applying to propositions about the values of physical quantities some general axioms governing partial truth for any kind of proposition. In Hamilton, Isham and Butterfield (1999), we developed the proposal for a still more general notion of context: a context was taken, not as a physical quantity, but as a commutative von Neumann algebra. In all three papers, an analogue for sieve-valued valuations of the famous *FUNC* condition, familiar from discussions of the Kochen-Specker theorem, played a central role.

The aim of this paper is to give another conceptual motivation for the proposal, which again relates to the idea of partial truth. Our previous papers discussed partial truth entirely within the framework of presheaves (even in the most general Sections of Butterfield and Isham (1999)). But presheaves provide just one kind of example, among many, of the mathematical notion of a topos. A topos is a sort of category. Very roughly, it is a category that behaves much like the category of sets; indeed, this category, which I will call $\mathcal{S}$, is itself a topos. The exact definition of a topos is complicated; but in fact, I will not need it in this paper. For my purposes, the main point is that in general – not just for presheaves – the notion of a topos provides a good framework for making sense of the idea of partial truth. Thus while $\mathcal{S}$ has just the two usual truth-values, $\{0, 1\}$, a topos in general has a larger collection of truth-values. And just as $\{0, 1\}$ is itself a set, i.e. an object in the category $\mathcal{S}$, so also in any topos the collection of truth-values is an object in the topos: it is called 'the subobject classifier', and written Ω.

This point – that in any topos, one can make good sense of partial truth – is well-known to logicians working in category theory and topos theory. But it is largely unknown among philosophers and physicists, even those sympathetic to the idea of somehow 'going beyond' the usual classical two truth-values. So it seems worthwhile to expound it; and all the more worthwhile to do so using two simple examples of toposes (viz. the category of endomaps, and the category of graphs) that provide simple analogues of the previous papers' proposal. I undertake this project in Sections 2, 3 and 4; which discuss, respectively, categories in general, toposes in general, and our proposal – to define partial truth-values in quantum theory and other physical theories, using toposes of presheaves.

I should stress at the outset four features of this framework for partial truth. All but the second are features that apply in any topos.

1. First, one common reason for being suspicious of notions of partial truth, and of many-valued logics, is the idea that the proposed notions are arbi-

trary: that other definitions, e.g. about how to define truth-functional connectives, would be just as well motivated by the logico-semantic phenomena or intuitions appealed to, as are the proposed definitions. But this reason does not apply here. For as we shall see, in any topos, the collection of generalized truth-values (the sub-object classifier Ω) is completely fixed by the structure of the topos; (and in general, there is no other topos with a similar but different Ω that one can argue to be as well motivated as the given one). In short, the many-valued logics that arise in toposes are *not* arbitrary.

2. In the toposes I shall be concerned with (viz. the topos of endomaps, the topos of graphs, and toposes of presheaves) the generalized truth-values (the elements of Ω) are natural in another sense: they are the answers to a natural 'multiple-choice question' about the objects in the topos. Besides, it will be a question whose answers exhibit a neat meshing between the various degrees or kinds of partial truth, and total truth; (where total truth can be roughly identified with classical truth, $1 \in \{0, 1\}$, and with the designated truth-value of traditional many-valued logics). For example, for graphs, the question will be, roughly speaking, 'What part of the given arc (i.e. arrow in a graph) lies within the given subgraph?'; and the answer will be closer to total truth, the more of the given arc is in the subgraph. Furthermore, in toposes of presheaves (the topic of our main proposal), this meshing is even stronger: the partial truth-values are completely controlled by the notion of total truth (in a precise sense discussed in Butterfield and Isham (1999)).

   So much by way of addressing general suspicion of notions of partial truth. The next two features are more formal; (and for reasons given at the start of Section 2, I will not discuss them in detail in this paper).

3. Any topos has notions of sum and product, which generalize the notions familiar from arithmetic and logic; with the sum and product obeying the familiar distributive law. So in relation to quantum theory, our proposal's retention of the distributive law sets us apart from the dominant tradition of quantum logic. (I will not here try to defend this departure from tradition; but only note that it has been argued, against this tradition, that the distributive law is indispensible to genuinely logical reasoning.)

4. On the other hand, our proposals do involve non-Boolean structure. More precisely: Any topos allows the definition of logical operations of 'and', 'or' and 'not' on its set of truth-values, in such a way that they obey an intuitionistic logic. For some toposes, like the category of sets, they also obey the other principles of classical logic; but *not* for our proposal's toposes of presheaves.

2. Category theory

In this Section and the next, I develop those aspects of the notion of a topos, that show how it provides a framework for discussing partial truth. More specifically, in Section 3 I will present the clause in the definition of a topos that requires the existence of the subobject classifier, Ω. To understand this clause, and how Ω functions as a collection of generalized truth-values, requires familiarity with some other notions of category theory: among them, the notions of monic, terminal object, point (also called 'global element') and generalized element. So in this Section, I first introduce the idea of a category and my examples of categories, and then present these notions (in the order listed).

Two warnings about the limitations of what follows. (1) A full understanding of 'how Ω functions' of course requires the full definition of a topos, which I cannot give here, for reasons of space. For the definition includes clauses requiring the existence for any pair of objects A,B in the category, of certain other objects, such as a product $A \times B$ and an exponential A^B; where these notions generalize the set-theoretic notions, and are characterized wholly in terms of the existence and uniqueness of certain arrows (morphisms) between the objects – but also take some space to state! But fortunately, we can see a good deal about 'how Ω functions', without these clauses. More specifically, they ensure that the generalized truth-values have an algebraic structure so as to obey an intuitionistic logic, including the distributive law; (as discussed in items 3. and 4. of Section 1). But even while neglecting this algebraic and logical structure, we will be able to see clearly how Ω encodes different degrees of partial truth.

(2) Space limitations also prevent my giving proofs, but they are all elementary and in the standard textbooks. I also cannot give many examples of categories (or diagrams). But I will take as examples, in addition to $\mathcal{S}$ and categories of presheaves, the category of endomaps and the category of graphs. These are good examples not just because they are toposes. As we shall see, their subobject classifiers (i) can be drawn vividly – for example, in the category of graphs, the object Ω is itself a graph; and (ii) provide simple analogues of the (perhaps off-puttingly complicated!) subobject classifier in a topos of presheaves.[2]

So I shall present in the following subsections: (1) the definition of a category, and my examples other than toposes of presheaves; (2) monics (presented as part of a discussion of inverses); (3) terminal objects, points and generalized elements.

2.1. Categories; examples

A category consists of a collection of *objects* and a collection of *arrows* (or *morphisms*), with the following three properties. (1) Each arrow f is associated with a pair of objects, known as its 'domain' (dom f) and the 'codomain' (cod f), and is written in the form $f : B \to A$ where $B = \mathrm{dom} f$ and $A = \mathrm{cod} f$. (2) Given two arrows $f : B \to A$ and $g : C \to B$ (so that the codomain of g is equal to the domain of f), there is a composite arrow $f \circ g : C \to A$; and this composition of arrows obeys the associative law. (3) Each object A has an identity arrow, $id_A : A \to A$, with the properties that for all $f : B \to A$ and all $g : A \to C$, $id_A \circ f = f$ and $g \circ id_A = g$.

I have already mentioned the prototype category (indeed, topos) $\mathcal{S}$, in which the objects are sets and the arrows are ordinary functions (set-maps) between them. In many categories, the objects are sets equipped with some type of additional structure, and the arrows are functions that preserve this structure. An obvious algebraic example is the category of groups, where an object is a group, and an arrow $f : G_1 \to G_2$ is a group homomorphism from G_1 to G_2. More generally, one often defines one category in terms of another; and in such a case, there is often only one obvious way of defining composition and identity arrows for the new category.

In addition to the category $\mathcal{S}$ of sets, I take as examples (both of which are toposes):

1. The category $\mathcal{E}$ of endomaps of sets. That is: starting with the category $\mathcal{S}$ of sets, with functions between sets as arrows, one defines:
   (i) an object of $\mathcal{E}$ is to be a set X equipped with an endomap α (*i.e.* α is an ordinary function on X, $\alpha : X \to X$); I write such an object as $(X;\alpha)$;
   (ii) an arrow of $\mathcal{E}$ is to be a function that preserves the endomap structure, *i.e.* an arrow from $(X;\alpha)$ to $(Y;\beta)$ is a function $f : X \to Y$ such that $f \circ \alpha = \beta \circ f$;
   and then one shows that with the obvious definitions of composition and of identity functions, the associative and identity laws hold.

2. The category $\mathcal{G}$ of (irreflexive, directed) graphs.
   (i) An object is a pair of sets, R and D, equipped with a pair of functions, $s : R \to D$ and $t : R \to D$, where R is called the set of *arcs* (or *arrows*) of the graph, D the set of *dots* of the graph, and s and t are called the *source* and *target* functions. So for an arc $r \in R$, $s(r) \in D$ is the dot that is the source of r, and $t(r)$ is its target. I shall write an object as $(s,t : R \to D)$.
   (ii) Intuitively, an arrow of $\mathcal{G}$ (not to be confused with an arc in a graph!) is to be a function that 'preserves the graph structure', mapping arcs to arcs and dots to dots in such a way as to preserve source and target relations between arcs and dots. That is, an arrow from $(s,t : R \to D)$ to $(s',t' : R' \to D')$ is a pair of functions $f_{Ar} : R \to R'$ and $f_{Do} : D \to D'$, for which both:

$$f_{Do} \circ s = s' \circ f_{Ar}; \text{ and } f_{Do} \circ t = t' \circ f_{Ar}. \quad (2.1)$$

And again one shows that with the obvious definitions of composition and of identity functions, the associative and identity laws hold.

2.2. Inverses, monics and epics

In any category, an arrow $f : A \to B$ is called *invertible* (or *an isomorphism*, or *iso*) iff there is an arrow $g : B \to A$ such that $g \circ f = id_A$ and $f \circ g = id_B$. Such a g is unique, and is called the *inverse* of f, written f^{-1}; and then A and B are called *isomorphic*. In $\mathcal{S}$, objects i.e. sets are isomorphic iff they have the same number of elements; in $\mathcal{E}$ and $\mathcal{G}$, objects are isomorphic iff they 'have the same diagram' (in a sense that can straightforwardly be made precise in terms of functions). If f has an inverse, then f satisfies:

$$\text{If } f \circ h = f \circ k, \text{then } h = k; \text{'left cancellation'}; \tag{2.2}$$

$$\text{If } h \circ f = k \circ f, \text{then } h = k; \text{'right cancellation'}. \tag{2.3}$$

I turn to the notion of a monic, and the corresponding ('dual') notion of an epic; these are defined in terms of the cancellation laws. By way of motivation, I will also note the connection to 'one-sided inverses', i.e. retractions and sections.[3]

An arrow $f : A \to B$ is called *injective for arrows from T* iff for any pair of arrows $x_1 : T \to A$ and $x_2 : T \to A$, if $f \circ x_1 = f \circ x_2$ then $x_1 = x_2$. If f is injective for arrows from T to A, for every T, we say f is *injective*, or a *monomorphism*, or a *monic*. In other words, f is called monic iff f obeys left-cancellation, Eq. (2.2).

In $\mathcal{S}$, this definition is equivalent to the usual meaning of 'injective' as the function being one-to-one. In $\mathcal{E}$, with objects $(X; \alpha)$, the monics from $(X; \alpha)$ to $(Y; \beta)$ are the arrows, i.e. functions from X to Y, that (i) are one-to-one as functions and (ii) have a range closed under β. (Requirement (ii) follows just from the definition of an arrow in $\mathcal{E}$.) Similarly in $\mathcal{G}$, with objects $(s, t : R \to D)$: a monic is a pair of functions, f_{Ar} on the arrows and f_{Do} on the dots, that are each one-to-one as functions; where again, the requirement that the pair preserve the structure of the graph that is the domain of the arrow follows just from the pair's being an arrow in $\mathcal{G}$. Thus the 'copy' of the domain-graph $A = (s, t : R \to D)$ that lies within the codomain-graph $B = (s', t' : R' \to D')$ can have more structure than A. It can have more arcs between 'a' pair of dots than A does, and it need not be a disconnected component of B.

A simple sufficient condition for being monic is given by the notion of a retraction. Given $f : A \to B$, a *retraction* for f is a 'left-inverse' of f, i.e. an arrow $r : B \to A$ such that $r \circ f = id_A$. It follows that if $f : A \to B$ has a retraction, then f is a monic.

In $\mathcal{S}$, the converse of this result is 'almost true' in that the only exception involves the empty set as domain. That is to say, if $f : A \to B$ is an injective function, and $A \neq \emptyset$, then f has a retraction, i.e. there are maps $r : B \to A$ such that $r \circ f = id_A$. (So the restriction to A non-empty is due to the fact that while there is a function, called the empty function, with domain $\emptyset$ and any codomain, there is no function with non-empty domain and $\emptyset$ as its codomain.)

But in a general category, the converse is not true: having a retraction is stronger than being monic. The simplest counterexample is a category with just two objects, say A and B, and one arrow between them, say from A to B (apart from the identity arrows on A and on B). This arrow is monic (for only id_A can be 'done before' it), but has no retraction. (Exercise! Think of a simple counterexample in $\mathcal{E}$ or in $\mathcal{G}$.)

The corresponding ('dual') definitions and result for surjectivity, rather than injectivity, are as follows. An arrow $f : A \to B$ is called *surjective for arrows to T* iff for any pair of arrows $x_1 : B \to T$ and $x_2 : B \to T$, if $x_1 \circ f = x_2 \circ f$ then $x_1 = x_2$. If f is surjective for arrows from B to T, for every T, we say f is *surjective*, or an *epimorphism* or *an epic*. In other words, f is called epic if f obeys right-cancellation, Eq (2.3). In $\mathcal{S}$, this definition is equivalent to the usual meaning of 'surjective'. (Exercise: Characterize the epics in $\mathcal{E}$ or in $\mathcal{G}$.)

A simple sufficient condition for being epic is given by the notion of a section. Given $f : A \to B$, a *section* for f is a 'right-inverse', i.e. a map $s : B \to A$ such that $f \circ s = id_B$. It follows that if $f : A \to B$ has a section, then f is an epic.

As before, the converse of this result is 'almost true' in $\mathcal{S}$, in that the only exception involves the empty set. But in a general category, the converse is not true: having a section is stronger than being epic. The simple example above will do: the category with just two objects, say A and B, and one arrow between them, say from A to B (apart from the identity arrows). This arrow is epic (for only id_B can be 'done after' it), but has no section.

Finally, on the relation between retractions and sections: If f has both a retraction r and a section s, then $r = s = f^{-1}$. But f can be both monic and epic, without being iso. Again, the simplest counterexample is the category with just two objects, say A and B, and one arrow between them, say from A to B.

2.3. Terminal objects, points and generalized elements

In this Subsection, I will first discuss a type of object and a type of arrow that may exist in a category: viz., terminal objects and points (also called global elements), illustrating them in $\mathcal{S}$, $\mathcal{E}$ and $\mathcal{G}$. This will lead to my last piece of stage-setting for Section 3's presentation of toposes, the notion of a generalized element; though as I mentioned, a full presentation of toposes would also require the notions of product and exponential (also called map object).

An object T is called a *terminal object* if for every object A there is exactly one arrow $f : A \to T$. We sometimes stress the uniqueness of the arrow using !, by writing $f : A \xrightarrow{!} T$. Any two terminal objects are isomorphic. We fix on one such object and write it as **1**. (Dually, an object S is called an *initial object* if for every object A there is exactly one arrow $f : S \to A$; initial objects are all isomorphic, and we fix on one such and write it as **0**.) An arrow $\mathbf{1} \to A$ is called a *point* of A; it is also called a *global element*. In general, an object A might have no points: we shall see examples in $\mathcal{E}$ and $\mathcal{G}$ in a moment – and more importantly for us, the Kochen-Specker theorem is the statement that a certain object has no points (cf. Section 4).

I apply these definitions to my examples of categories. In $\mathcal{S}$, each singleton set is a terminal object; we fix on one, often written $\{*\}$, as **1**. (Dually, the empty set $\emptyset$ is initial.) Every non-empty set A has points, and indeed the points of A give a 'listing' of the elements of A. Furthermore, **1** *separates* arbitrary arrows (i.e. functions) in the sense that: given two functions $f : A \to B$ and $g : A \to B$, if for every point $x : \mathbf{1} \to A$ of A, we have $f \circ x = g \circ x$, then $f = g$.

In $\mathcal{E}$, each singleton set $\{*\}$ equipped with the identity endomap $* \mapsto *$ is a terminal object; i.e. $(\{*\}; id_{\{*\}})$ is a terminal object. I will call this terminal object, unique upto isomorphism, 'the loop'. Then a point of an object $(X; \alpha)$ is an arrow from the loop; and so corresponds to a fixed point of $(X; \alpha)$, i.e. an $x \in X$ such that $\alpha(x) = x$. So an object $A = (X; \alpha)$ need not have any points, and if it does, the points of the object do not in general give a 'listing' of all its 'elements' (unlike the situation in $\mathcal{S}$).

In $\mathcal{G}$ the situation is similar to $\mathcal{E}$. Here, a graph with just one dot and one arc (from the dot as source to the dot as target) is terminal; we can write it as $(s, t : \{*\} \to \{*\})$, and I again call it 'the loop'. Then a point of an arbitrary graph $(s, t : R \to D)$ corresponds to any loop in the graph, i.e. an arc $r \in R$, and a dot $d \in D$ such that $s(r) = d = t(r)$. So as in $\mathcal{E}$, an object in $\mathcal{G}$ need not have any points, and if it does, the points do not in general give a 'listing' of all its 'elements'.

The idea that in general, the points of an object do not give a 'listing of all its elements' can be made more precise in terms of the idea, mentioned above for $\mathcal{S}$, of the terminal object separating arbitrary arrows. Explaining this will motivate the idea of an generalized element.

In general, terminal objects do not separate arbitrary arrows in the way that in $\mathcal{S}$, **1** separates arbitrary functions, i.e. by a 'canceling on the right' ('canceling when done first') for all points x. For example, in $\mathcal{E}$ (resp. $\mathcal{G}$), the terminal object **1**, i.e. the loop considered as a set-with-endomap (resp. considered as a graph), does not separate arbitrary arrows in $\mathcal{E}$ (resp. $\mathcal{G}$). The reason is that roughly speaking, points 'are' loops, and a set-with-endomap (resp. graph) can have plenty of structure apart from its loops, so that distinct arrows f and g can 'agree on all loops x' in the sense that for all $x, f \circ x = g \circ x$.

On the other hand, in many categories (including our examples, $\mathcal{E}$ and $\mathcal{G}$), there are one or two objects that *do* perform this separation of arbitrary arrows. In $\mathcal{E}$, the object in question 'is' the natural numbers with the successor map, i.e. $N := (Z^+, \sigma)$ with $Z^+ = \{1, 2, 3, \ldots\}$ and $\sigma(n) = n + 1$. That is to say, for any objects A, B in $\mathcal{E}$, and arrows $f : A \to B$ and $g : A \to B$: if for all arrows $x : N \to A$, we have $f \circ x = g \circ x$, then $f = g$. (Reason: the various x-images of N are all possible 'chains' in A, and f and g agreeing on their images of all possible chains implies that $f = g$).

In $\mathcal{G}$, the separation is done by a pair of (simple but not terminal) objects: the 'generic arc', *Arc* say, defined as having two dots and one arc from one to the other, and the 'naked dot', *Dot* say, defined as having one dot and no arcs. That is to say: for any objects (i.e. graphs) A, B in $\mathcal{G}$, and arrows $f : A \to B$ and $g : A \to B$: if (i) for all arrows $x : Arc \to A$, we have $f \circ x = g \circ x$ and also (ii) for all arrows $x : Dot \to A$, we have $f \circ x = g \circ x$; then $f = g$. (Reason: the various x-images of *Arc* and *Dot* code all the structure in A. For the component of an arrow $x : Arc \to A$ that describes x's action on dots, i.e. the 'dot-function' f_{Do}, can map two dots to one, so that *Arc* can be mapped into a loop in A.)

To sum up: the way that the various images of these special 'separating' objects – N (resp. *Arc* and *Dot*) – give a 'listing of the structure' of an arbitrary object A in the category $\mathcal{E}$ (resp. $\mathcal{G}$), generalizes the idea that in $\mathcal{S}$ the points of A give a 'listing' of all A's elements.

This suggests that we can think of an image of N (resp. *Arc* and *Dot*) as a kind of generalized (because structured) element of an arbitrary object A in $\mathcal{E}$ (resp. $\mathcal{G}$). And to avoid the vagueness of the word 'image', we can instead think of the *arrow itself*, i.e. $f : N \to A$ in $\mathcal{E}$ (resp. the arrows $f : Arc \to A$ and $g : Dot \to A$ in $\mathcal{G}$) as a generalized element of A.

We now generalize this idea to domains other than the special 'separating' objects, N (resp. *Arc* and *Dot*). That is to say, for any category we say that any arrow $f : B \to A$ is a *generalized element* of A. We also say f is a generalized element of A defined over B, or is a *B-element* of A, and we write $f \in_B A$. More vividly, we say B is the *stage of definition* of f, and that f is a *figure of shape B*, or a *B-figure*, in A. We often use the notation $f(x)$ in place of $f \circ x$ when x is a generalized element of the domain of f. It follows that in any category, the collection of all generalized elements of A separates arbitrary arrows to A in the above sense: given $f : B \to A$ and $g : B \to A$, if for all stages of definition C and C-elements of B (i.e. for all $x : C \to B$) we have $f \circ x = g \circ x$, then $f = g$.

The idea of a generalized element $f : B \to A$ of A, and its special case where f is monic, will be important in the definition of a subobject classifier.

3. Toposes and partial truth

I turn to developing one clause of the definition of a topos: the requirement that a topos contains a subobject classifier Ω. So far I have introduced Ω as a collection of generalized truth-values, generalizing the usual set $\{0, 1\}$. But from now on, the leading idea will be that Ω (together with some associated mathematical structure) generalizes the way that in set-theory characteristic functions classify whether an element x is in a given subset K of a set X. (This emphasis reflects Section 1's strategic decision to emphasise the general idea of partial truth (and so of parts), rather than logico-algebraic aspects of toposes.)

Thus we recall that for any set X, and any subset $K \subseteq X$, there is a characteristic function $\chi_K : X \to \{0, 1\}$, with $\chi_K(x) = 1$ or 0 according as $x \in K$ or $x \notin K$. Thus χ_K classifies the various x for the set-theoretically natural question (with only two possible answers!), "$x \in K$?". Furthermore, the structure of $\mathcal{S}$ secures the existence of the set of truth-values and the various functions χ_K: $\{0, 1\}$ is itself a set, i.e. an object in $\mathcal{S}$, and for each K, X with $K \subseteq X$, χ_K is an arrow from X to $\{0, 1\}$.

The main idea of the subobject classifier – as the name suggests – is that it is possible to formulate this 'classifying action' of the various χ_K in general category-theoretic terms, so as to give a fruitful generalization. I shall first generalize the idea of a subset of a given set, to the idea of a subobject of an object in an arbitrary category $\mathcal{C}$; and explain the notion of a pullback (Section 3.1). Then I shall give the definition of a subobject classifier (Section 3.2); and illustrate the definition in the toposes $\mathcal{E}$ and $\mathcal{G}$ (Section 3.3), and finally in toposes of presheaves (Section 3.4).

3.1. Monics as parts, pullbacks

In any category, one can define a categorial analogue of the set-theoretic idea of subset: it is called a 'subobject'. This notion is not only important for explaining subobject classifiers. It also provides a formal generalization of the notion of part as studied in mereology, so as to allow for 'structured parts'. I shall indicate this line of thought in this Subsection, but not pursue it in detail.

One starts from the idea that the subsets of a set X are in one-to-one correspondence with certain injective functions to X: a subset K of X corresponds to the injective (i.e., one-to-one) function $K \to X$ sending $x \in K$ to $x \in X$. Since category theory provides a generalization of injective functions, viz. the monic arrows (for short, monics), in any category one defines a *subobject* of any object X to be: a monic with codomain X. One also writes a monic with a 'hook-arrow', $k : K \hookrightarrow X$.

At first sight, this definition seems to be formally correct, but to have an informal defect; which I shall briefly discuss in turn. We immediately see that the definition is formally right, at least for the categories we have discussed,

$\mathcal{E}$ and $\mathcal{G}$. For we saw in Section 2.2's discussion of monics in these categories, that if $k : K \hookrightarrow X$ is a monic (in $\mathcal{E}$ or in $\mathcal{G}$) then the 'copy' of K inside X (the 'range' of the arrow k) 'looks exactly like' K itself. So for these categories at least, the definition of a subobject as a monic captures a notion of 'structured part'.

But on the other hand, this definition of subobject clearly ignores the underlying identity of the 'elements' that the objects in the category are comprised of. And it does so in $\mathcal{S}$, no less than in other categories such as $\mathcal{E}$ and $\mathcal{G}$. For a subset K of X in $\mathcal{S}$, it is the very same element x that is in K and in X: that is why it is natural to call K a part of X. But there is no such requirement for a monic: x in K can be sent by a monic k to a very different y in X. (Besides, K is not itself a monic and so *not* a subobject, according to this definition.) Of course, ignoring the underlying identity of basic elements is well-nigh universal in mathematics (and a springboard to structuralism in the philosophy of mathematics); and is all the more natural in category theory. But from the philosophical perspective of trying to generalize mereology or the notion of part, it seems a defect.

But it turns out that, despite this defect, the definition of a subobject as a monic does give a successful generalization of several formal aspects of mereology, such as the subobjects of an object having an inclusion relation between them. It is worth briefly indicating how this generalization proceeds, partly because it gives an example of the usefulness of Section 2.3's notion of a generalized element.

Given a monic $i : B \hookrightarrow A$, we say that a C-element $x \in_C A$, i.e. an arrow $x : C \to A$, is a *member* of i, and write $x \in i$, if x factors through i; that is, if there is some $h : C \to B$ such that $i \circ h = x$. This definition is natural (again, *modulo* the 'defect' above): it is clear that in $\mathcal{S}$, $x \in i$ iff $\mathrm{ran}(x) \subseteq \mathrm{ran}(i)$, and (exercise!) similar points hold in in $\mathcal{E}, \mathcal{G}$.

We now define an inclusion relation on monics to any one object. Given monics $i : A \hookrightarrow C$ and $j : B \hookrightarrow C$, we write $i \subseteq j$, and say that i is *included* in j, if i factors through j: that is, if there is an arrow $s : A \to B$ such that $j \circ s = i$. Again this definition is natural. It gives the intuitively right results in $\mathcal{S}, \mathcal{E}, \mathcal{G}$. And in any category, there are two nice elementary results: (1) Since j is monic, if there is any such s there is only one; and it is itself monic. So if i is included in j (i.e. $i \subseteq j$), it is included in only one way, and the inclusion s is itself a sub-object of the domain of j. (2) Given two subobjects of C, $i : A \hookrightarrow C$ and $j : B \hookrightarrow C$, $i \subseteq j$ iff every member of i (in the sense of the previous paragraph) is a member of j.

We then define: i is *equivalent* to j, $i \equiv j$, iff both $i \subseteq j$ and $j \subseteq i$. Again, this is mathematically natural. From $i \equiv j$ it follows that the domains of i and j are isomorphic; and result (2) implies that $i \equiv j$ iff i and j have exactly the same members. Besides, we can in general now recover the usual notion of a part (the notion respecting the 'underlying identity' of objects' elements – in the usual sense of 'element'!), by picking within an equivalence class of monics, the appropriate representative element (or rather its domain). We will see an example of this in Section 3.4's discussion of subobjects of presheaves.

Here is a final example of how generalized elements allow us to generalize mereological ideas. We can make precise the idea of 'that part of A on which a pair of parallel arrows $f : A \to B$ and $g : A \to B$ agree', as follows. Given a pair of parallel arrows, $f : A \to B$ and $g : A \to B$, we say that a sub-object $e : E \hookrightarrow A$ of A is an *equalizer* for f and g iff: for every C and every $x \in_C A$, x is a member of e iff $f \circ x = g \circ x$. (Or as we wrote more intuitively at the end of Section 2.3: iff $f(x) = g(x)$).

So much by way of formal generalization of the notion of part. My discussion of the subobject classifier needs one more preliminary notion: the notion of a pullback. We define a *pullback* of two arrows with common codomain, $f : A \to C$ and $g : B \to C$ to be an object P and arrows $p_1 : P \to A$ and $p_2 : P \to B$, such that:
(i) the 'square of arrows' is a commutative diagram in the usual sense, i.e. $f \circ p_1 = g \circ p_2$; and
(ii) for any T and any $h : T \to A$ and $k : T \to B$ such that the 'outer square' commutes, i.e. $f \circ h = g \circ k$, there is a unique $u : T \to P$ such that the entire diagram commutes, i.e. $p_1 \circ u = h$ and $p_2 \circ u = k$.
The maps p_1 and p_2 are called *projections*. Pullbacks are defined upto unique isomorphisms, which multiply one pair of projections to give the other.

Unfortunately, the best intuitive explanation of this rather abstract definition uses the notion of a product, $A \times B$ of two objects A and B; a notion I have not taken space to define. But here it suffices to make two comments. (1) In Section 3.2, the abstract definition will be enough for us. (2) (For readers who know the notion of product, i.e. the category-theoretic generalization of set-theory's cartesian product): If the product $A \times B$ exists, then a pullback P is a subobject of the product $A \times B$, in fact an equalizer. More precisely: suppose we are given $f : A \to C$ and $g : B \to C$, and that the product $A \times B$ exists, with projections (for the product) $p_1 : A \times B \to A, p_2 : A \times B \to B$. Then: (i) If there is an equalizer for the arrows, $f \circ p_1$ and $g \circ p_2$, then this equalizer is a pullback (with its projections being p_1, p_2). (ii) Conversely, if there is a pullback for f and g, then it has an arrow to $A \times B$ and that arrow is an equalizer.

3.2. The subobject classifier

Recall the main idea of a subobject classifier, Ω. It is an object in the topos, for which – just as for subsets of a given set X and characteristic functions in $\mathcal{S}$ – there is to be a one-to-one correspondence between subobjects of any given object X, and arrows from X to Ω.

Formally, we define: A *subobject-classifier* is an object Ω and a point (global element) $t : \mathbf{1} \to \Omega$ such that for any monic $k : K \hookrightarrow X$ there is a unique $\chi_k : X \to \Omega$ that makes the following square:

$$\begin{array}{ccc} K & \xrightarrow{!_K} & \mathbf{1} \\ \downarrow{\scriptstyle k} & & \downarrow{\scriptstyle t} \\ X & \xrightarrow{\chi_k} & \Omega \end{array} \tag{3.1}$$

into a pullback; where we have written $!_K$ for the unique arrow from K to $\mathbf{1}$. That is to say, we require:

(i) this square must commute, i.e. $K \xrightarrow{!_K} 1 \xrightarrow{t} \Omega$ and $K \xhookrightarrow{k} X \xrightarrow{\chi_k} \Omega$ are the same arrow;

(ii) for any T and any $h : T \to 1$ and $x : T \to X$ such that the 'outer square' commutes, i.e. $t \circ h = \chi_k \circ x$, there is a unique $u : T \to K$ such that the entire diagram commutes, i.e. $!_K \circ u = h$ and $k \circ u = x$.

The intuitive idea of this definition is readily expressed for the case where objects of the topos are structured sets, as follows.

(1) Think of Ω as a (structured) set of truth-values; think of $t : 1 \to \Omega$ as picking out a designated element, call it *true*, of Ω (intuitively, 'totally true').

(2) Writing the unique arrow from any object T to $\mathbf{1}$ as $!_T : T \to \mathbf{1}$, (1) means that for any T, the composite $t \circ !_T : T \to \Omega$ takes all of T to *true*. In particular, this is true for the case $T := K$.

(3) The fact that the square commutes then implies that χ_k takes all the elements of X that are 'in K', i.e. elements of X that are 'in the copy of K contained in X as the range of k', into the designated value *true* within Ω.

(4) The requirement that the square be a pullback relates to generalized elements of X. That is, since these generalized elements are just arrows $x : T \to X$, for arbitrary objects T, the requirement of being a pullback makes χ_k have the correct action on these elements. To be precise; first recall from Section 3.1 that we say a T-element x of X is a member of k, and write $x \in k$ iff x factors through k. And when the monic k is clear from the context, there is a common abuse of notation: we say that x is a member of the domain K, rather than the monic k, and we write $x \in K$, rather than $x \in k$.

Then the point is that our definition implies: the classifying arrow χ_k of any $k : K \hookrightarrow X$ is the unique arrow such that for any T-element x of X,

x is a member of K [properly speaking, of k] iff $\chi_k(x) := \chi_k \circ x = to!_T$. In short, $x \in K$ iff $\chi_k(x) = true$.

To sum up (1) to (4): the definition of the subobject classifier makes χ_k have the correct classifying action on elements of X, even on generalized elements of X, in the sense that for any such element x, $x \in K$ iff $\chi_k(x) = true$. (Besides, the definition requires that χ_k be the unique arrow with this correct action.)

I should stress three other general features of subobject classifiers, before turning to examples. These features are not obvious from the definition I have given, but follow from it together with the other clauses in the definition of a topos.

(A) Not only does Ω with its arrows χ_k correctly classify elements of X, as regards whether they are 'in K'. It also correctly classifies them, as regards whether they are 'partly in K'. That is, an element x of X, even a generalized element $x : T \to X$, gets sent by χ_k to the partial truth-value in Ω (a truth-value other than *true*) that correstly classifies the 'extent' to which x is in K. This will become clearer in Sections 3.3 and 3.4's examples.

(B) Ω is fixed by the structure of the topos concerned in the precise sense that, although the above definition of Ω characterizes Ω solely in terms of conditions on the topos' objects and arrows, Ω is provably unique (up to isomorphism).

(C) In any topos, Ω has a natural logical structure. More exactly, Ω has the internal structure of a Heyting algebra object: the algebraic structure appropriate for intuitionistic logic, mentioned in Section 1. Furthermore, in any topos, the collection of subobjects of any given object X is a complete Heyting algebra (a locale), written Ω^X. I shall not discuss this Heyting algebra structure in detail, but note only that its presence confirms my remarks 1., 3. and 4. in Section 1.

3.3. Ω in $\mathcal{E}$ and $\mathcal{G}$

I turn to displaying Ω in $\mathcal{E}$ and $\mathcal{G}$. As we shall see, the generalized truth-values (intuitively, the 'elements' of Ω) are the answers to a natural multiple-choice question about the objects in the topos, just as "$x \in K$?" is natural for sets.

In $\mathcal{E}$: Recall that a set X equipped with a given function $\alpha : X \to X$ is called an *endomap*, written $(X;\alpha)$; and the family of all endomaps forms a category – indeed, a topos – when one defines an arrow from $(X;\alpha)$ to $(Y;\beta)$ to be an ordinary set-function f between the underlying sets, from X to Y, that preserves the endomap structure, i.e. $f \circ \alpha = \beta \circ f$.

Applying the definition of a subobject, it turns out that a subobject of $(X;\alpha)$ is a subset of X that is closed under α, equipped with the restriction of α: i.e., a subobject is $(K, \alpha|_K)$, with $K \subseteq X$ and such that $\alpha(K) \subset K$. So a natural question, given $x \in X$ and a subendomap $(K, \alpha|_K)$, is: "How many iterations of α are needed to send x (or rather its descendant, $\alpha(x)$ or $\alpha^2(x)$ or $\alpha^3(x) \ldots$) into K?" The possible answers are '0 (i.e., $x \in K$)', '1', '2', ..., and 'infinity (i.e., the descendants never enter K)'; and if the answer for x is some natural number n (resp. 0, infinity), then the answer for $\alpha(x)$ is $n-1$ (resp. 0, infinity). So the possible answers can be presented in a diagram, as in Figure 1:

Figure 1. Ω for $\mathcal{E}$

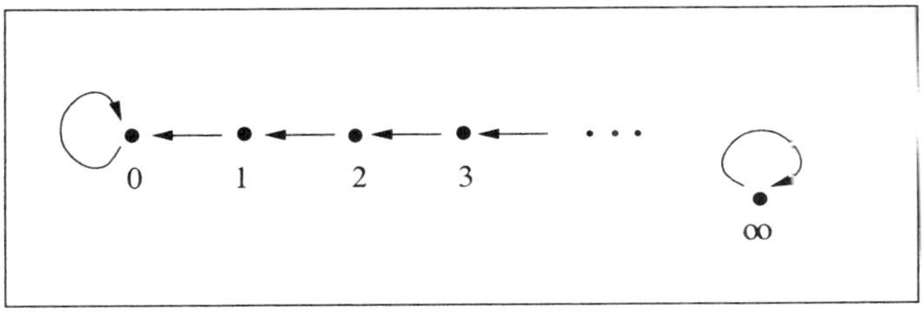

And it turns out that this is exactly the object Ω in the category of endomaps! To sum up this example: Ω is the collection of possible answers to "At what stage do you (strictly speaking: your descendant under α) enter K?" This notion of a *stage* will come to the fore in Sections 3.4 and 4.

I turn to $\mathcal{G}$. Recall that an object in $\mathcal{G}$ is a graph; we can think of a graph as a set of dots equipped with a (not necessarily symmetric) binary relation; where we think of the binary relation as a set of arcs among the dots. A subobject of a graph 'is' a subgraph; (more exactly, a corresponding monic with the given subgraph as 'range'). And we can think of a subgraph as a subset of the given graph's set of dots, equipped with a subset of the given graph's set of arcs.

So a natural question about the dots and arcs of a given graph X, in relation to a subgraph K of X, is: "Are you in K or not?". For a dot, there are only two possibilities: a dot in X is simply either in, or out, of a subgraph K. But for an arc in X, there are five possible relations to K!

(i) The arc can be wholly in K, in the sense that not only are the arc's source dot and target dot in K, but so is the arc itself.

(ii) The arc can be 'just out' of K in the sense that it is out, but its source and target dots are in.

(iii) The arc is out, and so is its source dot; but its target dot is in.

(iv) The arc is out, and so is its target dot; but its source dot is in.

(v) The arc can be 'wholly out' of K in that it is out, and so are its source and target dots.

These possibilities – two for dots, and five for arcs – can be pictured in a graph. And indeed it turns out that Ω is exactly this graph (upto isomorphism)! Figure 2 shows an example of how this Ω classifies the dots and arcs of a graph, as to whether they are in the subgraph K indicated by the dashed circle: the arcs of the given graph have Roman letters, and the arcs in Ω that are 'their truth-values' have the corresponding Greek letters.

Figure 2. A graph, a subgraph and Ω for $\mathcal{G}$

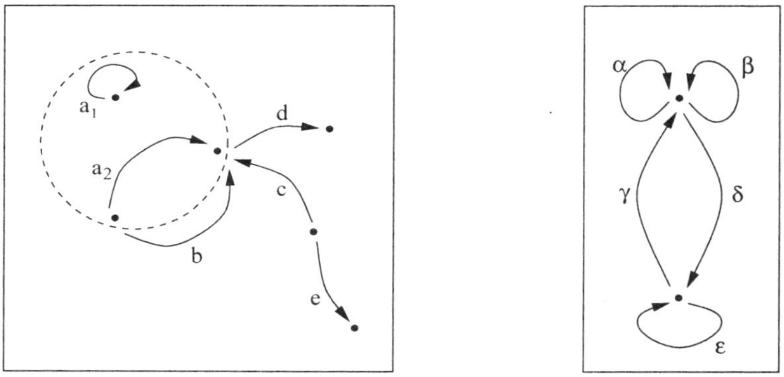

To sum up this example: Ω is the collection of possible answers to "What part of you is in K?" (where the source and target dots of a given arc count as part of it – but a dot has no proper parts!). This notion of a *part* of an element x being in a subobject will arise again in Section 4.

3.4. Toposes of presheaves

As discussed in Section 1, our proposal about partial truth in a physical theory uses a topos of presheaves. So to prepare for Section 4's discussion of that proposal, I now introduce presheaves.

We first need the idea of a covariant functor between a pair of categories $\mathcal{C}$ and $\mathcal{D}$. Intuitively, this is a arrow-preserving function from one category to the other. To be precise: A *covariant functor* $\mathbf{F}$ from a category $\mathcal{C}$ to a category $\mathcal{D}$ is a function that assigns

1. to each $\mathcal{C}$-object A, a $\mathcal{D}$-object $\mathbf{F}(A)$;

2. to each $\mathcal{C}$-arrow $f : A \to B$, a $\mathcal{D}$-arrow $\mathbf{F}(f) : \mathbf{F}(A) \to \mathbf{F}(B)$ such that: (i) $\mathbf{F}(\mathrm{id}_A) = \mathrm{id}_{\mathbf{F}(A)}$; and (ii) if $g : C \to A$, and $f : A \to B$ then $\mathbf{F}(f \circ g) = \mathbf{F}(f) \circ \mathbf{F}(g)$.

A *presheaf* on the category $\mathcal{C}$ is defined to be a covariant functor $\mathbf{X}$ from $\mathcal{C}$ to $\mathcal{S}$, the category of sets; (and $\mathcal{C}$ is then called the *base-category*). A presheaf is also called a *varying set*, since the membership of the assigned set $\mathbf{X}(A)$ varies from one object A in $\mathcal{C}$ to another.

Note that given a $\mathcal{C}$-arrow $f : A \to B$: if $\mathbf{X}(A) \neq \emptyset$, then $\mathbf{X}(B) \neq \emptyset$. So as we go 'downstream', the assigned set $\mathbf{X}(A)$ cannot 'disappear', becoming $\emptyset$: "once in, the descendants are forever in". But $\mathbf{X}$ can 'shrink' in that $\mathbf{X}(f)$ can be non-injective; and $\mathbf{X}$ can of course 'grow' (viz. if the codomain of $\mathbf{X}(f)$ is a superset of its range).

To make the collection of all presheaves on $\mathcal{C}$ into a category, we need to define what is meant by an arrow between two presheaves $\mathbf{X}$ and $\mathbf{Y}$. The intuitive idea is that an arrow from $\mathbf{X}$ to $\mathbf{Y}$ gives a 'picture' of $\mathbf{X}$ within $\mathbf{Y}$. Formally, we define an arrow to be a *natural transformation* $N : \mathbf{X} \to \mathbf{Y}$, by which is meant a family of functions (called the *components* of N) $N_A : \mathbf{X}(A) \to \mathbf{Y}(A)$, A an object in $\mathcal{C}$, such that if $f : A \to B$ is an arrow in $\mathcal{C}$, then the composite function $\mathbf{X}(A) \xrightarrow{N_A} \mathbf{Y}(A) \xrightarrow{\mathbf{Y}(f)} \mathbf{Y}(B)$ is equal to $\mathbf{X}(A) \xrightarrow{\mathbf{X}(f)} \mathbf{X}(B) \xrightarrow{N_B} \mathbf{Y}(B)$. In other words, we have the commutative diagram

$$\begin{array}{ccc} \mathbf{X}(A) & \xrightarrow{\mathbf{X}(f)} & \mathbf{X}(B) \\ \downarrow N_A & & \downarrow N_B \\ \mathbf{Y}(A) & \xrightarrow{\mathbf{Y}(f)} & \mathbf{Y}(B) \end{array} \tag{3.2}$$

The resulting category of presheaves on $\mathcal{C}$ is denoted $\mathcal{S}^{\mathcal{C}}$.

As in Section 3.1, a subobject of $\mathbf{X}$ is formally defined as a monic arrow (i.e. monic natural transformation) with codomain $\mathbf{X}$. But recalling our discussion of equivalent subobjects, we see that each equivalence class of subobjects of $\mathbf{X}$ has a member $i : \mathbf{K} \hookrightarrow \mathbf{X}$ that 'respects the underlying identity of elements': that is, i is such that for each A, the component map $i_A : \mathbf{K}(A) \to \mathbf{X}(A)$ is a subset embedding, i.e., $\mathbf{K}(A) \subseteq \mathbf{X}(A)$. Thus, if $f : A \to B$ is any arrow in $\mathcal{C}$, we get the analogue of the commutative diagram Eq. (3.2):

$$\begin{array}{ccc} \mathbf{K}(A) & \xrightarrow{\mathbf{K}(f)} & \mathbf{K}(B) \\ \downarrow i_A & & \downarrow i_B \\ \mathbf{X}(A) & \xrightarrow{\mathbf{X}(f)} & \mathbf{X}(B) \end{array} \tag{3.3}$$

but where now the vertical arrows are subset inclusions.

From now on, we fix on such 'identity-respecting' subobjects, for simplicity; (but without loss of generality, since every equivalence class of subobjects contains such). Then $\mathbf{K}(f)$ is the restriction of $\mathbf{X}(f)$ to $\mathbf{K}(A)$; and this implies remarks about "once in **K**, descendants are forever in", and about **K** 'shrinking' or 'growing' as we go 'downstream', corresponding to the remarks above about **X**. Thus the fact that $\mathbf{K}(f) = \mathbf{X}(f)\mid_{\mathbf{K}(A)}$ implies that for $x, y \in \mathbf{K}(A), \mathbf{K}(f)(x) = \mathbf{K}(f)(y)$ only if $\mathbf{X}(f)(x) = \mathbf{X}(f)(y)$. So **K** can only 'shrink' as we move from $\mathbf{K}(A)$ to $\mathbf{K}(B)$ to the extent allowed by **X** 'shrinking'. But **K** can 'grow' from $\mathbf{K}(A)$ to $\mathbf{K}(B)$, even if **X** does not; for $\mathbf{K}(B)$ can be a proper superset of $\mathrm{ran}(\mathbf{K}(f))$. This discussion of subobjects in $\mathcal{S}^{\mathcal{C}}$ is summed up in Figure 3.

Figure 3. A presheaf **X**, and a subobject **K** of **X**

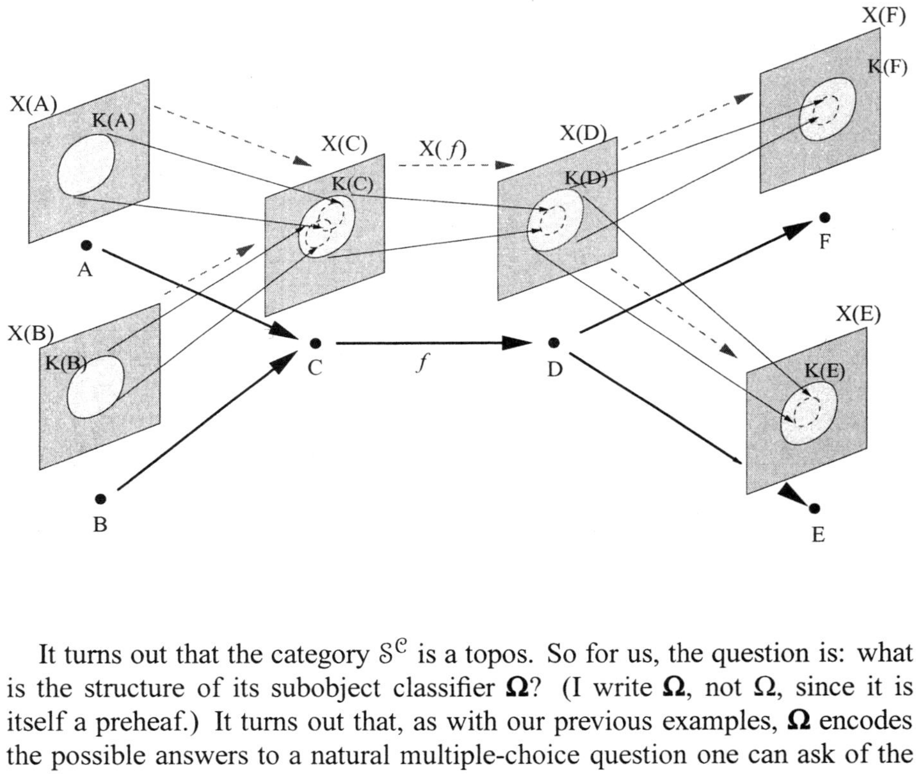

It turns out that the category $\mathcal{S}^{\mathcal{C}}$ is a topos. So for us, the question is: what is the structure of its subobject classifier $\mathbf{\Omega}$? (I write $\mathbf{\Omega}$, not Ω, since it is itself a preheaf.) It turns out that, as with our previous examples, $\mathbf{\Omega}$ encodes the possible answers to a natural multiple-choice question one can ask of the 'elements' of which objects in the topos are built. For presheaves, the question is asked of an element $x \in \mathbf{X}(A)$ at a 'stage' A in $\mathcal{C}$: "At which stages B in $\mathcal{C}$ are the 'descendants' of x, i.e. $\mathbf{K}(f)(x) = \mathbf{X}(f)(x)$, elements of $\mathbf{K}(B)$?" Note incidentally, that the topic, 'when do x's descendants under a certain function enter a certain set?', is reminiscent of Ω for the category of endomaps $\mathcal{E}$.

Given this question, and recalling that in a topos a subobject $\mathbf{K}$ of an object $\mathbf{X}$ corresponds to a characteristic arrow, call it $\chi^K : \mathbf{X} \to \mathbf{\Omega}$, we intuitively expect that an element $x \in \mathbf{X}(A)$ will be mapped by χ^K to the set of those B, with $f : A \to B$, at which $\mathbf{X}(f)(x) \in \mathbf{K}(B)$. So we expect the various partial truth-values to be various sets of objects in the base-category $\mathcal{C}$.

This expectation is essentially correct, but needs two qualifications.

(1) Since χ^K is an arrow in $\mathcal{S}^{\mathcal{C}}$, i.e. a natural transformation, it will consist of a component function $\chi^{\mathbf{K}}_A$ at each A in $\mathcal{C}$. So the objects A in the base-category $\mathcal{C}$ become *stages, contexts*, relative to which truth-values are ascribed; (in addition to their role as stages at which 'descendants' enter the set $\mathbf{K}(A)$).

(2) We need to allow for the possibility (not shown in Fig. 3) that there are two $\mathcal{C}$-arrows f, g from A to B (or even more than two). This means there can be distinct functions $\mathbf{X}(f)$ and $\mathbf{X}(g)$ from $\mathbf{X}(A)$ to $\mathbf{X}(B)$; and so for a given $x \in \mathbf{X}(A)$, we can have $\mathbf{X}(f)(x) \in \mathbf{K}(B)$ but $\mathbf{X}(g)(x) \notin \mathbf{K}(B)$. The most straightforward way that $\mathbf{\Omega}$ could encode these possibilities – i.e. assign different truth-values according to which of the functions $\mathbf{X}(f)$ and $\mathbf{X}(g)$ send x into $\mathbf{K}(B)$ – is for the truth-values to contain the $\mathcal{C}$-arrows themselves, f etc., rather than their codomains B etc. And indeed, this is exactly what $\mathbf{\Omega}$ does.

Besides, the feature that, going 'downstream' from A, once a descendant $\mathbf{X}(f)(x)$ of x gets into a subobject $\mathbf{K}$ of $\mathbf{X}$, all subsequent descendants are also in $\mathbf{K}$, implies that a set S of arrows that is a truth-value at A – i.e. a set of arrows with domain A, at whose codomain B $\mathbf{X}(f)(x) \in \mathbf{K}(B)$ – is 'closed under going downstream' in the sense that: if $f : A \to B$ belongs to S, and if $g : B \to C$ is any arrow, then $g \circ f : A \to C$ also belongs to S. A set of arrows each with domain A that has this property is called a *sieve* on A. So we expect the sieves on A to be the generalized truth-values at A; and since the subobject classifier $\mathbf{\Omega}$ is to be a presheaf on $\mathcal{C}$, we expect $\mathbf{\Omega}$'s assignment to each A in $\mathcal{C}$ to be the set of all sieves on A.

That leaves open how to define $\mathbf{\Omega}$'s action on $\mathcal{C}$-arrows f. But it turns out that a simple 'push-along' action, as in the following precise definition of $\mathbf{\Omega}$, works. The presheaf $\mathbf{\Omega} : \mathcal{C} \to \mathrm{Set}$ is defined by:

(i) For A in $\mathcal{C}$, $\mathbf{\Omega}(A)$ is to be the set of all sieves on A;

(ii) if $f : A \to B$, then $\mathbf{\Omega}(f) : \mathbf{\Omega}(A) \to \mathbf{\Omega}(B)$ is defined for all $S \in \mathbf{\Omega}(A)$ as

$$\mathbf{\Omega}(f)(S) := \{h : B \to C \mid h \circ f \in S\}. \tag{3.4}$$

We can now spell out how $\mathbf{\Omega}$ works as the subobject classifier for $\mathcal{S}^{\mathcal{C}}$. Let $\mathbf{K}$ be a subobject of $\mathbf{X}$. Then there is an associated *characteristic* arrow $\chi^{\mathbf{K}} : \mathbf{X} \to \mathbf{\Omega}$, whose 'component' $\chi^{\mathbf{K}}_A : \mathbf{X}(A) \to \mathbf{\Omega}(A)$ at each 'stage of truth' A in $\mathcal{C}$ is defined as

$$\chi^{\mathbf{K}}_A(x) := \{f : A \to B \mid \mathbf{X}(f)(x) \in \mathbf{K}(B)\} \tag{3.5}$$

for all $x \in \mathbf{X}(A)$. That the right hand side of Eq. (3.5) actually *is* a sieve on A follows from the definition of a subobject. Thus, in each 'branch' of $\mathcal{C}$ going 'downstream' from the stage A, $\chi_A^{\mathbf{K}}(x)$ picks out the first member B in that branch for which $\mathbf{X}(f)(x)$ lies in the subset $\mathbf{K}(B)$, and the commutative diagram Eq. (3.3) then guarantees that $\mathbf{X}(h \circ f)(x)$ will lie in $\mathbf{K}(C)$ for all $h : B \to C$. Thus each 'stage of truth' A in $\mathcal{C}$ serves as a possible context for an assignment to each $x \in \mathbf{X}(A)$ of a generalised truth-value.

There is a converse to Eq. (3.5): namely, each arrow $\chi : \mathbf{X} \to \mathbf{\Omega}$ (i.e., a natural transformation between the presheaves $\mathbf{X}$ and $\mathbf{\Omega}$) defines a subobject $\mathbf{K}^\chi$ of $\mathbf{X}$ via

$$\mathbf{K}^\chi(A) := \chi_A^{-1}\{1_{\mathbf{\Omega}(A)}\}. \tag{3.6}$$

at each stage of truth A. To sum up Eq. (3.5) and (3.6): we have a one-to-one correspondence between the subobjects of $\mathbf{X}$ and arrows $\chi : \mathbf{X} \to \mathbf{\Omega}$.

Two final remarks on $\mathbf{\Omega}$. (1) First, a logico-algebraic remark. The set $\mathbf{\Omega}(A)$ of sieves on any A in $\mathcal{C}$ has the structure of a Heyting algebra, under natural operations. I shall not define 'Heyting algebra' (it is a kind of distributive lattice), but only indicate how natural the logical operations in $\mathbf{\Omega}(A)$ are. The partial order is defined by $S_1 \leqslant S_2$ iff $S_1 \subseteq S_2$; so that the 1 of the algebra (the 'totally true' truth-value, *true*) is the set of all arrows with domain A (called the *principal sieve* on A), and the 0 of the algebra is the empty set of sieves. The logical connectives, 'and', 'or' are defined by:

$$S_1 \wedge S_2 := S_1 \cap S_2 \tag{3.7}$$

$$S_1 \vee S_2 := S_1 \cup S_2 \tag{3.8}$$

(2) I turn to points (global elements) of a presheaf. In $\mathcal{S}^{\mathcal{C}}$, a terminal object $\mathbf{1} : \mathcal{C} \to \mathcal{S}$ can be defined by: (i) $\mathbf{1}(A) := \{*\}$ at all stages A in $\mathcal{C}$; (ii) if $f : A \to B$ is an arrow in $\mathcal{C}$ then $\mathbf{1}(f) : \{*\} \to \{*\}$ is defined to be the function $* \mapsto *$. This is indeed a terminal object since, for any presheaf $\mathbf{X}$, there is a unique natural transformation $N : \mathbf{X} \to \mathbf{1}$ whose components $N_A : \mathbf{X}(A) \to \mathbf{1}(A) = \{*\}$ are the constant maps $x \mapsto *$ for all $x \in \mathbf{X}(A)$.

So a point (global element, also called *global section*) of a presheaf $\mathbf{X}$, i.e. an arrow $\gamma : \mathbf{1} \to \mathbf{X}$, corresponds to a choice of an element $\gamma_A \in \mathbf{X}(A)$ for each stage A in $\mathcal{C}$, such that, if $f : A \to B$, the 'matching condition'

$$\mathbf{X}(f)(\gamma_A) = \gamma_B \tag{3.9}$$

is satisfied. Some presheaves have no points: a 'globally matching' choice of element cannot be made. Examples include certain presheaves that arises naturally in any quantum theory: and the non-existence of such points is exactly the content of the Kochen-Specker theorem.

4. Presheaves of propositions and sieve-valued valuations in physics

Finally, I sketch our proposal applying the material in Section 3 to physical theories. As discussed in Section 1, this application is laid out in detail elsewhere; so here I will be very brief, emphasising only the connections with this paper's ideas of partial truth.

The fundamental ideas of the proposal are:

1. To make a category, $\mathcal{O}$ say, out of a theory's physical quantities. Each quantity is to be an object in the category, and there is to be an $\mathcal{O}$-arrow $f : A \to B$ iff $B = f(A)$ in the usual sense of the functional calculus of quantities (i.e. roughly the value of B is f of the value of A).[4]

2. To define a *coarse-graining presheaf* $\mathbf{G}$ on $\mathcal{O}$ as follows:

   (a) $\mathbf{G}$ assigns to each quantity A the set of propositions "$A \in \Delta$" (read as saying that the value of A lies in a Borel set of real numbers, $\Delta \subseteq \mathbb{R}$) for varying Δ; more exactly, $\mathbf{G}(A)$ is to be the set of these propositions' mathematical representatives, such as the algebra of A's spectral projectors on a quantum system's Hilbert space or the algebra of A's characteristic functions on a classical system's phase space; and

   (b) $\mathbf{G}$ assigns to each $\mathcal{O}$-arrow $f : A \to B$ the 'coarse-graining function' $\mathbf{G}(f) : \mathbf{G}(A) \to \mathbf{G}(B)$ defined by $\mathbf{G}(f) : A \in \Delta \mapsto f(A) \in f(\Delta)$.

3. To assign as the truth-value of the proposition $A \in \Delta$, relative to the context A, a sieve on A, written $\nu(A \in \Delta)$. These sieve-valued valuations can be required to have various natural properties, among them a generalization for the presheaf $\mathbf{G}$ of the main premise of the Kochen-Specker theorem: viz. the *FUNC* condition, that if $B = f(A)$, the real numbers r and r' that are the respective values of A and B are related by f, i.e. $r' = f(r)$.

4. To show that the states (both pure and mixed) of the physical theory (quantum or classical) define sieve-valued valuations enjoying such properties. I illustrate this with the case of a pure quantum state represented by a vector ψ in the Hilbert space $\mathcal{H}$. (The situation for a mixed quantum state is exactly analogous.) ψ defines a sieve on each bounded self-adjoint operator A on $\mathcal{H}$ (i.e. each A in $\mathcal{O}$) by the requirement that an $\mathcal{O}$-arrow $f : A \to B$ is in the sieve iff ψ is in the range of the spectral projector representing $f(A) \in f(\Delta)$ (i.e. the projector $\hat{E}[f(A) \in f(\Delta)]$, in a usual notation). That is, ψ defines a sieve by:

$$
\begin{aligned}
\nu^{\psi}(A \in \Delta) &:= \{f : A \to B \mid \hat{E}[B \in f(\Delta)]\psi = \psi\} \\
&= \{f : A \to B \mid \mathrm{Prob}(B \in f(\Delta); \psi) = 1\} \qquad (4.1)
\end{aligned}
$$

where $\text{Prob}(B \in f(\Delta); \psi)$ is the usual Born-rule probability that the result of a measurement of B will lie in $f(\Delta)$, given the state ψ.

This definition generalizes orthodox quantum theory's so-called eigenstate-eigenvalue link, in the sense that it requires, not that ψ be in the range of $\hat{E}[A \in \Delta]$, but only that ψ be in the range of the projector $\hat{E}[f(A) \in f(\Delta)]$; and this is a logically weaker requirement, since in the lattice $\mathcal{L}(\mathcal{H})$ of projectors on $\mathcal{H}$, $\hat{E}[A \in \Delta] \leqslant \hat{E}[f(A) \in f(\Delta)]$. Furthermore, one can check that the definition satisfies the presheaf version of *FUNC*, and also has other natural properties.

To sum up the proposal, as it applies to quantum theory: the new proposed truth-value of $A \in \Delta$ is to be determined by the set of quantities $f(A)$ for which the corresponding weaker proposition $f(A) \in f(\Delta)$ is true in the old (i.e., eigenstate-eigenvalue link) sense. To put it less exactly, but more memorably: the new proposed truth-value of a proposition is to be determined by the set of its consequences that are true in the old sense.[5]

From this summary, it is clear that this paper has added three motivations for this proposal. The first two are general. (1) I have given a general defence of partial truth as a notion that is made tractable and illuminating within a topos. (2) Section 3.4 motivated the definition of sieves, and thereby of $\mathbf{\Omega}$, in terms of a natural question one can ask about the elements x of an assigned set $\mathbf{X}(A)$: a question analogous to that encoded by the subobject classifier in $\mathcal{E}$.

(3) The third motivation relates to the last part of the summary above: the fact that the proposal defines the partial truth of $A \in \Delta$ in terms of which logically weaker propositions $f(A) \in f(\Delta)$ are wholly true. Since in philosophical logic it is natural (and common) to think of a proposition as the conjunction of its consequences, and to think of each consequence as part of the proposition, this feature of our proposal is intuitively appealing. For it implies that our proposal takes the partial truth of a proposition to be a matter of the whole truth of a part of the proposition. (And in this regard, the notion of partial truth in $\mathcal{S}^{\mathcal{O}}$ is like the notion of an arc being within a subgraph, within $\mathcal{G}$.)

To conclude: in this paper I have argued that topos theory provides a promising general framework for (i) making sense of partial truth, and (ii) developing a generalized mereology that considers a notion of 'structured part'. In particular, I have sketched how these notions apply to a recent proposal to use a topos of presheaves on a base-category of physical quantities, to define partial and contextual truth-values for propositions about the values of physical quantities in physical theories.[6]

Notes

1. Cf. Isham and Butterfield (1998, 2000), Butterfield and Isham (1999), Hamilton, Isham and Butterfield (2000).

2. Lawvere and Shanuel (1997) gives a very lucid and enjoyable introduction to categories and toposes, and discusses endomaps and graphs in detail.

3. But I will ignore the motivation for these notions in terms of determination and choice problems; i.e. the equivalence between an arrow having a retraction (resp. section) and determination (resp. choice) problems having a solution. For this, cf. e.g. Lawvere and Shanuel (1997).

4. Hamilton, Isham and Butterfield (2000) describe how the proposal carries over in all details to a more general notion of context, viz. as a commutative von Neumann algebra.

5. I should add that this summary of the proposal downplays some facts that are treated at length in the cited papers, including:

(i) The application to classical physical theories is perfectly analogous to the quantum theory case. But it is the interpretative problems of quantum theory – specifically, the difficulties of a classical realism, as shown by no-go theorems like the Kochen-Specker theorem – that motivate partial truth-values for propositions "$A \in \Delta$".

(ii) The Kochen-Specker theorem is equivalent to the non-existence of points (global elements) of certain presheaves on $\mathcal{O}$ (for quantum theory). This equivalence illuminates the Kochen-Specker theorem, and its various proofs (cf. Hamilton 2000). But it is not essential to our positive proposal. In particular, **G** is not one of the presheaves that lack points.

(iii) The presheaf generalization of *FUNC* enjoyed by the valuations ν is equivalent to ν specifying a subobject of **G**; i.e. to ν's assignment at each quantity A, $A \in \Delta \mapsto \nu(A \in \Delta)$, defining a natural transformation from **G** to Ω.

6. It is a pleasure to thank: the conference organizers and editors for all their work, audiences in Cracow, Chicago, London and Salzburg; Chris Isham and John Hamilton for many discussions of this material – and for invaluable computing help!

References

Butterfield, J., Isham, C.: (1999), 'A topos perspective on the Kochen-Specker theorem: II. Conceptual aspects, and classical analogues', *International Journal of Theoretical Physics* **38**, pp. 827–859. quant-ph/9808067.

Hamilton, J: (2000), 'An obstruction based approach to the Kochen-Specker theorem', submitted to *Journal of Physics A*, quant-ph/9912018.

Hamilton, J, Isham, C., Butterfield, J.: (2000), 'A topos perspective on the Kochen-Specker theorem: III. Von Neumann Algebras as the Base Category', forthcoming *International Journal of Theoretical Physics*, March 2000, quant-ph/9911020.

Isham, C., Butterfield, J.: (1998), 'A topos perspective on the Kochen-Specker theorem: I. Quantum states as generalised valuations', *International Journal of Theoretical Physics* **37**, pp. 2669–2733. quant-ph/980355.

Isham, C., Butterfield, J.: (2000), 'Some Possible Roles for Topos Theory in Quantum Theory and Quantum Gravity', forthcoming in *Foundations of Physics*, gr-qc/9910005.

Lawvere, W., Shanuel, S.: (1997), *Conceptual Mathematics: a First Introduction to Categories*, Cambridge University Press, Cambridge.

KANT, GÖDEL AND RELATIVITY*

Mauro Dorato
Dept. of Philosophy, Università degli Roma Tre,
Via Ostiense 234, 00146 Rome, Italy
tel. +39-06-54577523; fax +39-06-5457340
dorato@uniroma3.it

1. What is "the problem of time"?

Time is absolutely central in our inner experience, but at least since the foundation of modern physics, it also plays an important role in the description of the outer world. To the extent that there is a *physical* and a *mental* time, one can safely surmise that the problem of trying to establish whether and how they are related is one of the most fundamental issues in the philosophy of time. Are physical and mental time in conflict, or we can regard the former as some sort of appropriate extension to the outer world of the main features of the latter?

These questions, important as they are, already presuppose, however, that there are *two* sides of the problem, namely that *there is*, at a fundamental level of description, *a physical time*. Even independently of current physical speculations on the possibility that, within a quantum theory of gravity, spacetime could become a derived or an emergent notion, philosophers should ask questions that presuppose as little as possible preconceived answers. And given that the quintessential philosophical problem involves *what there is*, I take it that the main philosophical "problem of time" still involves a question that is at least as old as Aristotle, and that was clearly formulated by Kant:

Q_0: is time a subjective, purely ideal notion – *transcendental* in the sense of Kant – or is it rather part of the mind-independent, physical furniture of the universe?[1]

*I want to thank especially Jan Faye, Massimo Pauri, Mario Piazza, Robert Rynasiewicz and Steven Savitt for having read a previous version of the manuscript. Jeremy Butterfield, Elena Castellani, Marisa Dalla Chiara, Jesus Moisterin, Pieter Vermaas and, in general, the audience in Krakow, have raised interesting questions, that forced me to further clarify my thoughts. I obviously claim unique responsibility for the remaining mistakes.

P. Gärdenfors, J. Woleński and K. Kijania-Placek (eds.), In the Scope of Logic, Methodology and Philosophy of Science, Vol. I, 331-348.

In order to try to set the conceptual stage to begin answering this question, the paper is structured as follows: Section 2 offers some motivations as to why I take that Q_0 just *is* the crux of the matter in the philosophy of time. With no pretense of scholarly faithfulness to the original texts, Section 3 offers an interpretive sketch of Kant's views of time, in such a way as to distillate *three conditions* that are *together necessary* for the ideality of time in Kant's sense: (i) time must be *non-substantial*, and the resulting relationism must be constructed in such a way that both (ii) the difference between past and future and that (iii) between earlier and later than, must be *mind-dependent*. Following Gödel (1949), Section 4 and Section 5 bring to bear the special theory of relativity on (ii) and (iii).[2]

2. Time as the criterion of concrete existence and as *principium individuationis*: the neglect of Kant

Considering the complexity of the matter, it is not unwise to begin by reminding ourselves of humdrum, familiar truths, that seem, however, to have escaped our attention entirely. Despite the fact that in ordinary language we sometime say that 'we feel the passage of time', or that 'time flies', time by itself, even more than space, appears to common sense as a "non-object" or a "non-entity". Since the space and time of our experience look as though they are devoid of any perceivable qualities in the ordinary sense, it would seem legitimate to conclude that they are also devoid of any causal power. Given that the *properties* of objects or events can be identified with *their causal powers*, time has no causal power because it appear to have no properties. However, if time and space are causally inert, why can't we regard them as mere *nothing*, in the same sense in which empty space was, for the ancient atomists, the only instance of the parmenideian notion of "nothingness"?

There appears to be at least *two* important reasons to reply to the above question in the *negative*. The first is that time, together with space, is one of the main criteria for the reality of a *concrete* object or event, *physical or mental* as it may be. Except for the question of the existence of numbers – which, platonistically, could involve real but non-spatiotemporally extended, *abstract entities* – one can quite plausibly claim that a non-abstract entity x is real just in case x occupies a portion of space and time, i.e., just in case x *is in spacetime*. The second, related reason, is given by the fact that space and time yield a *principle of individuation*. While space distinguishes and separates physical objects and events that co-exist or exist at the same time, time is used to distinguish states or events belonging to the same substance, remaining in the same place or moving across space (see Howard 1997). In this sense, space and time, to use an expression that Auyang restricted to space, are "kind of dividers" (1995, p. 129): exactly as separation is space is sufficient to regard two *objects* as being *numerically* distinct, despite a perfect resemblance or even a *qualitative* identity as in quantum particles, separation in time is a necessary, individuating characteristic of *events*. It is not by chance that this conceptual aspect is reflected in some (pre-relativistic) theories of events, in which the Newtonian time of their occurrence is regarded as part of their definition, together with a substance and a property (see, for instance, Kim 1976).[3]

Keeping in mind these remarks, two rather striking questions naturally come to mind:

Q_1 If time (together with space) were a non-entity (unreal), how could it yield the most important criterion of the reality of things and events? Shouldn't this criterion imply, by itself, that time *is*, in some sense, real?

Q_2 How can we explain the fact that objects and events are objectively "separated" by space and time without also assuming that the latter, in some sense exist, either as "carrier" of the spatiotemporal relations in empty space, or as real, mind-independent relations exemplified by physical entities?

To my knowledge, questions like these have been neglected by the non-negligible, current philosophical literature on time: Nerlich (1994), who deserves the credit to have first raised Q_1 with respect to space, has unfortunately restricted his attention only to the latter. And since the inseparability of space and time implied by relativity is clearly not sufficient for the indistinguishability or even the identity of their properties, we must pay due attention to time.

As a first, tentative answer, we could follow a suggestion to be found in Kant's *Critique of Pure Reason*: couldn't it be that time, together with space, is an *a priori* intuition that sentient beings like ourselves necessarily (transcendentally) *presuppose* to organize and structure the sensation of the external and the internal world? Note that this suggestion would explain, at least *prima facie*, both why we presuppose time in every possible experience, internal and external, and why the spatio-temporal relations yielded by the transcendental subject – to which in the following I will just refer by the expression "the mind"[4] – provide structure and order (*the pure forms*) to our sensations. In this way, any concrete phenomenon, not matter how conceptualized in terms of the categories or concepts (substance or event, state or process), *always appear to be in space and time*. In sum, in order to provide an answer to Q_1 and Q_2, *prima facie* there is no need to assume that time is a mind-independent property of the physical universe.

It is interesting to note that with the notable exception of Ernst Cassirer (1920) and Kurt Gödel (1949) – who had claimed that the theory of relativity is a striking *confirmation* of Kant's claim that time is ideal[5] – virtually no philosopher in our century has tried to rely on a Kantian explanation for these puzzling questions.[6] Quite on the contrary, the general wisdom of contemporary philosophy of science has it that the Kantian philosophy of time has been superseded by the theory of relativity, in the same sense in which the latter has replaced Newton's theory of absolute space and time. With respect to time, in particular, we are told that since any global separation between past and future events can no longer be regarded as *absolute* – i.e., as independent, as Kant still thought, of particular inertial reference systems or matter distribution – Kant's doctrine of time must be abandoned.

In a word, relying on the widespread claim that Kant's *Critique of Pure Reason* was profoundly influenced by Newton's doctrine of absolute space and time as formulated in the *Principia*, the great majority of the philosophers of science of our century has simply ignored the possibility that an important aspect of Kant's *philosophical theory* of space and time – that advocating their transcendental nature – could be rescued from the abandonment of Newton's *physical theory* of absolute space and time. As evidence that at least Gödel's view on this subject differed from the received cliché, the following passage taken from an unpublished manuscript deserves being quoted: "the agreement described between certain consequences of modern physics and a doctrine that Kant set up 150 years ago in contradiction both to common sense and to the physicists and philosophers of his time, is greatly surprising, and it is hard to understand why so little attention is being paid to it in philosophical discussion of relativity theory" (Gödel 1990, vol. 2, p. 236).[7]

Obviously, philosophers that are possessed by the demon of realism would look at the two questions above from an opposite perspective. That time and space cannot exist in the same sense in which objects and event exist is clear from the simple remark that it does not make sense to say that spacetime is in spacetime. If localizability in spacetime is the rule by which we establish the existence of something, we cannot apply the rule to spacetime itself, for reasons that are even stronger than those that are usually presupposed by the claim that the standard meter in Paris is *not* one meter long. However, couldn't one grant that time exists *unlike* any ordinary object, and yet reject the Kantian view, according to which time and space are "characteristics not inherent in the things in themselves, but only in their relation to our sensibility" (Kant 1783, 284)?

3. Three logically independent senses in which time can be said to be real

In order to explore in more detail the plausibility of rejecting the Kantian explanations of Q_1 and Q_2, let us clarify why Kant's philosophy of time implies, *in modern parlance*, the mind-dependence of time. After all, it could be objected that Gödel's attribution to Kant of the claim that "time is ideal" is quite illegitimate. Especially in view of the famous distinction between phenomena and noumena, it could be replied that, according to Kant, time is unreal only insofar as it is referred to the noumena, or to the things in themselves, which are neither in space nor in time, but that, whenever it is referred to the world of phenomena, it is *real*. Isn't this the essential meaning of the famous Kantian view according to which *"time is empirically real but transcendentally ideal"*?[8]

Despite the undeniable fact that our current philosophical terminology can be superposed only partially with Kant's, I take it as uncontroversial that to the extent that it is a pure intuition, and is therefore *an a priori form* of our sensations and observations, time for Kant cannot be regarded as existing in a mind-independent way ("time, apart from the subject, is nothing in itself"). *To the extent that Kant's noumenal world can be identified, following Gödel, with whatever goes beyond the directly observable entities* – a world that, going beyond the phenomena, according to van Fraassen (1980) is and will remain forever inaccessible to our knowledge – the thesis that time is "transcendentally ideal" can have (at least) *three* related, "modern" interpretations:

(a) it may mean that time (i.e., spacetime) does *not* exist over and above physical entities, in particular those that are required to measure distances and temporal intervals.

(b) supposing that such a relationism is true, it may mean that any statement containing temporal relations referred to unobservable physical entities in van Fraassen's sense is reducible to *relations* between non-spatiotemporal, purely *physical* properties of events. If time can be "defined away", and it is therefore not a fundamental ingredient of the world as described by theoretical physics, we would have a situation like the one that is hypothesized, say, by the yet-to-be-built quantum theory of gravity, in which time is a purely "derived", "emergent" entity;

(c) it may mean that time as presupposed by theoretical physical models *cannot* be accounted for in terms of possible experiences of observers. In this case, *the time of our experience* can not be referred to the physical world at all, simply because it is *incompatible* with time as is conceived, say, within spacetime physics.

Clearly, the claim expressed in (a) is denied by a realist about time and spacetime: if time, regarded as the temporal aspect of spacetime, exists as a theoretical entity, then it exists over and above – i.e., independently of – the observable events and things located in it, *and therefore, in particular, independently of the knowing subject*. In this case, not only could the reductionist program not be completed, but what matters more, Kant's theory of time would be simply *false*, in the same sense in which Kant himself took the Newtonian thesis of the "existence of time as independent of anything external" as being incompatible with his view.

As for (b), which presupposes the truth of (a), it should be regarded as a condition of coherence of any mind-dependent, Kantian view of time: any difference in our experience must be accounted by a difference in the objective world: as Weyl used to say quoting Helmholtz, "a difference in the given perceptions always rests upon real conditions". In our case, the postulated, theoretical physical world, which we (somewhat arbitrarily) identify with the noumenal world, must contain *something* that is capable of giving rise to our experience of time, *via* the encounter of the physical world with our sensibility. This something could, for instance, be *causation*, physicalistically described.[9] For our purposes, it is not necessary to go into details in this context, which could be extremely relevant for the emergence of time in quantum gravity; let it suffice to say that if a Kantian program were to succeed, such a relation of causation should be characterizable in wholly non-temporal terms.

However, while substantivalism alone would be sufficient to reject Kant's philosophy of time (a possibility that Gödel did not consider), relationism by itself would not suffice to confirm it: Kant was no relationist either. In fact, even if we granted the relationist that time (spacetime) has no independent, substantial existence, but just amounts to certain temporal *relations* exemplified by physical entities of some sort, what matters more to evaluate Kant's theory of time is the issue whether such temporal relations, whatever they are, *are exemplified by physical events independently of observers*. This latter issue, in its turn, is tantamount to asking *whether the discriminations of the time of our experience can be coherently attributed to the physical world as described by our best spacetime theories*.

Clearly, such an issue calls into play the question raised in (c) above, which here will be given central emphasis, given that they also correspond to the way in which Gödel viewed the relationship between relativity theory and Kantian philosophy. He was in fact convinced that Kant was too skeptical about the possibility of getting to know the things in themselves, and thought that Einstein's contribution to physics had to be regarded as evidence that we have somehow been capable of throwing a partial glance at the noumenal world (see Wang 1995, p. 222). However, Gödel could argue in favor of some sort of progress in our knowledge of the noumenal world only to the extent that the time of our experience *cannot* be applied to the physical universe. Not by chance, Gödel's only quotation of Kant is directed at exactly this point "Those affections which we represent to ourselves as changes, in beings with other forms of cognition would give rise to a perception in which the idea of time, and therefore also of change, would not occur at all" (Gödel 1949, p. 558).[10] Gödel's claim – to be evaluated in the following – according to which the time of our experience cannot be accommodated by Einstein's theory of relativity becomes then of paramount importance, since it assumes a whole new dimension in the theory of knowledge. If my reading of Gödel is correct, such a claim becomes in fact a suggestive piece of evidence that Einstein's theory of relativity helped us to peek behind the veil of Maya of phenomena, and that such an overcoming of the bounds of our senses is indeed possible!

One way to gather evidence about the objectivity of the temporal discriminations of our experience ("those affections which we represent ourselves as changes") is to look at the way *we speak about time.* As the analytic philosophy of time of our century has made abundantly clear, there are in fact *two ways* of distinguishing events in time, the tenseless way of "earlier and later than", giving us the changeless order of *succession*, and the *tensed* way of past, present and future, giving us *transition and passage.* The former way calls into question the mind-dependence of the distinction between earlier and later, and therefore not only its reducibility to *asymmetrical physical processes* but, more in general, the vexed issued of the origin and nature of *de facto* irreversible physical processes (the so-called "arrow of time"). The latter way, by involving the issue of the privileged existence of a particular instant of time, the present, which in our experience seems to separate the fixity of the past from the openness of the future – calls into play the vexed problem of the objectivity of temporal becoming, that is, *the question whether the coming into existence at any instant of time of previously future facts and events is mind-dependent.*

Considering the by-now widely agreed-upon untranslatability of tenseless talk into tensed talk,[11] the thesis of the reality of time is then reduced to *three logically independent issues, namely the possibility of a substantial spacetime, the mind-independence of the distinction between earlier and later and the objectivity of the distinction between past, present and future.* Here it is important to remark that a discussion of the second and the third issue is required by the fact that Kant does not just deny the objectivity of time and space in themselves (i.e., their substantiality): this, as Gödel noted in his manuscript B2 (1990, p. 238) had already been accomplished by Leibniz. The true Kantian achievement, which in our century has been downplayed by every philosopher of space and time except Cassirer and Gödel, consists in the denial that temporal (and spatial) relations exists objectively, that is, that they are mind-independent.[12]

Apart from the issue of substantival versus relational time (spacetime) in the general theory of relativity – which here cannot be discussed[13] – I claim that we should regard time as transcendental *only if both* the distinction between earlier than and later than *and* that between past, present and future[14] is mind-dependent.

Now, while the latter distinction is usually regarded as mind-dependent, the former typically isn't,[15] something which makes the thesis of the mind-dependence of time (the Kantian view of time) particularly *strong*, since it implies that also the distinction between the two "directions" of time is relative to a temporal bias introduced by observers (see Price 1996). Furthermore, it should also be noted that although the reality of becoming would somehow affect the problem of the direction of time, the mind-dependence of the A series alone does not suffice to establish that time is ideal. It is in fact logically possible to regard the future as real as the past – thereby denying becoming – while claiming that time is "asymmetric" because of the existence of *de facto* or nomologically irreversible processes. However, the distinction between past present and future is so essential to our experience that we discovered that the special theory of relativity cannot in principle represent it, Kant's theory of time would receive a significant confirmation by modern physics in the sense conveyed by (c) above.

4. The "now" and Minkowski spacetime

In what follows, I will assume that trying to establish whether the distinction between past, present and future has an objective, physical counterpart – and is therefore *mind-independent* – means trying to establish whether such a distinction is *definable* in terms of *invariant* structure of Minkowski spacetime as is standardly presented in physics textbooks.

Such a choice appears preferable for two reasons. The first is given by the fact that the requirement of definability in terms of invariant relations ensures that the candidate becoming relation be invariant for all possible observers. In our context, such an invariance suffices to ensure the intersubjective validity of the becoming relation, something that should be required by any theory of objective becoming. The second reason is given by the remark that attempts at introducing additional structure in Minkowski spacetime, as in Rakic (1997), essentially amount to change the physical theory only in order to make it compatible with one's pet metaphysical hypothesis. Such a move should be resisted because, given a sufficient amount of tinkering with a physical theory – even if, as in Rakic's case, the tinkering consists in the addition of empirically superfluous structure which does not lead to any contradiction with the known phenomena – one can make a physical theory compatible with *any* metaphysical hypothesis. In this case, however, it is clear that the project of studying the compatibility between the time of our experience and the time of physics would be *trivialized*.

Our task is rather to find a geometrical structure S within Minkowski spacetime with respect to which one *could make* the distinction between what is already definite and fixed (the present and perhaps the past) and what is open (the future) required by objective becoming. Such an S should in any case be regarded as a merely *necessary condition* for the existence of a distinction between the actual (the present) and the merely possible (the future), with the tacit understanding that a physical theory cannot be deemed capable *in principle* to explain what the coming into being in the present required by becoming really amounts to. Clearly, *it is not physics' business to tell us which instant of time is present*, and this remarks holds for Newtonian spacetime as well as for relativity. However, if the postulation of *S* were to lead to inconsistencies with the structure of Minkowski spacetime, we would have discovered an inconsistency between the time of experience and the time of relativity theory, a fact that would lend support to Kant's view of time.

In a recent paper read at the PSA98 meeting, Savitt has produced new and cogent arguments against the claim that the notion of "the present", as it is commonly understood in our experience of time, can be implemented in Minkowski spacetime (Savitt 2000). While in our experience only what occurs in the present properly speaking exists – let us call this view *presentism*, tensely or tenselessly construed – for Savitt "there is no time like the present" in Minkowski spacetime, because there is no geometrical structure capable of capturing the relevant metaphysical intuitions.

Savitt considers various options that have been proposed in the literature, among which (1) the single spatiotemporal point or the "here-now" (Stein 1991), (2) events *on* (not in) the backward light cone (Godfrey-Smith 1979), (3) the spacelike-related region with respect to a point (Weingard 1972) and (4) the set of events that are orthogonal to the world line of an inertial observer (an hyperplane of simultaneity) (Putnam 1967). He finds counterintuitive features with each of these proposals, like the *non-achronality* of the second and the third option,[16] or the *arbitrariness* of choosing between one "here-now" over another for option (1), or between two different hypersurfaces of simultaneity *relative to* two observers in *different* inertial frames for option (4). Which of the two points or the two hypersurfaces is the bearer of reality in a theory that, like Einstein's relativity, does not consider any single "perspective" as privileged?[17] As to (1), if it is understood as implying that each event is real from its own perspective but no event is privileged, it is hard to tell the difference between this "pluralistic solipsism", as Stein called it (1968), and the view that all events are real, except that in the latter view, *any* event is real as of *any* other event, while in the former, reality functions as an indexical like in Lewis' indexical theory of actuality for possible worlds.

Shouldn't we conclude that a Kantian view of the distinction between past and future is confirmed, as Gödel had it, by the special theory of relativity? Before yielding to this verdict, let us give the realist about becoming the right to reply. Her best strategy may consist in trying to account for our experience of time by using the physical principles embodied in Minkowski spacetime. By making due allowance to the necessary idealization that is always present in any physical model, the most obvious choice to make for the physical counterpart of the present in Minkowski spacetime is *an appropriately chosen seg-*

ment of a worldline. It is obvious that the enterprise of trying to model the present in Minkowski spacetime is doomed to fail from the start if we don't have a more precise idea of the structure of our experience. In this respect, the notion of a pointlike event is completely *inappropriate* to model the psychological present, since the latter, if anything, has a certain, content-dependent, but *finite (non-pointlike) temporal duration*. This fact is responsible for one of the objections that Savitt raised to the "here-now view of reality" in relativistic spacetimes, involving two observers (Carol and Ted) who are very close to each other in the *same* inertial frame. Combining the stipulation that the present is a spatiotemporally non-extended point with the view that only what is present is real, our two observers cannot recognize their reciprocal existence, (2000, p. S568) simply because they are located in causally disconnected regions of spacetime. Since if they are close enough, as we are supposing, in real life they *do* share the same present, the point-like event is clearly an inadequate choice for the present of our experience. Can we remedy this situation?

As we are about to see, however, the gain in faithfulness to our experience that could be achieved by considering a segment of a worldline as "the present" is paid in terms of non-invariance. The length of the experienced time along any two segments of two *distinct* worldlines in Minkowski spacetime – representing the different spatiotemporal carrier of Carol and Ted – is worldline-dependent, as the twin paradox notoriously shows.[18] Since this dilemma can serve as an introduction to the only two rigorous results in the literature, to be evaluated later, it is worth delving into it in some more detail.

Without buying into the controversial notion of the "specious present", it is plausible to suppose that if the temporal interval separating two flashes of light presented to a normal subject diminishes more and more, below a certain minimal threshold the signals will be *perceived* as simultaneous. Let us then assume, in accord with various empirical findings (Giulio 1995, p. 151), that if the two flashes of light, presented as a double pulse stimulation to the rod-free foveal region of the retina, are temporally separated by less than approximately 15 milliseconds (14 to 19 thousands of a second), the two events are judged as being *one*. In this temporal interval, however, a light ray covers approximately 4500 kilometers. Imagine a sphere whose center is in the middle of the room where you are sitting and whose radius is exactly 4500 km. Think first of a flash of light emitted on the surface of the sphere, and then consider another event, coinciding with our turning on the light bulb of the projector right in the center of the sphere: the two events would be perceived as simultaneous, despite their distance. This holds, a fortiori, for any flash emitted within the sphere reaching our retinas within the temporal threshold given above.

This fact has remarkable consequences. The only reason why we naturally believe that the present extends at a distance is given by the fact that our capacity of discriminating two signals as temporally successive allows light to travel back and forth many times between ourselves and the objects around us (see Stein 1991). In the environment we have evolved, we never had to deal with objects that are hidden behind the visual horizon, and the latter extends far less than 4500 km, and is therefore well within the sphere of what we could call psychological simultaneity. This is why, in the sense of a "naive physics" implanted in our brains through the evolution, *we have evolved with the "natural" belief that there exists a universe of objects or events that is objectively simultaneously with our perception even if at a distance.* However, from a physical point of view, a cosmic now *need not* exist, and the metric of Minkowski spacetime – by dictating the existence of events that are *not causally connectable* with any event p, no matter how small the neighborhood centered in p is – clearly *bans* such a temporal structure.

Going back to Savitt's criticism of option (1), it should be now clear why it forces us to switch from Stein's point-like present (the here-now) to *an appropriately long segment of a worldline* ($ct = 4500$ Km), a modification that is capable of explaining why, provided they are "close enough", as it is always the case in our experience, Ted has information about Carol and viceversa within a single act of perception. The light signals have had enough time to travel back and forth between parts of their worldlines on their respective light-cones, a fact that has interesting consequence also for the origin of our concept of *space.* From the above remarks, it follows that the physical counterparts of a set of *subjectively* simultaneous events do not objectively co-exist with each other, *even though in our experience they appear to be simultaneous.* In this sense, space as the order of coexistence is a construction of the perceiving subject in a sense envisaged by Kant, exactly as a picture of the night sky shows objects in its center and on the sides of as if they were simultaneous, while in reality the light that reaches us from the sides took much longer.

In a word, my objection to Stein's assumption that the psychological present is modeled by a point, therefore, is that such an assumption is not true to the facts of our experience that he himself has so clearly and deeply stated. Probably the reason for Stein's avoidance of the identification between our psychological present and a segment of worldline is the loss of invariance. The trouble is, however, that if "we stick to the point", as Stein does, we cannot avoid the difficulties illustrated above about Ted and Carol not sharing their present even if they are very close to each other!

Be that as it may, in order to present my other objections to Stein's theorem – which defines becoming in terms of the relation of past causal connectibility (1991) – let us charitably set this dilemma aside, by stipulating that it can be solved by invoking the approximate nature of models.

5. Becoming and Minkowski spacetime

Once we agree that the present of our experience is to be modeled by a physical point, two of the premises of Stein's theorem are easily justified by the explanatory scheme given above. Clearly, any event that might possibly affect our present experience a along the worldline of our body λ should be regarded *as real with respect to a*. Otherwise, how could it be causally affecting us now? It follows that all events in the causal past of any event a should be regarded as having become as of a. Furthermore, any event b in the causal past of a such that b carries the mark of some previous event c in its causal past, makes it the case that also c is real with respect to a. This gives us transitivity, and, together with our first remark, Stein's first two premises:

1. A becoming relation B is a transitive, reflexive relation holding among spacetime events;

2. If any event a is a cause of b, the former has become (is real) as of the latter, that is, aBb. The other two premises are:

3. the relation B is invariant under automorphisms of Minkowski spacetime preserving the time orientation of Minkowski spacetime;

4. for any event e, there are events that as of e are unreal, or have not become.

While the third premise fulfills the invariance requirement postulated at the beginning of the fourth Section,[19] the fourth premise is justified by the fact that if we want to know whether becoming is compatible with the structure of Minkowski spacetime, we need to assume that it is possible, so that at least some events are unreal as of any event of the spacetime. On the basis of these four premises, Stein has proved that for any pair of events a and b, a has become as of b just in case a is in the causal past of b.

Clifton and Hogarth (1995) have later extended Stein's result to the relation of past *chronological* connectibility (becoming along a worldline), in such a way that the latter, together with Stein's past causal connectibility, have been shown to be the only becoming relations that are definable in terms of Minkowski spacetime. This result in particular implies that if a single event, spacelike related with an event p, were regarded as being real with respect to p, then all events would have to be regarded real as of any other event, a result which gives us an additional argument to reject Weingard's and Putnam's proposals to identify "the now" with a spacelike related region (see 4. above).

Unfortunately, there are three major conceptual problems with Stein's and Clifton's and Hogarth's technical results (SCH from now on). The first involves the EPR paradox and Bell's non-local correlations, which implies, against a consequence of SCH's becoming, that spacelike related events *are* mutually real. The second is the difficulty of interpreting *becoming along worldlines* as a *bona fide physical* becoming, while the third calls into question the origin and nature of the asymmetry of causation, tacitly assumed by Stein and Clifton-Hogarth in their theorems (see the second premise above). As I have discussed the first problem elsewhere (Dorato 1996), here I will conclude my discussion by broaching the second and the third objection in turn.

The fact that the present is reduced to a point shows that in relativity we can't have a global becoming, that is, a becoming in time, but only a local becoming along worldlines, each providing a chronological perspective of the whole spacetime. Here is a defense of this view by one of its few supporters: "What we need not, and I think cannot, conclude is that there is some neutrally describable space-time reality which each of the chronological perspective is a perspective *of.* There is no identity of instants at a distance, and no neutral position from which the actual state of the world could be viewed *sub specie aeternitatis* – as was possible in the classical picture" (Godfrey-Smith 1978, p. 242).

This quotation may perhaps reply to some of the difficulties raised by Savitt (1998), concerning the fact that the difference between such pluralistic solipsism and the so-called "block view", in which spacetime is completely laid out, is dubious. Perspectivism just argues that there is no perspective-independent description of the universe, even though all perspectives are equally admissible, just because reality itself is relational and perspectival: "perspective is one of the components of reality. Far from being its deformation, it is rather its principle of organization" (Ortega & Gasset 1923/1961, p. 91). The block view of the universe argues, on the contrary, that Godfrey Smith's chronological perspectives are purely *mental*, a by-product of the existence of sentient organisms that need to act in a timely way, and whose experienced time agrees locally with the structure of spacetime, which, however, is timeless and, in some sense, the *sum* of all perspectives, as a Leibnitzian God regarded as "the monad of all the monads".

To a certain extent, this conflict reflects in part two broadly construed theories of the aims of our knowledge of the universe. Independently on one's view of the matter, however, it must be admitted that science progressed by trying to "discount" the perspectival effects due to our special position in the universe.[20] Moreover, to the extent that the block-view theorist is correct in arguing that such a "perspectivism", in order to make sense at all, needs *conscious observers*, or at least unconsciously perceiving monads located in every point of spacetime, the strategy of the realist about becoming would be completely useless. Becoming would in fact turn out to be mind-dependent, or at least would require a moderately animistic metaphysics of a Leibnizian sort. The Kantian position on the distinction between past and future would be vindicated.

On the other hand, the realist can still reply that the special theory of relativity does not need observers for its formulation. However, even granting that observers are only *possible* occupants of timelike curves, what sense does it make to consider a physical event as the entity with respect to which the *"chronological perspective"* above is referred to? Such a physical event *a* would have to be regarded as bearing in itself all the traces of its causal past, plus the dispositions to transmit the causal influence to the future, but the word perspective, associated to an event described physically, would have nothing but a metaphorical meaning. In any case, associating to the chronological perspective of an event in Minkowski spacetime the sum of the causal influence of its past light cone – an influence whose nature, given a sufficient amount of determinism, is coded in the properties, dispositions, or causal powers of the event itself – still presupposes a *temporally asymmetric relation of causal connectibility*, an assumption that also SCH have *tacitly* made in order to derive their result.

This gives us the third, and more telling objection: on what grounds are we entitled to assume that causation (causal connectibility) is temporally asymmetric? If our aim is merely to show the *compatibility* of becoming and STR, we are free to postpone this worry, as Stein correctly did. But if we are trying to find out whether becoming is a real feature of the universe, such a worry should become central.

Stein argues that since "the interpretation of the Einstein-Minkowski structure as spatio-temporal depends critically upon the principle that the state at any point *a* is subject to influence by the states at all points in the "causal past" of *a*, it follows that "the states at all such points should be supposed to be definite as of *a*" (Stein 1991, 149). These quotations patently show that Stein is assuming that causation has temporally asymmetric and ontological implications: *if* becoming is real, and if *a* causes *b*, the former event must be definite (real) as of the latter's viewpoint. The converse, however, does not hold, otherwise also events in the future light cone should be regarded as real. But if we are assuming that becoming is mind-independent, from the temporal viewpoint of its cause, each of its possible effects should be regarded as unreal.

Clearly, attributing the arrow of causation such an *ontological status* requires a justification in *physicalistic* terms, in contrast to those theories claiming that the arrow of causation depends only on the existence of conscious agents. If the origin of the arrow of causation were subjective (Healey 1983, Price 1996), *relativistic becoming – depending as it does on a temporally asymmetric causation – would end up being subjective as well.*[21] Consequently, there appears to be an important link between the two main issues of the philosophy of time hinted above that needs to be explored, and that calls into question the relationship between becoming and the arrow of time (earlier and later), the third sense of "ideal/real" required by a Kantian approach to time. One of the merits of SCH is actually have brought to the fore such a relationship.

Since I have discussed the issue in another paper,[22] here I will simply state the main difficulties. Causation is the fundamental asymmetry of our experience, and clearly depends on the fact that we can act only for the sake of the future. Other asymmetries of our experience, like that of traces, counterfactual dependence and explanation, depend on that of causation: if backward causation were a common phenomenon, we would have traces of the future and counterfactual dependence of present events on the future rather than on the past, as well as explanation of present events by future causes. Unfortunately for the causal realist, and despite the important role of the asymmetry of asymmetry in our experience, any philosophical analysis that tries to *reduce* such an asymmetry to some physical process, or to an exchange of conserved physical quantities, must rely on *physical laws of conservation* (see Dowe 1992, Heathcote 1989, Salmon 1994). Since the latter are time reversal invariant, Stein's assumption of an asymmetry in the causal relation, at least in physicalistic terms, *is unjustified, unless one can relate it to* de facto *asymmetric phenomena in time*. However, such an association of the direction of causation with entropy growth or positive energy (Faye 1997) would be *extrinsic*: there is nothing in the concept of cause or effect that is in any way linked to physical attributes that make an event into a cause or an effect.

Of course, one could reply that the justification of a direction of causation is in our experience: we cannot affect the past, we can influence the future, so that the latter must be conceptualized as "empty" or open. But if we are trying to decide whether such an intuition is objective, we cannot rely on a metaphysical elaboration of common sense. It is only physics that can tell us whether causation is *objectively* asymmetric as required by Stein's assumption for becoming: if the only justification for the asymmetry of causation comes from metaphysics, for our purposes we would be moving in a circle.

In view of these difficulties, I think one can safely conclude that any form of becoming – even the *tenseless*, relational form advocated by Stein, which has been regarded as a watering down of its stronger, tensed version (Callender 1997) – cannot be reconciled with the special theory of relativity. In any of the choices presented above, the kind of time that one would recover by trying to define becoming in Minkowski spacetime would have so little in common with experienced time that Gödel's claim about the inapplicability of the time of our experience to a mind-independent world seems correct. It is at least ironical to remark that Kant's view of time seems vindicated by that very theory (relativity) that many 20th century philosophers regarded as its official burier.

Notes

1. Physicists currently engaged in building a quantum theory of gravity are beginning to suggest that "time emerges from the observer/observed relation or just is such a relation" (Rovelli 1997, p. 217).

2. The question (iii) will not be discussed in full, but only in its relation with (ii).

3. Correcting Kim's definition to make it relativistically invariant simply entails the addition of a spatial location to the 'proper' time of occurrence, where proper has the technical sense of relativity.

4. Even though this rendering could be accused by Kantian scholars of being too"psychologistic", here I am only concerned to study the problem of the mind-dependence of the temporal relations and not the conceptual-logical question of the condition of possibility of our knowledge in general.

5. For Gödel's theory of time, see Yourgrau (1991), Savitt (1994), Wang (1995) and Earman (1995, ch. 6).

6. A brilliant, contemporary exception is Pauri (2000).

7. This quotation is taken from manuscript B2, written between 1946 and 1949, and therefore as a preparation to Gödel's (1949a.)

8. "Our observations prove the empirical reality of time, that is, its objective validity with respect to any object that can ever be given to our senses" (Kant 1787, 2 Section, B 6, my trans.; second edition of the *Critique of Pure Reason*). In this context, ideal means that "time in itself, apart from the subject, is nothing". (*ibid.*)

9. Interestingly, there is some evidence that Gödel regarded causation as a truly, objectively existing relation, giving rise to our experience of succession and change: "The real idea behind time is causation: the time structure of the world is just its causal structure." (Wang 1995, p. 229).

10. The passage quoted by Gödel is in the 1787 edition of the *Critique of Pure Reason*, p. 54.

11. See Mellor (1981) and Faye (1989), who defend a tenseless ontology and are therefore tenseless theorists.

12. In Kant's parlance, as Gödel reports it in his unpublished manuscript *C*1, the three theses presented above correspond to the claims that "time is neither something 'existing in itself', nor 'a characteristic or ordering inhering in the objects', but only a characteristic inherent in the relation of the objects to something else" (1990, p. 247).

13. Rynasiewicz (1996) offers some good reasons in favor of my ignoring this side of the issue: to the extent that the general theory of relativity fails to distinguish between empty space and matter, the latter theory cannot be invoked to settle the dispute between relationists and substantivalist, which was meaningful only in the 17th century.

14. Since McTaggart (1908), the series of events going from past, to present to future is known as the A series, while the B series is the series of events ordered by 'earlier and later than'.

15. Grünbaum (1973), for instance, does not regard the distinction between earlier and later than as being mind-dependent, though he is notoriously an advocate of the mind-dependence of nowness.

16. A region of spacetime R is achronal if and only if any two events belonging to R are connected only by spacelike curves.

17. See also Dorato (1995, ch. 10).

18. The proper time along a geodesic connecting two point, as is well known, maximizes the length.

19. The temporal orientation is needed to coherently distinguish the two directions of time.

20. The Copernican revolution, and the belief that other galaxies are not part of our own Milky Way are two famous examples.

21. Note that also Clifton and Hogarth (1995), despite their suggestion of a worldline-dependent becoming, need to rely on asymmetric causation. If, as they write, our having lived through certain events is a sufficient condition to regard them as real, an event can be judged to be earlier than another only if the first is remembered while the second is perceived. In other words, one has to have a trace of the first event while one perceives the second as being later than it. And the trace asymmetry presupposes the asymmetry of causation.

22. See Dorato (2000).

References

Auyang, Sunny: (1995), *How is Quantum Field Theory Possible?*, Oxford University Press, Oxford.

Callender, Craig: (1997), 'Review of M. Dorato, 'Time and Reality', Clueb, Bologna', *British Journal for the Philosophy of Science* **48**, pp. 117–120.

Cassirer, Ernst: (1920), *Zur Einstein'schen Relativitätstheorie*, Bruno Cassirer Verlag, Berlin.

Clifton, Robert, Hogarth, Mark: (1995), 'The Definability of Objective Becoming in Minkowski Spacetime', *Synthese* **103**, pp. 355–387.

Dorato, Mauro: (1995), *Time and Reality*, Clueb, Bologna.

Dorato, Mauro: (1996), 'On Becoming, Relativity and Nonseparability', *Philosophy of Science* **64**, pp. 585–604.

Dorato, Mauro: (2000), 'Becoming and the Arrow of Causation', *PSA 98*, *Philosophy of Science*, Supplement to vol. 67, pp. S523–S534.

Dowe, Phil: (1992), 'Wesley Salmon's Process Theory of Causality and the Conserved Quantity Theory', *Philosophy of Science* **59**, pp. 195–216.

Earman, John: (1974), 'An Attempt to add a little Direction to The Problem of the Direction of Time', *Philosophy of Science* **41**, pp. 15–47.

Earman, John: (1995), *Bangs, Crunches, Whimpers and Shrieks. Singularities and Acausalities in Relativistic Spacetimes*, Oxford University Press, Oxford.

Faye, Jan: (1989), *The Reality of the Future*, Odense University Press, Odense.

Faye, Jan: (1997), 'Causation, Reversibility and the Direction of Time', in J. Faye, U. Scheffler, M. Urchs (eds.), *Perspectives on Time*, Boston Studies in the Philosophy of Science, vol. 189, Kluwer, Dordrecht, pp. 237–266.

Giulio, Ludovico: (1995), 'La microstruttura del presente', in E. Agazzi (ed.), *Il tempo nella scienza e nella filosofia, Guida*, Napoli, pp. 147–166.

Gödel, Kurt: (1949), 'A Remark About the Relationship between Relativity Theory and Idealistic Philosophy', in P.A. Schilpp (ed.), *Albert Einstein: Philosopher-Scientist*, LaSalle IL, Open Court, pp. 557–562, reprinted with corrections and additions in Gödel (1990).

Gödel, Kurt: (1990), *Collected Works*, S. Feferman et. al. (eds.), Vol. 2, Oxford University Press, Oxford.

Godfrey-Smith, William: (1991), 'Special Relativity and the Present', *Philosophical Studies* **36**, pp. 233–244.

Grünbaum, Adolf: (1973), *Philosophical problems of space and time*, 2nd enlarged edition, Reidel, Dordrecht.

Heatchcote, Adrian: (1989), 'A Theory of Causality: Causality = Interaction (as defined by a suitable field theory)', *Erkenntnis* **31**, pp. 77–108.

Howard, Don: (1997), 'A Peek behind The Veil of Maya. Einstein, Schopenhauer and the Historical Background of the Conception of Space as a Ground for the Individuation of Physical Systems', in J. Earman, J. Norton (eds.), *The Cosmos of Science*, University of Pittsburgh Press, Pittsburgh, pp. 87–150.

Kant, Immanuel: (1781), *Kritik der reinen Vernunft*, in *Kants Gesammelte Schriften*, vol. 3, Deutsche Akademie der Wissenschaft, Berlin-Leipzig, 1900–1902.

Kant, Immanuel: (1783), 'Prolegomena zu jeder künftiger Metaphysik die sich als Wissenschaft auftreten wird', in *Kants Gesammelte Schriften*, vol. 4, Deutsche Akademie der Wissenschaft, Berlin-Leipzig, 1900–1902.

Kim, Jaegwon: (1976), 'Events as property exemplification', in M. Brand, D. Walton (eds.), *Action Theory*, Reidel, Dordrecht, pp. 198–215.

McTaggart, John: (1908), 'The unreality of time', *Mind* **68**, pp. 457–474.

Mellor, Hugh: (1981), *Real Time*, Cambridge University Press, Cambridge.

Nerlich, Graham: (1994), *The Shape of Space*, Cambridge University Press, 2nd edition, Cambridge.

Ortega & Gassett, J.: (1923/1961), *The Modern Theme*, New York, Harper.

Pauri, Massimo: (2000), 'Leibniz, Kant and the Quantum: a provocative point of view about observation, space-time, and the mind-body issue', in E. Agazzi, M. Pauri (eds.), *The Reality of the Unobservable*, Reidel, Dordrecht, forthcoming.

Price, Huw: (1996), *Time's Arrow and Archimedes' Point*, Oxford University Press, Oxford.

Putnam, Hilary: (1967), 'Time and Physical Geometry', *Journal of Philosophy* **64**, pp. 240–247.

Rakic, Nataša: (1997), 'Past, present and future and special relativity', *British Journal for the Philosophy of Science* **48**, pp. 257–280.

Rovelli, Carlo: (1997), 'Half-way through the Woods: Contemporary Research in Space and Time', in J. Earman, J. Norton (eds.), *The Cosmos of Science*, University of Pittsburgh Press, Pittsburgh, pp. 180–224.

Rynasiewicz, Robert: (1996), 'Absolute versus Relational Space-Time: an Outmoded Debate?', *The Journal of Philosophy* **43**, 1, pp. 279–306.

Salmon, Wesley: (1994), 'Causality without counterfactuals', *Philosophy of Science* **61**, pp. 297–312.

Savitt, Steven: (1994), 'The Replacement of Time', *Australasian Journal of Philosophy* **72**, pp. 463–474.

Savitt, Steven: (2000), 'There is no time like the present (in Minkowski spacetime)', *PSA* 98, *Philosophy of Science*, Supplement to vol. 67, pp. S563–S574.

Stein, Howard: (1991), 'On Relativity Theory and Openness of the Future,' *Philosophy of Science* **58**, pp. 147–167.

Shapere, Dudley: (199?), 'The origin of time', manuscript.

Van Fraassen, Bas: (1980), *The Scientific Image*, Oxford University Press, Oxford.

Wang, Hao: (1995), 'Time in Philosophy and in Physics from Kant and Einstein to Gödel', *Synthese* **102**, pp. 215–234.

Weingard, Robert: (1972), 'Relativity and the Reality of Past and Future Events', *British Journal for the Philosophy of Science* **23**, pp. 199–121.

Yourgrau, Palle: (1991), *The Disappearance of Time*, Cambridge University Press, Cambridge.

SYMMETRIES: PHYSICS OR METAPHYSICS?

Margaret Morrison
Philosophy Dept.
University of Toronto
Toronto, Canada

1. Outlining the problem

Speaking about the unsuccessful attempts to find magnetic monopoles, Dirac once remarked that "[...] pretty mathematics by itself is not an adequate reason for nature to have made use of a theory." Pretty mathematics is often associated with symmetry. Perhaps the simplest way to think about symmetry in physics is to use Weyl's idea: a thing is symmetrical if one can subject it to a certain operation and it appears exactly the same after the operation. This applies not just to physical objects that undergo operations such as rotation and translation in space but also laws, which are transformed in certain ways that leave their form unchanged. Symmetry is studied mathematically using group theory where a group G is defined as a set of elements or operations for which there is a multiplication law with the following three properties:

1. Associativity $(g_1g_2)g_3 = g_1(g_2g_3)$
2. An identity element e such that for every $g \in G$, $eg = ge = g$
3. An inverse element g^{-1} such that for every $g \in G$, $g^{-1}g = gg^{-1} = g$

Groups of transformations that leave the Hamiltonian or Lagrangian invariant are called symmetry groups. Some symmetries are considered to be directly physical such as the reflections and rotations for molecules and translations in crystal lattices. Others are more abstract and involve changing properties as in the case of symmetries associated with gauge theories and the organization of nuclear and sub nuclear processes.

The phase transformations of a theory like quantum electrodynamics are one parameter transformations and form what is called a one dimensional abelian group (meaning that any two transformations commute) known as the $U(1)$ group of a $U(1)$ gauge symmetry. As I mentioned at the beginning the mathematical description of all symmetry operations are dealt with in group theory which uses algebra to describe operations like rotations. These symmetry

P. Gärdenfors, J. Woleński and K. Kijania-Placek (eds.), In the Scope of Logic, Methodology and Philosophy of Science, Vol. I, 349-367.

groups, however, are more than simply mathematizations of certain kinds of transformations since in the non-abelian case (non-commutative transformations) it is the mathematical structure of the symmetry group that determines the structure of the gauge fields and therefore, indirectly, the form of the interactions. In these more complicated situations there are several wave functions or fields transforming together as in the case of $SU(2)$ and $SU(3)$ transformations that involve unitary matrices acting on multiplets.[1] These symmetries are internal symmetries and are typically associated with families of identical particles. In each case the conserved quantities are simply the quantum numbers that label the members of the multiplets (such as isospin and colour) together with operators that induce transitions from one member of a multiplet to another. Hence the operators correspond on the one hand to the conserved dynamical variables (isospin etc.) and to the group of transformations of the symmetry group of the multiplets on the other.

For example, the conserved quantity isospin allows the proton and neutron to be considered as two states of the same particle. Here, a local gauge invariance means that although we can, in one location, label the proton as the up state of isospin $1/2$ and the neutron as the down state, the up state need not be the same at another location. Because this assignment of isospin values is arbitrary at each point some form of connection or rule is required to coordinate the various choices at different positions. In electromagnetism phase values are connected by the potential A_μ, hence an analogous isospin (connection) potential field is required. But, because the $SU(2)$ symmetry group which governs isospin is also the group that governs rotations in a three dimensional space, the potential field associated with this group is very different from the $U(1)$ phase group of electromagnetism. In the $SU(2)$ case the "phase" is replaced by a local variable that specifies the direction of the isospin.

The use of gauge theory applies not only to weak and strong interactions, governed by $SU(2)$ and $SU(3)$, but forms the foundation for all the fundamental forces (including gravity). The theory of Lie groups is the background group theory necessary for gauge interactions. These Lie groups can be treated as abstract entities in order to clarify their structure but they also occur naturally as transformation groups, that is, as representations – rotations of Euclidean 3-space, Lorentz transformations, and so on. Gauge theories, which determine the form of interactions, are based on or can be constructed from rigid internal groups using a gauge principle (or minimal principle). These rigid internal groups are space-time independent compact Lie groups that are used to classify fields, particles, and other physical quantities such as charge, currents, and masses associated with them.[2] The notion of a classification here refers to the assignment of some finite multiplet of such quantities (e.g. the field $\Psi_a(x)$) to a unitary representation $U(g)$ of the group which yields the following transformation[3]

$$\Psi(x) \rightarrow U(g)\Psi x.$$

The Lagrangian that describes the interaction of the fields with each other and with the electromagnetic field is invariant with respect to two different kinds of gauge transformations. The first type simply expresses the fact that the Lagrangian is electrically neutral and symmetric with respect to the rigid symmetry group $U(1)$ with parameter θ, (i.e. $\Phi \rightarrow e^{i\theta Q}$). The second kind of gauge invariance is the type originally discussed by Weyl and states that the Lagrangian is invariant with respect to a symmetry group that is not rigid but varies from point to point in configuration space $\theta \rightarrow \theta(x)$. Here the introduction of the electromagnetic field is necessary in order to keep the derivatives of the matter fields covariant. The process of letting $\theta \rightarrow \theta(x)$ is referred to as "gauging the $U(1)$ group" while the former process of letting $L(x) \rightarrow L_{em}(x)$ is known as "gauging $L(x)$", the matter Lagrangian, with respect to $U(1)$. Once the charges of the matter fields are assigned the form of the interaction of the electromagnetic (EM) field with matter is completely determined. In that sense symmetry can be said to determine the dynamics for the EM field. But such is not the case with the electroweak theory which requires the $SU(2)$ x $U(1)$ group. Here the form of the field interactions are governed by the non-abelian structure of the group but more is required to complete the physical dynamics. Symmetry breaking must be introduced into the theory as well.

Given the central role that abstract symmetries play in contemporary physics, questions about their ontological status become tremendously important. For example, anyone familiar with symmetries knows there exists a connection between symmetries and conserved quantities, a connection first established by Emmy Noether in 1918. But what does this tell us about whether symmetries themselves are part of the physical framework of our world? Noether's theorem states that for every symmetry of the laws of physics there is a corresponding conservation law/conserved quantity.[4] However, this relationship, in itself, is silent on the ontological question since the reference here is to symmetries of laws; and if we think of laws as simply abstract descriptions, then nothing *physical* about symmetries in nature follows from the connection between laws and conserved quantities. However, the connection needs to be further explicated in terms of the mathematics of our physical theories and the objects it purports to describe. That is, given our theories are formulated in the language of mathematics, do the symmetries of the mathematics correspond to symmetries in the world.

Questions about the correspondence between mathematical and physical objects is complicated by the fact that spontaneous symmetry breaking (SSB) or hidden symmetries are also a crucial feature of modern physics. So, the issue isn't simply whether the symmetry present in the mathematical form of physical laws is present in the world, but rather how should we understand the physical implications of symmetrical equations whose solutions are asymmetrical. Given that we have empirical evidence for the lack of symmetry, do we have a corresponding degree of evidence for the physical existence of symmetry? If so, what does this evidence consist in? Answers to these questions require that we look at some examples of symmetries and their link to the physical dynamics of particular systems. What I want to claim, on the basis of these kinds of examples, is that a physical interpretation of symmetries (the idea that symmetries are an inherent feature of the physical world) cannot be legitimated using empirical arguments from contemporary physics. Hence, any role they play over and above a mathematical one is strictly metaphysical.

When thinking about the relation between the mathematical and physical aspects of symmetries a useful distinction to keep in mind is the one between symmetries of laws (of nature) and symmetries of objects; a distinction that is traceable to Sophus Lie (1893) and Pierre Curie (1894). Lie was concerned mainly with symmetry in the equations used to represent laws of nature and saw in their invariance the origin of the conservation principles. Curie was interested in providing explanations of magneto and electro-mechanical effects, and hence in the symmetries of physical states or the solutions to the equations.[5]

Wigner (1931, 1979), whose work on symmetry methods in quantum physics earned him a Nobel Prize, further emphasised the importance of the distinction, claiming that symmetry principles apply only to the laws of nature rather than the physical states themselves. Indeed, one of the points I want to emphasise below is the way the comparison between the symmetries of laws and states can function as the conceptual basis for the notion of spontaneous symmetry breaking. The immediate concern, however, is the following: given that we know, in certain circumstances, the world exhibits non-symmetrical behaviour; are we similarly justified in thinking that underlying this is an inherent symmetry that is simply hidden or spontaneously broken. That is, should we assume that the natural state of the universe is the symmetry described by the Lagrangian or Hamiltonian?

One might want to claim that an argument relying on SSB does not represent a fair assessment of the issue, putting the case for the physical status of symmetries at a distinct disadvantage. Instead, an evaluation of the physical status of symmetries should be based on instances akin to the electromagnetic case where there is an immediate connection between the $U(1)$ symmetry group and the electromagnetic field. The difficulty is that the gauge formulation of electromagnetism contributes nothing to the theory's physical content that was not already in place. Ideally what we want is an example where symmetry provides us with some information about a physical system that could not be recovered from the phenomenology of the physics or other aspects of theoretical structure. That is, a situation where symmetry functions in an explanatory way, as a "stand in" for physical explanations. Such a case might provide increased evidence for a physical interpretation of symmetries.

One such instance is the way in which the Poincaré group (or inhomogeneous Lorentz group) can be used to yield a description of the purely kinematical properties like spin for quantum relativistic systems. Spin is a somewhat complicated case because there is a definite sense in which its physical nature is not fully understood. Yet, many would claim that group theory does provide the requisite framework for understanding the physical aspects of spin. On such a view there is no room for questioning the physical reality of symmetries, it simply collapses the distinction between the physical structure of the world and the mathematics used to describe it. Yet, I want to claim that even in this context it becomes apparent that symmetry cannot deliver on its promise to provide a truly *physical* explanation. Before discussing the spin example let us look more closely at SSB to see how it bears on questions about the physical status of symmetries.

2. Symmetry, causes and predictive power

Some remarks found in Pierre Curie's work (1894) provide a useful context in which to discuss the example of spontaneous symmetry breaking.[6] Although this phenomenon was not known at the time Curie claimed that in order for certain phenomena to occur it was necessary that some symmetry elements be missing, that "asymmetry is what creates the phenomena". His example was the Weidemann effect (the longitudinal magnetization of a uniform iron cylinder), which requires the absence of certain rotation and reflection symmetries.[7] He also specified two other principles, that the symmetry elements of causes must be found in their effects and that when effects show a particular asymmetry that asymmetry must be found in the causes that give rise to them. These principles are important not because they set the agenda for thinking about how symmetries need to be conceived empirically, but because they expose the ways in which symmetries can function as a metaphysical category or constraint. In order to show this we need to make explicit the relation between spontaneous symmetry breaking and the second and third of Curie's principles relating causes and effects.

Those familiar with quantum field theory know that the spontaneous symmetry breaking needed to introduce boson masses into the electroweak theory was an idea borrowed from superconductivity. The idea of a spontaneously broken symmetry (SSB) needs to be distinguished from that of a broken or approximate symmetry. In the latter case the symmetry is approximately true of the laws of nature or true for some interaction laws but not others. For example, isospin symmetry holds for weak interactions but it violated by electromagnetic forces. Here the term broken means that there are terms in the

Hamiltonian that violate symmetry; neither the Hamiltonian nor the physical world exhibit symmetry. In the case of hidden symmetries or SSB we have symmetrical laws governing asymmetrical phenomena. That is, the solutions to the equations of motion (the physical states) have less symmetry than the equations themselves. Here the symmetry is not really broken because the equations of motion remain symmetrical.

We can see then why some might think that the straightforward idea of a broken symmetry is not inherently problematic from a philosophical point of view. We simply say that in some cases the symmetry holds and in some cases not. The spontaneously broken symmetry however, presents us with a different concern. The empirical evidence points to an asymmetrical physical state while at the same time we want to affirm the existence of a hidden symmetry based on the form of the equations of motion. But what exactly is the relation between the symmetrical equations and the asymmetrical solutions? Should we think of asymmetry as somehow causally connected to symmetry? If so then what does this fact reveal about the relations specified by Curie's principles? And, if no causal relation exists between the equations and the physical states, how can we justify the ontological and epistemological status of hidden symmetry as a fundamental feature of the empirical world? Details from the history of the electroweak theory serve to clarify the issue.

Attempts to understand weak interactions by Schwinger in the 1950's and later Glashow and Weinberg in the 1960's were based on the idea that leptons carried a "weak" isospin analogous the strongly interacting particles discussed by Yang and Mills (1954). Consequently weak interactions could be considered invariant under rotations in the isospin space, and the corresponding Lagrangian would be invariant under the group of isospin rotations characterized by the $SU(2)$ group. On analogy with electromagnetism where the force is mediated by photon exchange, weak decay was thought to be mediated by the exchange of a spin-1 quantum called the W boson.[8] What this suggested was that weak and electromagnetic processes could be described using the same kind of theory. The most serious problem with the W particle was that its mass prevented the theory from being renormalizable.[9] Although local symmetry required the introduction of gauge particles like the W to maintain an invariant Lagrangian, gauge invariance allows only massless particles; so the W would effectively break the symmetry. Hence, one obtained an asymmetric solution (massive bosons) from a symmetric theory (gauge theory).

In order for the theory to work it had to be possible for the gauge particles to acquire a mass in a way that would preserve gauge invariance. Moreover, in order for electromagnetism and the weak force to be describable using the same theory some account of the coupling of massive and massless particles was crucial. The answer to these problems was provided by a mechanism known as spontaneous symmetry breaking.

It was Steven Weinberg (1967) who suggested was that one could understand the mass problem and the coupling differences by supposing that the symmetries relating the two interactions were exact symmetries of the Lagrangian that were somehow broken by the vacuum. As I mentioned above, these ideas originated much earlier and were motivated by work done in solid-state physics on superconductivity. There it was common to use quasi-particles that could not be directly mapped onto fundamental observable fields, to explain certain types of experimental results. Traditionally in field theory a direct correspondence was assumed between physical particles and terms in the Lagrangian. However, this idea was challenged on the grounds that when one takes interaction terms into account the correspondence between fields and observable particles might be broken. In other words, one might not be able to simply read-off from the Lagrangian an array of physical, observable particles – more might be involved.

A characteristic feature of superconductivity was that the superconductive ground state broke the electromagnetic gauge invariance. In their work in the 1930's on the macroscopic theory of superconductivity Landau and Ginzburg (1937) showed that SSB occurs whenever different phases have different symmetry. They also introduced the notion of a characteristic order parameter that takes nonzero values in the asymmetrical phase and vanishes elsewhere. An effective wave function ψ of the superconducting electrons was introduced as a characteristic function. In cases where ψ does not vanish SSB has occurred and the superconducting state characterized by ψ is asymmetrical.

Bardeen, Cooper and Schrieffer (1957) in their microscopic theory interpreted ψ as an energy gap. The gap was produced in the spectrum just above the ground state and was thought to result from a microscopic phenomenon known as Cooper pairing. The pairs of electrons have opposite momenta and spins with the paring brought about by an attractive phonon-mediated interaction between electrons. The gap is the amount of energy required to break the paring. So, the SSB is, in effect, the energy gap which is caused by the Cooper paring.

The translation of these ideas to particle physics by Nambu and Jona-Lasinio (1961) was based on an analogy between the physical vacuum of a quantum field theory and the ground state of an interacting many-body system – both of which are states of minimum energy. In quantum field theory the state of minimum energy need not be one in which all of the fields have zero average value. Because fluctuations are always possible it may be the case that in the true ground state of the theory some field will have a non-zero equilibrium (expectation) value, producing an asymmetrical situation. One such example is the ferromagnet where the basic interactions between the constituents of the system are rotationally symmetrical even though its ground state configuration is not.[10]

Although this was an attractive idea it had one major drawback. Goldstone, a high energy physicist, had shown that cases of spontaneous symmetry breaking involve the production of a massless, spinless particle, now referred to as the Goldstone boson. For each degree of freedom in which the symmetry is spontaneously broken, a massless scalar field always appears. Since no such particles were thought capable of existing it was difficult to see how the idea of spontaneous symmetry breaking could be applied to field theory. But, once again a clue was forthcoming from the field of superconductivity.

In that context symmetry breaking does not give rise to Goldstone excitations, a result that is directly traceable to the presence of the long-range Coulomb interaction between electrons (Anderson 1958). P.W. Anderson who showed that the introduction of a long-range field like the electromagnetic field might serve to eliminate massless particles from the theory provided the first indication of a similar effect in relativistic theories in 1963. Basically, what happens is that in case of local symmetries that are spontaneously broken the zero-mass Goldstone bosons can combine with the massless gauge bosons to form massive vector bosons; thereby removing the undesirable feature of massless particles as well as the threat to gauge invariance due to the need to insert masses by hand. This result was fully developed for relativistic field theories by Peter Higgs (1964a&b, 1966) and others who showed that Goldstone bosons could be eliminated by a gauge transformation so that they no longer appeared as physical particles. Instead they emerge as the longitudinal components of the vector bosons.

The Higgs model is based on the idea that even the vacuum state can fail to exhibit the full symmetry of the laws of physics. If the vacuum state is unique, meaning that there is only one state of lowest energy, then it must exhibit the full symmetry of the laws. However, the vacuum may be a degenerate (non-unique) state such that for each unsymmetrical vacuum state there are others of the same minimal energy that are related to the first by various symmetry transformations that preserve the invariance of physical laws. The phenomena observed within the framework of this unsymmetrical vacuum state will exhibit the broken symmetry even in the way that the physical laws appear to operate. The analog of the vacuum state for the universe is the ground state of the ferromagnet. Although there is no evidence that the vacuum state for the electroweak theory is degenerate it can be made so by the introduction of the Higgs mechanism, which involves the artificial insertion of an additional field. This field has a definite but arbitrary orientation in the isospin vector space as well as properties that would make the vacuum state degenerate. And, not only is the symmetry breaking Higgs mechanism able to rid the theory of the Goldstone boson but in doing so solves the mass problem for vector bosons while leaving the theory gauge invariant.

The Higgs field (or its associated particle the Higgs boson) is really a complex $SU(2)$ doublet

$$\begin{pmatrix} \Phi^+ \\ \Phi^- \end{pmatrix}$$

consisting of four real fields that are required to transform the massless gauge fields into massive ones. So, in addition to the four scalar quanta we also have four vector gauge quanta (three from $SU(2)$ and one from $U(1)$). A massless gauge boson, like the photon, has two orthogonal spin components transverse to the direction of motion while massive gauge bosons have three including a longitudinal component in the direction of motion. Three of the four Higgs fields are absorbed by the W^{+-} and Z^0 and appear as the longitudinal polarization in each of the three quanta. Since longitudinal polarization in a vector field is equivalent to mass we obtain three massive gauge quanta with charges $+, -$ and 0. The fourth vector quantum remains massless and is identified as the photon. Of the four original scalar bosons only the Higgs boson remains. The four gauge fields and the Higgs field undergo unique, nonlinear interactions with each other and because the Higgs field is not affected by the vector bosons it should be observable as a particle in its own right.

The Higgs field breaks the symmetry of the vacuum by having a preferred direction in space (like the ferromagnet). When the W and Z particles acquire masses the charges I_3 and Y (weak isospin and hypercharge) associated with the $SU(2) \times U(1)$ symmetry are not conserved in the weak interactions because the vacuum absorbs these quantum numbers. The electric charge Q associated with the $U(1)$ symmetry of electromagnetism remains conserved. The gauge theory associated with the electroweak unification predicts the existence of four gauge quanta, a neutral photon like object sometimes referred to as the X^0 or (B) associated with the $U(1)$ symmetry as well as a weak isospin triplet W^{+-} and W^0 associated with $SU(2)$. As a result of the Higgs symmetry breaking mechanisms the X^0 and the W^0 are mixed so that the neutral particles one sees in nature are really two different linear combinations of these two. One of the neutral particles, the Z^0 has a mass while the other, the photon, is massless. Since the masses of the W^{+-} and the Z^0 are governed by the structure of the Higgs field they do not affect the basic gauge invariance of the theory. And, the so-called "weakness" of the weak interaction which is mediated by the W^{+-} and the Z^0 is understood as a consequence of the masses of these particles.

From this story one can see that the Curie principles governing the relationship between asymmetric causes and effects are not actually violated in the case of spontaneous symmetry breaking. In both the superconductivity and the electroweak cases there is a separate and independent causal story that accounts for SSB. Unlike the case of broken symmetries where one has asymmetrical terms in both the Hamiltonian and its solutions, in SSB an additional mechanism is required in order to causally explain how the process takes place while retaining a gauge invariant Lagrangian. However, with electroweak theory more is required than simply the introduction of the Higgs mechanism; additional assumptions about the vacuum as a plenum form the theoretical foundation for SSB. But, I want to claim that the very core of these assumptions is inherently metaphysical.

Why metaphysical? Primarily because they are not informed by empirical evidence relating to properties of the vacuum and are adopted solely to legitimise other theoretical assumptions; that is, assumptions about the vacuum as plenum are crucial for the possibility of a degenerate vacuum. And, as Brown and Cao (1991) point out, the introduction of SSB into quantum field theory (QFT) would not have been possible save for the *conceptual* shift from vacuum as void to vacuum as plenum. Indeed the history of physics from Descartes through to Newton, Maxwell, Einstein and Dirac has been marked by changing conceptions about how the vacuum ought to be characterised; decisions based not so much on evidential necessity but on philosophical expediency.

In the context of QFT it was Heisenberg (1958) who first introduced the idea of a degenerate vacuum to account for internal quantum numbers such as isospin and strangeness and the breaking of isospin symmetry in electromagnetism. No significant results emerged from Heisenberg's work but in 1960 Nambu reintroduced the idea into particle physics. He viewed the vacuum as a plenum packed with many virtual degrees of freedom, an idea that allowed him to accept the notion of a degenerate vacuum. The work of Goldstone and finally Higgs completed the picture. It is because the explanation of SSB involves not just the existence of the Higgs particle (field) as the mechanism responsible for making the vacuum degenerate, but also the connection between degeneracy and the *assumption* of the vacuum as plenum that its conceptual basis is primarily metaphysical. To that extent hypotheses about hidden symmetries cannot be supported using empirical arguments. Moreover, the causal mechanisms responsible for SSB are traceable to theoretical assumptions that bear no direct relation to the *hidden symmetry* in the Lagrangian. In other words, there is no direct causal link between the symmetrical equations and their asymmetrical solutions.

It is possible, however, to give a different interpretation to the causal structure of this example. Because SSB is required to generate particle masses and thereby guarantee the gauge invariance of the Lagrangian one could claim that there *is* a direct causal link between asymmetry and symmetry; that asymmetry is in fact the causal basis that allows for the preservation of symmetry – a result that would constitute a clear violation of the Curie principles. This is significant, not because of the importance of the principles themselves as a framework for defining appropriate constraints on symmetry; but rather because both the causal and acausal interpretations point to the rather tenuous character of hidden symmetry.

In the acausal case we interpret the role of the Higgs mechanism as an extraneous causal factor responsible for SSB. Hence we preserve the structure of the Curie framework – asymmetrical states are not causally linked to symmetrical equations. Instead they result from an intervening condition – the presence of the Higgs field. But, on that interpretation one is at a loss to provide a *physical* justification for hidden symmetry. Why postulate a hidden symmetry at all? If, on the other hand, we interpret the symmetry breaking at lower energy levels as the causal mechanism that is in some odd way responsible for symmetry preservation at higher ones, the Curie principles are violated. We are then also left in the difficult situation of having to explain why there should be a causal connection between broken symmetry and a gauge invariant Lagrangian; why one should be a necessary condition for the other. In either case we are led to question the physical significance of the symmetry present in the Lagrangian.

An obvious answer, of course, is that gauge invariance is required for renormalizability, for the theory to give physically meaningful predictions. Once 't Hooft (1971) had shown that renormalizability was not affected by what occurs at the scale where symmetry is broken, the effect was to disentangle the symmetry from the dynamics. A new renormalization scheme was introduced that discarded short distance effects through a procedure that made no explicit reference to the physical parameters at that level. Consequently there was no longer any dependence of physical results (i.e. the results due to renormalization) on the physics that operates at short distances. Renormalizibility as a constraint on acceptable field theories requires that we preserve the symmetry in the Lagrangian. But does that warrant any conclusions about the physical status of the symmetry? To invoke Wigner's distinction what we have here is a symmetry of laws (equations) rather than a symmetry of physical states (solutions).

One might want to argue that Wigner's distinction is in some sense artificial or too restrictive, that only by considering both laws and states together was it possible to embark on the path that led to SSB. However, if we attempt to link in a causal way the symmetry of equations describing the physical system with the restricted symmetry of its physical states we are left with the situation described above. Put differently, if we explain the asymmetry of the physical states independently of the symmetrical equations via the Higgs mechanism then the physical status of that symmetry cannot be independently justified. But, if we attempt to link the two by claiming that the asymmetry is a necessary condition for the symmetry we are similarly at a loss as to explain any causal role for the symmetrical equations in the physical dynamics. Either way the symmetry cannot be extended beyond its mathematical function.

My goal has been to use the Curie principles as a framework for displaying how hidden symmetry functions in a metaphysical way in the context of physical theory. Curie himself had other reasons for adopting restrictions on the relations between symmetrical effects and their causes; reasons that relate to contemporary uses of symmetry, specifically predictive power. In electrodynamics, for example, gauge invariance can be used to predict the existence of the electromagnetic field. However, when the direct connection between what Curie considered causes and effects, or between physical states and equations, breaks down so too does the predictability afforded by the symmetry. In Curie's case the concern was to determine the type of symmetries possessed by the field and its quantities.

For instance, in the case of a magnetic field caused by the current in a circular coil, and in electromagnetic induction where a magnetic field occurs in the cause, we are led to posit an intrinsic symmetry for the field that corresponds to the group whose elements are the transformations that can be performed on a cylinder rotating about its axis. The idea is that on the basis of experimental observations the principles allow us to determine symmetries in causes and/or effects.[11] In the electroweak example we saw that predictability provided by a gauge invariant Lagrangian was at odds with the physics of weak interactions, thereby creating a rather intricate net of causal relations. The electroweak theory itself had significant predictive power (e.g. weak neutral currents and W and Z bosons) but our concern here is the predictive power of the $SU(2)$ x $U(1)$ gauge group/theory. The gauge invariance of the Lagrangian predicts fields/particles with mass zero in direct contradiction with the phenomenology of the physics. Hence, it was only by introducing SSB that a reconciliation was possible between the physics and the mathematics.

These issues aside, the notion of predictability used by Curie has an interesting twist. Recall that one of his fundamental claims was that asymmetry rather than symmetry was responsible for creating phenomena. For example, if the physical world is invariant with respect to a particular transformation then this fact could never be discovered because there would be no way of distinguishing two points that were equivalent; the situation at one is identical to that at the other. How then are claims about invariance upheld?

The answer is that invariance principles determine the equivalence of coordinate systems only to a limited extent, the extent to which our descriptions can be separated into two categories – initial conditions and laws of nature.[12] It is indeed possible that two situations, equivalent with respect to certain invariance principles, may be different given different initial conditions. This, in effect, is what allows us to distinguish otherwise identical situations. Hence, if we think of laws of nature governing physical systems, then given the same initial conditions we can say that certain invariances hold provided the two systems evolve in the same way. What this implies is that even in the most straightforward cases invariance or symmetry principles apply to laws of nature but only provided certain initial conditions are present. Hence, even as meta-laws their status is not absolute in the sense of providing the ultimate structural features of the world. While this in itself is not an argument against the physical status of symmetries, it does show that symmetry principles, functioning as higher order laws, also require a context of application in order for any physical implications to be realised.[13]

As I mentioned in section one, it is still an option for the defender of symmetries to argue that SSB presents a special and somewhat problematic case for the physical status of symmetries, that there are other contexts where the case is quite straightforward.[14] Undeniably there are instances where there seems to be definitive connections between particular physical states and symmetries; for example, where states of elementary particles constitute representations of symmetry groups. The situation I have in mind is the relationship between symmetry and spin. Although spin is in some sense a well-understood characteristic of certain particles, to give a full explanation of its physical nature is a less than straightforward task. Not only is there a good deal of ambiguity surrounding its connection with relativity but our physical understanding of spin seems to be in terms of its group theoretical properties. Because no one denies that spin is physically real its connection with symmetry principles becomes especially important.

3. The causal link restored: symmetry as physical explanation

In the 1930s Wigner showed how one could use the Poincaré group (or the inhomogeneous Lorentz group) to get a description of the purely kinematical properties like spin for quantum relativistic systems.[15] Each relativistic wave function corresponded to some unitary representation of the Poincaré group and as such one could sometimes say that an elementary particle is associated (or identified) with a unitary irreducible representation of the group.[16] Given that definition, an elementary particle can then be characterised by its mass and spin. Spin turns out to be simply a group invariant characterising the unitary representation of the relativity group associated with the wave equation. Consequently, one thinks of spin not as the physical rotation associated with a particle, but rather as a symmetry, a way of mathematically stating that a system can undergo a certain rotation.

But, even before the work of Wigner one could understand spin as a group theoretical property; it emerged out of Dirac's (1928) formulation of the wave equation. The positive energy solutions for the equation provided a description of a free spin $1/2$ particle that was consistent with the requirements of special relativity. Spin seemed to appear as a consequence of the transformation law of the solutions under rotation, which in turn were determined by the transformation law under homogenous Lorentz transformations. In other words, spin seemed intimately connected to the relativistic invariance of the Dirac equation.

In fact, one can think of the structure and properties of any quantum field as dictated by the representation of the homogenous Lorentz group under which it transforms. A representation is defined as the set of matrices $D(\Lambda)$ that satisfy the group multiplication law

$$D(\bar{\Lambda})D(\Lambda) = D(\bar{\Lambda}\Lambda)$$

One of the most general irreducible representations of the Lorentz group is called a "spinor" which is really the spin $1/2$ representation of the group. In order to calculate the behaviour of the Dirac equation

$$(i\gamma^{\mu}\delta_{\mu} - M)\psi = 0$$

under the Lorentz group we need to define how spinors transform under some representation $S(\Lambda)$ of that group. Once this is done one then needs to construct invariants under the group. To show this one defines a new field

$$\bar{\psi} \equiv \psi^{t}\gamma^{0}$$

which, under a Lorentz transformation, obeys the conditions required to form invariants. Hence we find that $\bar{\psi}\gamma^{\mu}\psi$ is a genuine vector under the Lorentz group which is tantamount to claiming that the γ^{μ} matrices transform as vectors under the spinor representation of the group. And, because γ^{μ} transforms as a vector, the Lagrangian corresponding to the Dirac equation

$$L = \bar{\psi}(i\gamma^{\mu}\delta_{\mu} - M)\psi$$

is invariant under the Lorentz group.[17] As far as group theory is concerned then the wave function ψ carries the representation $(1/2,0)(0,1/2)$ of the homogenous Lorentz group.

The question of whether this group theoretical account gives us a physical understanding of spin is not unlike the question of whether spin is a relativistic effect. It depends on the context in which the question is raised. If one is satisfied that a particle is simply a unitary irreducible representation of, say, the inhomogeneous Lorentz group, then a similar explanation of spin will no doubt be acceptable. But, from a philosophical point of view this is rather unsatisfying, in part because it forces us to a view of the physical world that is aptly characterized in a remark of Steven Weinberg's which states that at the deepest level all we find are symmetries and responses to symmetries; matter itself dissolves, and the universe itself is revealed as one long reducible representation of the symmetry group of nature. In another context Weinberg (1993) similarly remarks that "Broken symmetry is a very Platonic notion: the reality we observe in our laboratories is only an imperfect reflection of a deeper and more beautiful reality, the reality of the equations that display all the symmetries of the theory".

But notice that in each of these contexts the symmetries take on a metaphysical status that is wholly *distinct* from the physical aspects of the world. And, in neither case do they provide what we typically expect from metaphysics – an explanation of some puzzling *physical* phenomena. Once again the connection between the symmetries and any corresponding physical states is lacking. Instead we are asked to accept the symmetries as fundamentally real while assigning the "material" world as subsidiary role, thereby obviating the need to justify symmetries on the basis of their physical properties or implications.

4. Conclusions

We can, however, think about the matter in another way. One might want to sanction the importance of symmetries as part of the mathematical structure of physics without being a mathematical realist about groups or a materialist about the symmetries themselves. Such a view distinguishes between the "physical" content of a theory and its mathematical formulation. Whether aspects of the latter have a physical status or refer to physically real things will depend entirely on their connection with features of particular systems. For example should we consider the gauge field predicted in electromagnetism real? Yes, because it turns out to be just the electromagnetic field. So if one is willing to sanction the latter then so too the former.

However, the Lagrangian of a system is *not* a physical property, rather it redescribes physical information and to that extent nothing follows about symmetries in the Lagrangian directly corresponding to physical symmetries in nature. For instance, rotational and translational symmetries in the Lagrangian are compatible with the absence of any geometrical symmetries in a physical system and in some cases with the non-conservation of angular and linear momentum. A similar situation applies to the Hamiltonian. Moreover, the deduction of conservation laws from the symmetry in the Lagrangian, although a valid mathematical proof is not an *explanation* of the conservation laws or why the particular quantities are conserved. Here again the metaphysics of symmetries fails to satisfy the demands of explanation.

What happens in the physical world and indeed the laws of nature themselves are not completely determined by the symmetries. But the situation is perhaps worse than this. Because of their elusive character many of the links between symmetries and the constitution of the physical world cannot be uniquely specified either by physical investigation or even philosophical speculation. At best symmetries have the status of a hypothesis established by a hypothetico-deductive argument; but no conclusions about their physical nature should follow from that.

Notes

1. A multiplet is a family of particles that share a particular quantum number. For example, the pion and all its charged states have isospin 1.

2. A gauge group is a Lie group whose parameters are space-time dependent.

3. For more see O'Raigeartah (1986).

4. Although not part of her original theorem, there is some question as to whether the reverse is also true.

5. Radicati (1987) notes these points and provides a discussion of early work on symmetry and spontaneous symmetry breaking.

6. Brown and Cao (1991) take issue with the claim that SSB violates the Curie principles. Although I agree with the main thrust of their argument I want to draw attention to some another ways of thinking about the relationship between the role of the Curie principles and symmetries.

7. For a complete discussion of this effect see Radicati (1987).

8. There were however significant differences between the photon and the W boson. The boson had to exist in at least two charged states W^+ and W^- since all weak interactions involved the exchange of electric charge between the currents J_μ and J'_μ . In addition the boson had to be extremely massive in order to explain the short range of the weak force. Also if neutral current phenomena were to be explained there needed to be a neutral boson as well, a particle commonly known as the Z^0.

9. In perturbation theory the probability of an event is given by the sum of contributions representing increasingly complex Feynman diagrams. Each contribution should, for all intents and purposes, be a dimensionless number. The mass factor in the W boson propagator leads to an ever-increasing number of infinite contributions to the perturbation series, which cannot be reabsorbed, into redefinition of the masses and couplings. Hence it cannot provide physically meaningful answers.

10. That is, the Hamiltonian describing the spin-spin interaction is rotationally invariant while below the ferromagnetic transition temperature the ground state is not; instead a particular direction in space is singled out as the preferred one due to the alignment of the spins. This situation (where the ground state configuration does not display the symmetry of the Hamiltonian) is referred to as a symmetry that is spontaneously broken. Technically, the symmetry is simply "hidden" since there are no non-symmetric terms added to the Hamiltonian. In a case where, for example, an external magnetic field term is added, resulting in the selection of one specific direction, rotational invariance is explicitly broken. But, in the case where any direction is equally good the symmetry still exists, it is merely hidden by the association of one of the possible directions with the ground state of the ferromagnet.

11. For a detailed discussion of the ways in which Curie's principle can be used see Chalmers (1970).

12. This point is also discussed in Houtappel, Van Dam, and Wigner (1965).

13. In addition to the specific link between symmetry and predictability discussed by Curie, the connection between symmetries and conserved quantities described by Noether's theorem also relates symmetry, laws of nature and predictability.

Noether's theorem states that when a symmetry group governs a particular phenomenon and the appropriate Lagrangian does not change under the group transformations then a particular quantity is conserved. Hence, for every continuous symmetry of a Lagrangian there is a quantity that is conserved by its dynamics. The issue here, however, is not just predictability but also whether the relationship between symmetries and conserved quantities is evidence for the physical interpretation of symmetries. Without going into too much detail two points suggest not. The first is simply the fact that the symmetry in question is a symmetry of the Lagrangian rather than a physical symmetry per se, and nothing about the former implies the latter. The second concerns the issue of whether it is possible to conclude, conversely, that for every conservation law there is a corresponding symmetry. One might want to argue that if the presence of a conserved quantity in nature implies a symmetry then that is sufficient for claiming that the symmetry must also have a physical status. However, work by Rosen and Freundlich (1978) has shown that for a certain set of assumptions that forms an essential ingredient of any classical Lagrangian theory, symmetry transformation groups need not be associated with conservations and allows conservations not to be associated with transformation groups.

14. I have assumed for the sake of argument that the SSB produced by the Higgs mechanism is indeed a legitimate hypothesis and tried to show how it is embodies certain metaphysical assumptions. However, one could adopt a more skeptical position, claiming that its introduction is ad hoc; thereby calling into question the legitimacy of the $SU(2)$ x $U(1)$ and the $SU(3)$ x $SU(2)$ x $U(1)$ symmetries themselves. Because the Higgs mechanism was introduced to rid the theory of Goldstone bosons one might claim this simply reintroduces the problem at another level – one set of particles/fields are introduced simply to solve a problem created by another. The other related issue concerns the structure of the symmetry group itself. Even if one were to accept symmetries as part of the physical constitution of the world there may be reasons for not accepting the symmetry groups of the standard model as the right sort of structure. The idea behind grand unified theories (GUTs) is that there ought to be one large group G of which $SU(3)$ x $SU(2)$ x $U(1)$ is merely a subgroup. This is due to the fact that the algebra underlying the latter contains three irreducible pieces with a different coupling constant for each. In that sense one assumes that the real symmetry has a much more elegant structure than the one currently thought to be in play.

15. The Poincaré group is simply an enlargement of the Lorentz group that results from adding translations.

16. An irreducible representation is one that cannot be split up into smaller pieces, each of which would transform under a smaller representation of the same group. All the basic fields of physics transform as irreducible representations of the Lorentz and Poincaré groups. The complete set of finite dimensional representations of the rotation group O (2) or the orthogonal group comes in two classes, the tensors and spinors.

17. This representation of the Lorentz group is not unitary because the generators are not all represented by Hermitian matrices. They do satisfy a pseudounitarity relation. See Weinberg(1995).

References

Anderson, P.W.: (1958), 'Random Phase Approximation in the Theory of Superconductivity', *Physical Review* **112**, pp. 1900–1916.

Anderson, P.W.: (1963), 'Plasmons, Gauge Invariance and Mass', *Physical Review* **130**, pp. 439–442.

Bardeen, J., Cooper, L.N., Schrieffer, J.R.: (1957), 'Theory of Superconductivity', *Phys. Rev.* **108**, pp. 1175–1204.

Brown, L., Cao, T.: (1991), 'Spontaneous breakdown of symmetry: Its rediscovery and integration into quantum field theory', *Historical Studies in the Physical Sciences* **21**, pp. 211–35.

Chalmers, A.: (1970), 'Curie's Principle', *British Journal for the Philosophy of Science* **21**, p. 133.

Curie, P.: (1894), 'Sur la symmetrie des phenomenes physiques: symetrie d'un champ electrique et d'un champ magnetique', *Journal de Physique*, 3e serie, pp. 393–415.

Dirac, P.A.M.: (1928), 'The Quantum Thoery of the Electron', *Proceedings of the Royal Society of London* A117, pp. 610–624; A118, pp. 351–361.

Glashow, S.L.: (1961a), 'Partial Symmetries of Weak Interactions', *Nuclear Physics* **22**, pp. 579–588.

Glashow, S., Gell-Mann, M.: (1961), 'Gauge Theories of Vector Particles', *Annals of Physics* **15**, pp. 437–460.

Goldstone, J.: (1961), 'Field Theories with "Superconductor Solutions" ', *Nuovo Cimento* **19**, p. 154.

Goldstone, J., Salam, A., Weinberg, S.: (1962), 'Broken Symmetries', *Physical Review* **127**, p. 965.

Heisenberg, W.: (1958), 'Research on the non-linear spinor theory with indefinite metric in Hilbert space', International Conference on High Energy Physics, CERN, *Proceedings* Geneva, pp. 119-122.

Higgs, P.W.: (1964a), 'Broken Symmetries, Massless Particles and Gauge Fields', *Physics Letters* **12**, pp. 132-133.

Higgs, P.W.: (1964b), 'Broken Symmetries and Masses of Gauge Bosons', *Physical Review Letters* **13**, pp. 508–509.

Higgs, P.W.: (1966), 'Spontaneous Symmetry Breaking Without Massless Bosons', *Physical Review* **145**, pp. 1156–1163.

Houtappel, R.M.F., Van Dam, H., Wigner, E.: (1965), 'The conceptual basis and use of the geometric invariance principles', *Reviews of Modern Physics* **37**, pp. 595–632.

Landau, L., Ginzburg, V.L.: (1937), 'On the theory of superconductivity' in L. Landau, 'On the theory of phase transitions', *Physikalische Zeitschrift der Sowjetunion II*, pp. 26–47, 545–555.

Lie, S., Engel, F.: (1893), *Theorie der transformationsgruppen*, 3 Vols., Leipzig.

Moriyasu, K.: (1983), *An Elementary Primer for Gauge Theory*, World Scientific, Singapore.

Nambu, Y.: (1960), 'A 'superconductor' model of elementary particles and its consequences', in Midwest conference on Theoretical Physics *Proceedings*, Purdue University.

Nambu, Y., Jona-Lasinio, G.: (1961), 'Dynamical Model of Elementary Particles Based on and Analogy with Superconductivity', *Physical Review* **122**, pp. 345–358 and **124**, p. 246.

Noether, E.: (1918), *Nachr. kgl. Ges. Wiss.*, Gottigen 235.

O'Raifeartaigh, L.: (1986), *Group Structure of Gauge Theories*, CUP, Cambridge.

Poincaré, H.: (1898), 'On the Foundations of Geometry', *The Monist* **9**, p. 1.

Radicati, L.A.: (1987), 'Remarks on the early development of the notion of symmetry breaking', in M.G. Doncel et. al. (eds.), *Symmetries in Physics (1600–1980)*, Belletera, pp. 197–207.

Rosen, J., Freundlich, Y.: (1978), 'Symmetry and Conservation', *American Journal of Physics* **46**, pp. 1030–1041.

Salam, A., Ward, J.C.: (1964), 'Electromagnetic and Weak Interactions', *Phys. Letters* **13**, pp. 168–171.

Schwinger, J.: (1957), 'A Theory of Fundamental Interactions', *Annals of Physics* **2**, pp. 407–434.

Weinberg, S.: (1967), 'A Model of Leptons', *Physical Review Letters* **19**, pp. 1264–1266.

Weinberg, S.: (1974), 'Recent Progress in Gauge Theories of Weak, Electromagnetic and Strong Interactions', *Reviews of Modern Physics* **46**, pp. 255–277.

Weinberg, S.: (1980), 'Conceptual foundations of the unified theory of weak and electromagnetic interactions', *Reviews of Modern Physics* **52**, pp. 515–523.

Weinberg, S.: (1993), *Dreams of a Final Theory*, Pantheon, New York.

Weinberg, S.: (1995), *Quantum Theory of Fields*, Vol.1, CUP, Cambridge.

Wigner, E.: (1931), *Gruppentheorie und ihre Anwendung auf die Quantenmechanik*, Braunschweig.

Wigner, E.: (1979), *Symmetries and Reflections*, Woodbridge, New York.

Yang, C.N., Mills, R.: (1954), 'Conservation of isotopic spin and isotopic gauge invariance', *Physical Review* **96**, p. 191.

Table of Contents Volume II

XVI HISTORY OF LOGIC, METHODOLOGY, AND PHILOSOPHY OF SCIENCE

XVII PHILOSOPHICAL QUESTIONS RAISED BY THE HISTORY AND SOCIOLOGY OF SCIENCE

SPECIAL SYMPOSIUM:

A HUNDRED YEARS OF THE PHILOSOPHY OF SCIENCE

SPECIAL SYMPOSIUM:

COGNITIVE SCIENCE MEETS PHILOSOPHY OF SCIENCE

CLOSING LECTURE

SPECIAL LECTURE

SYNTHESE LIBRARY

1. J. M. Bochénski, *A Precis of Mathematical Logic.* Translated from French and German by O. Bird. 1959 ISBN 90-277-0073-7
2. P. Guiraud, *Problèmes et méthodes de la statistique linguistique.* 1959 ISBN 90-277-0025-7
3. H. Freudenthal (ed.), *The Concept and the Role of the Model in Mathematics and Natural and Social Sciences.* 1961 ISBN 90-277-0017-6
4. E. W. Beth, *Formal Methods.* An Introduction to Symbolic Logic and to the Study of Effective Operations in Arithmetic and Logic. 1962 ISBN 90-277-0069-9
5. B. H. Kazemier and D. Vuysje (eds.), *Logic and Language.* Studies dedicated to Professor Rudolf Carnap on the Occasion of His 70th Birthday. 1962 ISBN 90-277-0019-2
6. M. W. Wartofsky (ed.), *Proceedings of the Boston Colloquium for the Philosophy of Science, 1961–1962.* [Boston Studies in the Philosophy of Science, Vol. I] 1963 ISBN 90-277-0021-4
7. A. A. Zinov'ev, *Philosophical Problems of Many-valued Logic.* A revised edition, edited and translated (from Russian) by G. Küng and D.D. Comey. 1963 ISBN 90-277-0091-5
8. G. Gurvitch, *The Spectrum of Social Time.* Translated from French and edited by M. Korenbaum and P. Bosserman. 1964 ISBN 90-277-0006-0
9. P. Lorenzen, *Formal Logic.* Translated from German by F.J. Crosson. 1965 ISBN 90-277-0080-X
10. R. S. Cohen and M. W. Wartofsky (eds.), *Proceedings of the Boston Colloquium for the Philosophy of Science, 1962–1964.* In Honor of Philipp Frank. [Boston Studies in the Philosophy of Science, Vol. II] 1965 ISBN 90-277-9004-0
11. E. W. Beth, *Mathematical Thought.* An Introduction to the Philosophy of Mathematics. 1965 ISBN 90-277-0070-2
12. E. W. Beth and J. Piaget, *Mathematical Epistemology and Psychology.* Translated from French by W. Mays. 1966 ISBN 90-277-0071-0
13. G. Küng, *Ontology and the Logistic Analysis of Language.* An Enquiry into the Contemporary Views on Universals. Revised ed., translated from German. 1967 ISBN 90-277-0028-1
14. R. S. Cohen and M. W. Wartofsky (eds.), *Proceedings of the Boston Colloquium for the Philosophy of Sciences, 1964–1966.* In Memory of Norwood Russell Hanson. [Boston Studies in the Philosophy of Science, Vol. III] 1967 ISBN 90-277-0013-3
15. C. D. Broad, *Induction, Probability, and Causation.* Selected Papers. 1968 ISBN 90-277-0012-5
16. G. Patzig, *Aristotle's Theory of the Syllogism.* A Logical-philosophical Study of *Book A* of the *Prior Analytics.* Translated from German by J. Barnes. 1968 ISBN 90-277-0030-3
17. N. Rescher, *Topics in Philosophical Logic.* 1968 ISBN 90-277-0084-2
18. R. S. Cohen and M. W. Wartofsky (eds.), *Proceedings of the Boston Colloquium for the Philosophy of Science, 1966–1968, Part I.* [Boston Studies in the Philosophy of Science, Vol. IV] 1969 ISBN 90-277-0014-1
19. R. S. Cohen and M. W. Wartofsky (eds.), *Proceedings of the Boston Colloquium for the Philosophy of Science, 1966–1968, Part II.* [Boston Studies in the Philosophy of Science, Vol. V] 1969 ISBN 90-277-0015-X
20. J. W. Davis, D. J. Hockney and W. K. Wilson (eds.), *Philosophical Logic.* 1969 ISBN 90-277-0075-3
21. D. Davidson and J. Hintikka (eds.), *Words and Objections.* Essays on the Work of W. V. Quine. 1969, rev. ed. 1975 ISBN 90-277-0074-5; Pb 90-277-0602-6
22. P. Suppes, *Studies in the Methodology and Foundations of Science. Selected Papers from 1951 to 1969.* 1969 ISBN 90-277-0020-6
23. J. Hintikka, *Models for Modalities.* Selected Essays. 1969 ISBN 90-277-0078-8; Pb 90-277-0598-4

24. N. Rescher *et al.* (eds.), *Essays in Honor of Carl G. Hempel.* A Tribute on the Occasion of His 65th Birthday. 1969 ISBN 90-277-0085-0
25. P. V. Tavanec (ed.), *Problems of the Logic of Scientific Knowledge.* Translated from Russian. 1970 ISBN 90-277-0087-7
26. M. Swain (ed.), *Induction, Acceptance, and Rational Belief.* 1970 ISBN 90-277-0086-9
27. R. S. Cohen and R. J. Seeger (eds.), *Ernst Mach: Physicist and Philosopher.* [Boston Studies in the Philosophy of Science, Vol. VI]. 1970 ISBN 90-277-0016-8
28. J. Hintikka and P. Suppes, *Information and Inference.* 1970 ISBN 90-277-0155-5
29. K. Lambert, *Philosophical Problems in Logic.* Some Recent Developments. 1970 ISBN 90-277-0079-6
30. R. A. Eberle, *Nominalistic Systems.* 1970 ISBN 90-277-0161-X
31. P. Weingartner and G. Zecha (eds.), *Induction, Physics, and Ethics.* 1970 ISBN 90-277-0158-X
32. E. W. Beth, *Aspects of Modern Logic.* Translated from Dutch. 1970 ISBN 90-277-0173-3
33. R. Hilpinen (ed.), *Deontic Logic.* Introductory and Systematic Readings. 1971 *See also* No. 152. ISBN Pb (1981 rev.) 90-277-1302-2
34. J.-L. Krivine, *Introduction to Axiomatic Set Theory.* Translated from French. 1971 ISBN 90-277-0169-5; Pb 90-277-0411-2
35. J. D. Sneed, *The Logical Structure of Mathematical Physics.* 2nd rev. ed., 1979 ISBN 90-277-1056-2; Pb 90-277-1059-7
36. C. R. Kordig, *The Justification of Scientific Change.* 1971 ISBN 90-277-0181-4; Pb 90-277-0475-9
37. M. Čapek, *Bergson and Modern Physics.* A Reinterpretation and Re-evaluation. [Boston Studies in the Philosophy of Science, Vol. VII] 1971 ISBN 90-277-0186-5
38. N. R. Hanson, *What I Do Not Believe, and Other Essays.* Ed. by S. Toulmin and H. Woolf. 1971 ISBN 90-277-0191-1
39. R. C. Buck and R. S. Cohen (eds.), *PSA 1970.* Proceedings of the Second Biennial Meeting of the Philosophy of Science Association, Boston, Fall 1970. In Memory of Rudolf Carnap. [Boston Studies in the Philosophy of Science, Vol. VIII] 1971 ISBN 90-277-0187-3; Pb 90-277-0309-4
40. D. Davidson and G. Harman (eds.), *Semantics of Natural Language.* 1972 ISBN 90-277-0304-3; Pb 90-277-0310-8
41. Y. Bar-Hillel (ed.), *Pragmatics of Natural Languages.* 1971 ISBN 90-277-0194-6; Pb 90-277-0599-2
42. S. Stenlund, *Combinators, γ Terms and Proof Theory.* 1972 ISBN 90-277-0305-1
43. M. Strauss, *Modern Physics and Its Philosophy.* Selected Paper in the Logic, History, and Philosophy of Science. 1972 ISBN 90-277-0230-6
44. M. Bunge, *Method, Model and Matter.* 1973 ISBN 90-277-0252-7
45. M. Bunge, *Philosophy of Physics.* 1973 ISBN 90-277-0253-5
46. A. A. Zinov'ev, *Foundations of the Logical Theory of Scientific Knowledge (Complex Logic).* Revised and enlarged English edition with an appendix by G. A. Smirnov, E. A. Sidorenka, A. M. Fedina and L. A. Bobrova. [Boston Studies in the Philosophy of Science, Vol. IX] 1973 ISBN 90-277-0193-8; Pb 90-277-0324-8
47. L. Tondl, *Scientific Procedures.* A Contribution concerning the Methodological Problems of Scientific Concepts and Scientific Explanation. Translated from Czech by D. Short. Edited by R.S. Cohen and M.W. Wartofsky. [Boston Studies in the Philosophy of Science, Vol. X] 1973 ISBN 90-277-0147-4; Pb 90-277-0323-X
48. N. R. Hanson, *Constellations and Conjectures.* 1973 ISBN 90-277-0192-X

49. K. J. J. Hintikka, J. M. E. Moravcsik and P. Suppes (eds.), *Approaches to Natural Language.* 1973 ISBN 90-277-0220-9; Pb 90-277-0233-0
50. M. Bunge (ed.), *Exact Philosophy.* Problems, Tools and Goals. 1973 ISBN 90-277-0251-9
51. R. J. Bogdan and I. Niiniluoto (eds.), *Logic, Language and Probability.* 1973 ISBN 90-277-0312-4
52. G. Pearce and P. Maynard (eds.), *Conceptual Change.* 1973 ISBN 90-277-0287-X; Pb 90-277-0339-6
53. I. Niiniluoto and R. Tuomela, *Theoretical Concepts and Hypothetico-inductive Inference.* 1973 ISBN 90-277-0343-4
54. R. Fraissé, *Course of Mathematical Logic* – Volume 1: *Relation and Logical Formula.* Translated from French. 1973 ISBN 90-277-0268-3; Pb 90-277-0403-1 (For *Volume 2* see under No. 69).
55. A. Grünbaum, *Philosophical Problems of Space and Time.* Edited by R.S. Cohen and M.W. Wartofsky. 2nd enlarged ed. [Boston Studies in the Philosophy of Science, Vol. XII] 1973 ISBN 90-277-0357-4; Pb 90-277-0358-2
56. P. Suppes (ed.), *Space, Time and Geometry.* 1973 ISBN 90-277-0386-8; Pb 90-277-0442-2
57. H. Kelsen, *Essays in Legal and Moral Philosophy.* Selected and introduced by O. Weinberger. Translated from German by P. Heath. 1973 ISBN 90-277-0388-4
58. R. J. Seeger and R. S. Cohen (eds.), *Philosophical Foundations of Science.* [Boston Studies in the Philosophy of Science, Vol. XI] 1974 ISBN 90-277-0390-6; Pb 90-277-0376-0
59. R. S. Cohen and M. W. Wartofsky (eds.), *Logical and Epistemological Studies in Contemporary Physics.* [Boston Studies in the Philosophy of Science, Vol. XIII] 1973 ISBN 90-277-0391-4; Pb 90-277-0377-9
60. R. S. Cohen and M. W. Wartofsky (eds.), *Methodological and Historical Essays in the Natural and Social Sciences. Proceedings of the Boston Colloquium for the Philosophy of Science, 1969–1972.* [Boston Studies in the Philosophy of Science, Vol. XIV] 1974 ISBN 90-277-0392-2; Pb 90-277-0378-7
61. R. S. Cohen, J. J. Stachel and M. W. Wartofsky (eds.), *For Dirk Struik. Scientific, Historical and Political Essays.* [Boston Studies in the Philosophy of Science, Vol. XV] 1974 ISBN 90-277-0393-0; Pb 90-277-0379-5
62. K. Ajdukiewicz, *Pragmatic Logic.* Translated from Polish by O. Wojtasiewicz. 1974 ISBN 90-277-0326-4
63. S. Stenlund (ed.), *Logical Theory and Semantic Analysis.* Essays dedicated to Stig Kanger on His 50th Birthday. 1974 ISBN 90-277-0438-4
64. K. F. Schaffner and R. S. Cohen (eds.), *PSA 1972. Proceedings of the Third Biennial Meeting of the Philosophy of Science Association.* [Boston Studies in the Philosophy of Science, Vol. XX] 1974 ISBN 90-277-0408-2; Pb 90-277-0409-0
65. H. E. Kyburg, Jr., *The Logical Foundations of Statistical Inference.* 1974 ISBN 90-277-0330-2; Pb 90-277-0430-9
66. M. Grene, *The Understanding of Nature.* Essays in the Philosophy of Biology. [Boston Studies in the Philosophy of Science, Vol. XXIII] 1974 ISBN 90-277-0462-7; Pb 90-277-0463-5
67. J. M. Broekman, *Structuralism: Moscow, Prague, Paris.* Translated from German. 1974 ISBN 90-277-0478-3
68. N. Geschwind, *Selected Papers on Language and the Brain.* [Boston Studies in the Philosophy of Science, Vol. XVI] 1974 ISBN 90-277-0262-4; Pb 90-277-0263-2
69. R. Fraissé, *Course of Mathematical Logic* – Volume 2: *Model Theory.* Translated from French. 1974 ISBN 90-277-0269-1; Pb 90-277-0510-0 (For *Volume 1* see under No. 54)

70. A. Grzegorczyk, *An Outline of Mathematical Logic.* Fundamental Results and Notions explained with all Details. Translated from Polish. 1974 ISBN 90-277-0359-0; Pb 90-277-0447-3
71. F. von Kutschera, *Philosophy of Language.* 1975 ISBN 90-277-0591-7
72. J. Manninen and R. Tuomela (eds.), *Essays on Explanation and Understanding.* Studies in the Foundations of Humanities and Social Sciences. 1976 ISBN 90-277-0592-5
73. J. Hintikka (ed.), *Rudolf Carnap, Logical Empiricist.* Materials and Perspectives. 1975 ISBN 90-277-0583-6
74. M. Čapek (ed.), *The Concepts of Space and Time.* Their Structure and Their Development. [Boston Studies in the Philosophy of Science, Vol. XXII] 1976 ISBN 90-277-0355-8; Pb 90-277-0375-2
75. J. Hintikka and U. Remes, *The Method of Analysis.* Its Geometrical Origin and Its General Significance. [Boston Studies in the Philosophy of Science, Vol. XXV] 1974 ISBN 90-277-0532-1; Pb 90-277-0543-7
76. J. E. Murdoch and E. D. Sylla (eds.), *The Cultural Context of Medieval Learning.* [Boston Studies in the Philosophy of Science, Vol. XXVI] 1975 ISBN 90-277-0560-7; Pb 90-277-0587-9
77. S. Amsterdamski, *Between Experience and Metaphysics.* Philosophical Problems of the Evolution of Science. [Boston Studies in the Philosophy of Science, Vol. XXXV] 1975 ISBN 90-277-0568-2; Pb 90-277-0580-1
78. P. Suppes (ed.), *Logic and Probability in Quantum Mechanics.* 1976 ISBN 90-277-0570-4; Pb 90-277-1200-X
79. H. von Helmholtz: *Epistemological Writings. The Paul Hertz / Moritz Schlick Centenary Edition of 1921 with Notes and Commentary by the Editors.* Newly translated from German by M. F. Lowe. Edited, with an Introduction and Bibliography, by R. S. Cohen and Y. Elkana. [Boston Studies in the Philosophy of Science, Vol. XXXVII] 1975 ISBN 90-277-0290-X; Pb 90-277-0582-8
80. J. Agassi, *Science in Flux.* [Boston Studies in the Philosophy of Science, Vol. XXVIII] 1975 ISBN 90-277-0584-4; Pb 90-277-0612-2
81. S. G. Harding (ed.), *Can Theories Be Refuted?* Essays on the Duhem-Quine Thesis. 1976 ISBN 90-277-0629-8; Pb 90-277-0630-1
82. S. Nowak, *Methodology of Sociological Research.* General Problems. 1977 ISBN 90-277-0486-4
83. J. Piaget, J.-B. Grize, A. Szeminśska and V. Bang, *Epistemology and Psychology of Functions.* Translated from French. 1977 ISBN 90-277-0804-5
84. M. Grene and E. Mendelsohn (eds.), *Topics in the Philosophy of Biology.* [Boston Studies in the Philosophy of Science, Vol. XXVII] 1976 ISBN 90-277-0595-X; Pb 90-277-0596-8
85. E. Fischbein, *The Intuitive Sources of Probabilistic Thinking in Children.* 1975 ISBN 90-277-0626-3; Pb 90-277-1190-9
86. E. W. Adams, *The Logic of Conditionals.* An Application of Probability to Deductive Logic. 1975 ISBN 90-277-0631-X
87. M. Przełęcki and R. Wójcicki (eds.), *Twenty-Five Years of Logical Methodology in Poland.* Translated from Polish. 1976 ISBN 90-277-0601-8
88. J. Topolski, *The Methodology of History.* Translated from Polish by O. Wojtasiewicz. 1976 ISBN 90-277-0550-X
89. A. Kasher (ed.), *Language in Focus: Foundations, Methods and Systems.* Essays dedicated to Yehoshua Bar-Hillel. [Boston Studies in the Philosophy of Science, Vol. XLIII] 1976 ISBN 90-277-0644-1; Pb 90-277-0645-X

90. J. Hintikka, *The Intentions of Intentionality and Other New Models for Modalities.* 1975 ISBN 90-277-0633-6; Pb 90-277-0634-4
91. W. Stegmüller, *Collected Papers on Epistemology, Philosophy of Science and History of Philosophy.* 2 Volumes. 1977 Set ISBN 90-277-0767-7
92. D. M. Gabbay, *Investigations in Modal and Tense Logics with Applications to Problems in Philosophy and Linguistics.* 1976 ISBN 90-277-0656-5
93. R. J. Bogdan, *Local Induction.* 1976 ISBN 90-277-0649-2
94. S. Nowak, *Understanding and Prediction.* Essays in the Methodology of Social and Behavioral Theories. 1976 ISBN 90-277-0558-5; Pb 90-277-1199-2
95. P. Mittelstaedt, *Philosophical Problems of Modern Physics.* [Boston Studies in the Philosophy of Science, Vol. XVIII] 1976 ISBN 90-277-0285-3; Pb 90-277-0506-2
96. G. Holton and W. A. Blanpied (eds.), *Science and Its Public: The Changing Relationship.* [Boston Studies in the Philosophy of Science, Vol. XXXIII] 1976 ISBN 90-277-0657-3; Pb 90-277-0658-1
97. M. Brand and D. Walton (eds.), *Action Theory.* 1976 ISBN 90-277-0671-9
98. P. Gochet, *Outline of a Nominalist Theory of Propositions.* An Essay in the Theory of Meaning and in the Philosophy of Logic. 1980 ISBN 90-277-1031-7
99. R. S. Cohen, P. K. Feyerabend, and M. W. Wartofsky (eds.), *Essays in Memory of Imre Lakatos.* [Boston Studies in the Philosophy of Science, Vol. XXXIX] 1976 ISBN 90-277-0654-9; Pb 90-277-0655-7
100. R. S. Cohen and J. J. Stachel (eds.), *Selected Papers of Léon Rosenfield.* [Boston Studies in the Philosophy of Science, Vol. XXI] 1979 ISBN 90-277-0651-4; Pb 90-277-0652-2
101. R. S. Cohen, C. A. Hooker, A. C. Michalos and J. W. van Evra (eds.), *PSA 1974. Proceedings of the 1974 Biennial Meeting of the Philosophy of Science Association.* [Boston Studies in the Philosophy of Science, Vol. XXXII] 1976 ISBN 90-277-0647-6; Pb 90-277-0648-4
102. Y. Fried and J. Agassi, *Paranoia.* A Study in Diagnosis. [Boston Studies in the Philosophy of Science, Vol. L] 1976 ISBN 90-277-0704-9; Pb 90-277-0705-7
103. M. Przełęcki, K. Szaniawski and R. Wójcicki (eds.), *Formal Methods in the Methodology of Empirical Sciences.* 1976 ISBN 90-277-0698-0
104. J. M. Vickers, *Belief and Probability.* 1976 ISBN 90-277-0744-8
105. K. H. Wolff, *Surrender and Catch.* Experience and Inquiry Today. [Boston Studies in the Philosophy of Science, Vol. LI] 1976 ISBN 90-277-0758-8; Pb 90-277-0765-0
106. K. Kosík, *Dialectics of the Concrete.* A Study on Problems of Man and World. [Boston Studies in the Philosophy of Science, Vol. LII] 1976 ISBN 90-277-0761-8; Pb 90-277-0764-2
107. N. Goodman, *The Structure of Appearance.* 3rd ed. with an Introduction by G. Hellman. [Boston Studies in the Philosophy of Science, Vol. LIII] 1977 ISBN 90-277-0773-1; Pb 90-277-0774-X
108. K. Ajdukiewicz, *The Scientific World-Perspective and Other Essays, 1931-1963.* Translated from Polish. Edited and with an Introduction by J. Giedymin. 1978 ISBN 90-277-0527-5
109. R. L. Causey, *Unity of Science.* 1977 ISBN 90-277-0779-0
110. R. E. Grandy, *Advanced Logic for Applications.* 1977 ISBN 90-277-0781-2
111. R. P. McArthur, *Tense Logic.* 1976 ISBN 90-277-0697-2
112. L. Lindahl, *Position and Change.* A Study in Law and Logic. Translated from Swedish by P. Needham. 1977 ISBN 90-277-0787-1
113. R. Tuomela, *Dispositions.* 1978 ISBN 90-277-0810-X
114. H. A. Simon, *Models of Discovery and Other Topics in the Methods of Science.* [Boston Studies in the Philosophy of Science, Vol. LIV] 1977 ISBN 90-277-0812-6; Pb 90-277-0858-4

115. R. D. Rosenkrantz, *Inference, Method and Decision.* Towards a Bayesian Philosophy of Science. 1977 ISBN 90-277-0817-7; Pb 90-277-0818-5
116. R. Tuomela, *Human Action and Its Explanation.* A Study on the Philosophical Foundations of Psychology. 1977 ISBN 90-277-0824-X
117. M. Lazerowitz, *The Language of Philosophy.* Freud and Wittgenstein. [Boston Studies in the Philosophy of Science, Vol. LV] 1977 ISBN 90-277-0826-6; Pb 90-277-0862-2
118. Not published 119. J. Pelc (ed.), *Semiotics in Poland, 1894–1969.* Translated from Polish. 1979 ISBN 90-277-0811-8
120. I. Pörn, *Action Theory and Social Science.* Some Formal Models. 1977 ISBN 90-277-0846-0
121. J. Margolis, *Persons and Mind.* The Prospects of Nonreductive Materialism. [Boston Studies in the Philosophy of Science, Vol. LVII] 1977 ISBN 90-277-0854-1; Pb 90-277-0863-0
122. J. Hintikka, I. Niiniluoto, and E. Saarinen (eds.), *Essays on Mathematical and Philosophical Logic.* 1979 ISBN 90-277-0879-7
123. T. A. F. Kuipers, *Studies in Inductive Probability and Rational Expectation.* 1978 ISBN 90-277-0882-7
124. E. Saarinen, R. Hilpinen, I. Niiniluoto and M. P. Hintikka (eds.), *Essays in Honour of Jaakko Hintikka on the Occasion of His 50th Birthday.* 1979 ISBN 90-277-0916-5
125. G. Radnitzky and G. Andersson (eds.), *Progress and Rationality in Science.* [Boston Studies in the Philosophy of Science, Vol. LVIII] 1978 ISBN 90-277-0921-1; Pb 90-277-0922-X
126. P. Mittelstaedt, *Quantum Logic.* 1978 ISBN 90-277-0925-4
127. K. A. Bowen, *Model Theory for Modal Logic.* Kripke Models for Modal Predicate Calculi. 1979 ISBN 90-277-0929-7
128. H. A. Bursen, *Dismantling the Memory Machine.* A Philosophical Investigation of Machine Theories of Memory. 1978 ISBN 90-277-0933-5
129. M. W. Wartofsky, *Models.* Representation and the Scientific Understanding. [Boston Studies in the Philosophy of Science, Vol. XLVIII] 1979 ISBN 90-277-0736-7; Pb 90-277-0947-5
130. D. Ihde, *Technics and Praxis.* A Philosophy of Technology. [Boston Studies in the Philosophy of Science, Vol. XXIV] 1979 ISBN 90-277-0953-X; Pb 90-277-0954-8
131. J. J. Wiatr (ed.), *Polish Essays in the Methodology of the Social Sciences.* [Boston Studies in the Philosophy of Science, Vol. XXIX] 1979 ISBN 90-277-0723-5; Pb 90-277-0956-4
132. W. C. Salmon (ed.), *Hans Reichenbach: Logical Empiricist.* 1979 ISBN 90-277-0958-0
133. P. Bieri, R.-P. Horstmann and L. Krüger (eds.), *Transcendental Arguments in Science.* Essays in Epistemology. 1979 ISBN 90-277-0963-7; Pb 90-277-0964-5
134. M. Marković and G. Petrović (eds.), *Praxis.* Yugoslav Essays in the Philosophy and Methodology of the Social Sciences. [Boston Studies in the Philosophy of Science, Vol. XXXVI] 1979 ISBN 90-277-0727-8; Pb 90-277-0968-8
135. R. Wójcicki, *Topics in the Formal Methodology of Empirical Sciences.* Translated from Polish. 1979 ISBN 90-277-1004-X
136. G. Radnitzky and G. Andersson (eds.), *The Structure and Development of Science.* [Boston Studies in the Philosophy of Science, Vol. LIX] 1979 ISBN 90-277-0994-7; Pb 90-277-0995-5
137. J. C. Webb, *Mechanism, Mentalism and Metamathematics.* An Essay on Finitism. 1980 ISBN 90-277-1046-5
138. D. F. Gustafson and B. L. Tapscott (eds.), *Body, Mind and Method.* Essays in Honor of Virgil C. Aldrich. 1979 ISBN 90-277-1013-9
139. L. Nowak, *The Structure of Idealization.* Towards a Systematic Interpretation of the Marxian Idea of Science. 1980 ISBN 90-277-1014-7

140. C. Perelman, *The New Rhetoric and the Humanities.* Essays on Rhetoric and Its Applications. Translated from French and German. With an Introduction by H. Zyskind. 1979
ISBN 90-277-1018-X; Pb 90-277-1019-8
141. W. Rabinowicz, *Universalizability.* A Study in Morals and Metaphysics. 1979
ISBN 90-277-1020-2
142. C. Perelman, *Justice, Law and Argument.* Essays on Moral and Legal Reasoning. Translated from French and German. With an Introduction by H.J. Berman. 1980
ISBN 90-277-1089-9; Pb 90-277-1090-2
143. S. Kanger and S. Öhman (eds.), *Philosophy and Grammar.* Papers on the Occasion of the Quincentennial of Uppsala University. 1981 ISBN 90-277-1091-0
144. T. Pawlowski, *Concept Formation in the Humanities and the Social Sciences.* 1980
ISBN 90-277-1096-1
145. J. Hintikka, D. Gruender and E. Agazzi (eds.), *Theory Change, Ancient Axiomatics and Galileo's Methodology.* Proceedings of the 1978 Pisa Conference on the History and Philosophy of Science, Volume I. 1981 ISBN 90-277-1126-7
146. J. Hintikka, D. Gruender and E. Agazzi (eds.), *Probabilistic Thinking, Thermodynamics, and the Interaction of the History and Philosophy of Science.* Proceedings of the 1978 Pisa Conference on the History and Philosophy of Science, Volume II. 1981 ISBN 90-277-1127-5
147. U. Mönnich (ed.), *Aspects of Philosophical Logic.* Some Logical Forays into Central Notions of Linguistics and Philosophy. 1981 ISBN 90-277-1201-8
148. D. M. Gabbay, *Semantical Investigations in Heyting's Intuitionistic Logic.* 1981
ISBN 90-277-1202-6
149. E. Agazzi (ed.), *Modern Logic – A Survey.* Historical, Philosophical, and Mathematical Aspects of Modern Logic and Its Applications. 1981 ISBN 90-277-1137-2
150. A. F. Parker-Rhodes, *The Theory of Indistinguishables.* A Search for Explanatory Principles below the Level of Physics. 1981 ISBN 90-277-1214-X
151. J. C. Pitt, *Pictures, Images, and Conceptual Change.* An Analysis of Wilfrid Sellars' Philosophy of Science. 1981 ISBN 90-277-1276-X; Pb 90-277-1277-8
152. R. Hilpinen (ed.), *New Studies in Deontic Logic.* Norms, Actions, and the Foundations of Ethics. 1981 ISBN 90-277-1278-6; Pb 90-277-1346-4
153. C. Dilworth, *Scientific Progress.* A Study Concerning the Nature of the Relation between Successive Scientific Theories. 3rd rev. ed., 1994 ISBN 0-7923-2487-0; Pb 0-7923-2488-9
154. D. Woodruff Smith and R. McIntyre, *Husserl and Intentionality.* A Study of Mind, Meaning, and Language. 1982 ISBN 90-277-1392-8; Pb 90-277-1730-3
155. R. J. Nelson, *The Logic of Mind.* 2nd. ed., 1989 ISBN 90-277-2819-4; Pb 90-277-2822-4
156. J. F. A. K. van Benthem, *The Logic of Time.* A Model-Theoretic Investigation into the Varieties of Temporal Ontology, and Temporal Discourse. 1983; 2nd ed., 1991 ISBN 0-7923-1081-0
157. R. Swinburne (ed.), *Space, Time and Causality.* 1983 ISBN 90-277-1437-1
158. E. T. Jaynes, *Papers on Probability, Statistics and Statistical Physics.* Ed. by R. D. Rozenkrantz. 1983 ISBN 90-277-1448-7; Pb (1989) 0-7923-0213-3
159. T. Chapman, *Time: A Philosophical Analysis.* 1982 ISBN 90-277-1465-7
160. E. N. Zalta, *Abstract Objects.* An Introduction to Axiomatic Metaphysics. 1983
ISBN 90-277-1474-6
161. S. Harding and M. B. Hintikka (eds.), *Discovering Reality.* Feminist Perspectives on Epistemology, Metaphysics, Methodology, and Philosophy of Science. 1983
ISBN 90-277-1496-7; Pb 90-277-1538-6
162. M. A. Stewart (ed.), *Law, Morality and Rights.* 1983 ISBN 90-277-1519-X

163. D. Mayr and G. Süssmann (eds.), *Space, Time, and Mechanics.* Basic Structures of a Physical Theory. 1983 ISBN 90-277-1525-4
164. D. Gabbay and F. Guenthner (eds.), *Handbook of Philosophical Logic.* Vol. I: Elements of Classical Logic. 1983 ISBN 90-277-1542-4
165. D. Gabbay and F. Guenthner (eds.), *Handbook of Philosophical Logic.* Vol. II: Extensions of Classical Logic. 1984 ISBN 90-277-1604-8
166. D. Gabbay and F. Guenthner (eds.), *Handbook of Philosophical Logic.* Vol. III: Alternative to Classical Logic. 1986 ISBN 90-277-1605-6
167. D. Gabbay and F. Guenthner (eds.), *Handbook of Philosophical Logic.* Vol. IV: Topics in the Philosophy of Language. 1989 ISBN 90-277-1606-4
168. A. J. I. Jones, *Communication and Meaning.* An Essay in Applied Modal Logic. 1983 ISBN 90-277-1543-2
169. M. Fitting, *Proof Methods for Modal and Intuitionistic Logics.* 1983 ISBN 90-277-1573-4
170. J. Margolis, *Culture and Cultural Entities.* Toward a New Unity of Science. 1984 ISBN 90-277-1574-2
171. R. Tuomela, *A Theory of Social Action.* 1984 ISBN 90-277-1703-6
172. J. J. E. Gracia, E. Rabossi, E. Villanueva and M. Dascal (eds.), *Philosophical Analysis in Latin America.* 1984 ISBN 90-277-1749-4
173. P. Ziff, *Epistemic Analysis.* A Coherence Theory of Knowledge. 1984 ISBN 90-277-1751-7
174. P. Ziff, *Antiaesthetics.* An Appreciation of the Cow with the Subtile Nose. 1984 ISBN 90-277-1773-7
175. W. Balzer, D. A. Pearce, and H.-J. Schmidt (eds.), *Reduction in Science.* Structure, Examples, Philosophical Problems. 1984 ISBN 90-277-1811-3
176. A. Peczenik, L. Lindahl and B. van Roermund (eds.), *Theory of Legal Science.* Proceedings of the Conference on Legal Theory and Philosophy of Science (Lund, Sweden, December 1983). 1984 ISBN 90-277-1834-2
177. I. Niiniluoto, *Is Science Progressive?* 1984 ISBN 90-277-1835-0
178. B. K. Matilal and J. L. Shaw (eds.), *Analytical Philosophy in Comparative Perspective.* Exploratory Essays in Current Theories and Classical Indian Theories of Meaning and Reference. 1985 ISBN 90-277-1870-9
179. P. Kroes, *Time: Its Structure and Role in Physical Theories.* 1985 ISBN 90-277-1894-6
180. J. H. Fetzer, *Sociobiology and Epistemology.* 1985 ISBN 90-277-2005-3; Pb 90-277-2006-1
181. L. Haaparanta and J. Hintikka (eds.), *Frege Synthesized.* Essays on the Philosophical and Foundational Work of Gottlob Frege. 1986 ISBN 90-277-2126-2
182. M. Detlefsen, *Hilbert's Program.* An Essay on Mathematical Instrumentalism. 1986 ISBN 90-277-2151-3
183. J. L. Golden and J. J. Pilotta (eds.), *Practical Reasoning in Human Affairs.* Studies in Honor of Chaim Perelman. 1986 ISBN 90-277-2255-2
184. H. Zandvoort, *Models of Scientific Development and the Case of Nuclear Magnetic Resonance.* 1986 ISBN 90-277-2351-6
185. I. Niiniluoto, *Truthlikeness.* 1987 ISBN 90-277-2354-0
186. W. Balzer, C. U. Moulines and J. D. Sneed, *An Architectonic for Science.* The Structuralist Program. 1987 ISBN 90-277-2403-2
187. D. Pearce, *Roads to Commensurability.* 1987 ISBN 90-277-2414-8
188. L. M. Vaina (ed.), *Matters of Intelligence.* Conceptual Structures in Cognitive Neuroscience. 1987 ISBN 90-277-2460-1

189. H. Siegel, *Relativism Refuted.* A Critique of Contemporary Epistemological Relativism. 1987 ISBN 90-277-2469-5
190. W. Callebaut and R. Pinxten, *Evolutionary Epistemology.* A Multiparadigm Program, with a Complete Evolutionary Epistemology Bibliograph. 1987 ISBN 90-277-2582-9
191. J. Kmita, *Problems in Historical Epistemology.* 1988 ISBN 90-277-2199-8
192. J. H. Fetzer (ed.), *Probability and Causality.* Essays in Honor of Wesley C. Salmon, with an Annotated Bibliography. 1988 ISBN 90-277-2507-8; Pb 1-5560-8052-2
193. A. Donovan, L. Laudan and R. Laudan (eds.), *Scrutinizing Science.* Empirical Studies of Scientific Change. 1988 ISBN 90-277-2608-6
194. H.R. Otto and J.A. Tuedio (eds.), *Perspectives on Mind.* 1988 ISBN 90-277-2640-X
195. D. Batens and J.P. van Bendegem (eds.), *Theory and Experiment.* Recent Insights and New Perspectives on Their Relation. 1988 ISBN 90-277-2645-0
196. J. Österberg, *Self and Others.* A Study of Ethical Egoism. 1988 ISBN 90-277-2648-5
197. D.H. Helman (ed.), *Analogical Reasoning.* Perspectives of Artificial Intelligence, Cognitive Science, and Philosophy. 1988 ISBN 90-277-2711-2
198. J. Woleński, *Logic and Philosophy in the Lvov-Warsaw School.* 1989 ISBN 90-277-2749-X
199. R. Wójcicki, *Theory of Logical Calculi.* Basic Theory of Consequence Operations. 1988 ISBN 90-277-2785-6
200. J. Hintikka and M.B. Hintikka, *The Logic of Epistemology and the Epistemology of Logic.* Selected Essays. 1989 ISBN 0-7923-0040-8; Pb 0-7923-0041-6
201. E. Agazzi (ed.), *Probability in the Sciences.* 1988 ISBN 90-277-2808-9
202. M. Meyer (ed.), *From Metaphysics to Rhetoric.* 1989 ISBN 90-277-2814-3
203. R.L. Tieszen, *Mathematical Intuition.* Phenomenology and Mathematical Knowledge. 1989 ISBN 0-7923-0131-5
204. A. Melnick, *Space, Time, and Thought in Kant.* 1989 ISBN 0-7923-0135-8
205. D.W. Smith, *The Circle of Acquaintance.* Perception, Consciousness, and Empathy. 1989 ISBN 0-7923-0252-4
206. M.H. Salmon (ed.), *The Philosophy of Logical Mechanism.* Essays in Honor of Arthur W. Burks. With his Responses, and with a Bibliography of Burk's Work. 1990 ISBN 0-7923-0325-3
207. M. Kusch, *Language as Calculus vs. Language as Universal Medium.* A Study in Husserl, Heidegger, and Gadamer. 1989 ISBN 0-7923-0333-4
208. T.C. Meyering, *Historical Roots of Cognitive Science.* The Rise of a Cognitive Theory of Perception from Antiquity to the Nineteenth Century. 1989 ISBN 0-7923-0349-0
209. P. Kosso, *Observability and Observation in Physical Science.* 1989 ISBN 0-7923-0389-X
210. J. Kmita, *Essays on the Theory of Scientific Cognition.* 1990 ISBN 0-7923-0441-1
211. W. Sieg (ed.), *Acting and Reflecting.* The Interdisciplinary Turn in Philosophy. 1990 ISBN 0-7923-0512-4
212. J. Karpiński, *Causality in Sociological Research.* 1990 ISBN 0-7923-0546-9
213. H.A. Lewis (ed.), *Peter Geach: Philosophical Encounters.* 1991 ISBN 0-7923-0823-9
214. M. Ter Hark, *Beyond the Inner and the Outer.* Wittgenstein's Philosophy of Psychology. 1990 ISBN 0-7923-0850-6
215. M. Gosselin, *Nominalism and Contemporary Nominalism.* Ontological and Epistemological Implications of the Work of W.V.O. Quine and of N. Goodman. 1990 ISBN 0-7923-0904-9
216. J.H. Fetzer, D. Shatz and G. Schlesinger (eds.), *Definitions and Definability.* Philosophical Perspectives. 1991 ISBN 0-7923-1046-2
217. E. Agazzi and A. Cordero (eds.), *Philosophy and the Origin and Evolution of the Universe.* 1991 ISBN 0-7923-1322-4

218. M. Kusch, *Foucault's Strata and Fields.* An Investigation into Archaeological and Genealogical Science Studies. 1991 ISBN 0-7923-1462-X
219. C.J. Posy, *Kant's Philosophy of Mathematics.* Modern Essays. 1992 ISBN 0-7923-1495-6
220. G. Van de Vijver, *New Perspectives on Cybernetics.* Self-Organization, Autonomy and Connectionism. 1992 ISBN 0-7923-1519-7
221. J.C. Nyíri, *Tradition and Individuality.* Essays. 1992 ISBN 0-7923-1566-9
222. R. Howell, *Kant's Transcendental Deduction.* An Analysis of Main Themes in His Critical Philosophy. 1992 ISBN 0-7923-1571-5
223. A. García de la Sienra, *The Logical Foundations of the Marxian Theory of Value.* 1992 ISBN 0-7923-1778-5
224. D.S. Shwayder, *Statement and Referent.* An Inquiry into the Foundations of Our Conceptual Order. 1992 ISBN 0-7923-1803-X
225. M. Rosen, *Problems of the Hegelian Dialectic.* Dialectic Reconstructed as a Logic of Human Reality. 1993 ISBN 0-7923-2047-6
226. P. Suppes, *Models and Methods in the Philosophy of Science: Selected Essays.* 1993 ISBN 0-7923-2211-8
227. R. M. Dancy (ed.), *Kant and Critique: New Essays in Honor of W. H. Werkmeister.* 1993 ISBN 0-7923-2244-4
228. J. Woleński (ed.), *Philosophical Logic in Poland.* 1993 ISBN 0-7923-2293-2
229. M. De Rijke (ed.), *Diamonds and Defaults.* Studies in Pure and Applied Intensional Logic. 1993 ISBN 0-7923-2342-4
230. B.K. Matilal and A. Chakrabarti (eds.), *Knowing from Words.* Western and Indian Philosophical Analysis of Understanding and Testimony. 1994 ISBN 0-7923-2345-9
231. S.A. Kleiner, *The Logic of Discovery.* A Theory of the Rationality of Scientific Research. 1993 ISBN 0-7923-2371-8
232. R. Festa, *Optimum Inductive Methods.* A Study in Inductive Probability, Bayesian Statistics, and Verisimilitude. 1993 ISBN 0-7923-2460-9
233. P. Humphreys (ed.), *Patrick Suppes: Scientific Philosopher.* Vol. 1: Probability and Probabilistic Causality. 1994 ISBN 0-7923-2552-4
234. P. Humphreys (ed.), *Patrick Suppes: Scientific Philosopher.* Vol. 2: Philosophy of Physics, Theory Structure, and Measurement Theory. 1994 ISBN 0-7923-2553-2
235. P. Humphreys (ed.), *Patrick Suppes: Scientific Philosopher.* Vol. 3: Language, Logic, and Psychology. 1994 ISBN 0-7923-2862-0
Set ISBN (Vols 233–235) 0-7923-2554-0
236. D. Prawitz and D. Westerståhl (eds.), *Logic and Philosophy of Science in Uppsala.* Papers from the 9th International Congress of Logic, Methodology, and Philosophy of Science. 1994 ISBN 0-7923-2702-0
237. L. Haaparanta (ed.), *Mind, Meaning and Mathematics.* Essays on the Philosophical Views of Husserl and Frege. 1994 ISBN 0-7923-2703-9
238. J. Hintikka (ed.), *Aspects of Metaphor.* 1994 ISBN 0-7923-2786-1
239. B. McGuinness and G. Oliveri (eds.), *The Philosophy of Michael Dummett.* With Replies from Michael Dummett. 1994 ISBN 0-7923-2804-3
240. D. Jamieson (ed.), *Language, Mind, and Art.* Essays in Appreciation and Analysis, In Honor of Paul Ziff. 1994 ISBN 0-7923-2810-8
241. G. Preyer, F. Siebelt and A. Ulfig (eds.), *Language, Mind and Epistemology.* On Donald Davidson's Philosophy. 1994 ISBN 0-7923-2811-6
242. P. Ehrlich (ed.), *Real Numbers, Generalizations of the Reals, and Theories of Continua.* 1994 ISBN 0-7923-2689-X

243. G. Debrock and M. Hulswit (eds.), *Living Doubt*. Essays concerning the epistemology of Charles Sanders Peirce. 1994 ISBN 0-7923-2898-1
244. J. Srzednicki, *To Know or Not to Know*. Beyond Realism and Anti-Realism. 1994 ISBN 0-7923-2909-0
245. R. Egidi (ed.), *Wittgenstein: Mind and Language*. 1995 ISBN 0-7923-3171-0
246. A. Hyslop, *Other Minds*. 1995 ISBN 0-7923-3245-8
247. L. Pólos and M. Masuch (eds.), *Applied Logic: How, What and Why*. Logical Approaches to Natural Language. 1995 ISBN 0-7923-3432-9
248. M. Krynicki, M. Mostowski and L.M. Szczerba (eds.), *Quantifiers: Logics, Models and Computation*. Volume One: Surveys. 1995 ISBN 0-7923-3448-5
249. M. Krynicki, M. Mostowski and L.M. Szczerba (eds.), *Quantifiers: Logics, Models and Computation*. Volume Two: Contributions. 1995 ISBN 0-7923-3449-3
Set ISBN (Vols 248 + 249) 0-7923-3450-7
250. R.A. Watson, *Representational Ideas from Plato to Patricia Churchland*. 1995 ISBN 0-7923-3453-1
251. J. Hintikka (ed.), *From Dedekind to Gödel*. Essays on the Development of the Foundations of Mathematics. 1995 ISBN 0-7923-3484-1
252. A. Wiśniewski, *The Posing of Questions*. Logical Foundations of Erotetic Inferences. 1995 ISBN 0-7923-3637-2
253. J. Peregrin, *Doing Worlds with Words*. Formal Semantics without Formal Metaphysics. 1995 ISBN 0-7923-3742-5
254. I.A. Kieseppä, *Truthlikeness for Multidimensional, Quantitative Cognitive Problems*. 1996 ISBN 0-7923-4005-1
255. P. Hugly and C. Sayward: *Intensionality and Truth*. An Essay on the Philosophy of A.N. Prior. 1996 ISBN 0-7923-4119-8
256. L. Hankinson Nelson and J. Nelson (eds.): *Feminism, Science, and the Philosophy of Science*. 1997 ISBN 0-7923-4162-7
257. P.I. Bystrov and V.N. Sadovsky (eds.): *Philosophical Logic and Logical Philosophy*. Essays in Honour of Vladimir A. Smirnov. 1996 ISBN 0-7923-4270-4
258. Å.E. Andersson and N-E. Sahlin (eds.): *The Complexity of Creativity*. 1996 ISBN 0-7923-4346-8
259. M.L. Dalla Chiara, K. Doets, D. Mundici and J. van Benthem (eds.): *Logic and Scientific Methods*. Volume One of the Tenth International Congress of Logic, Methodology and Philosophy of Science, Florence, August 1995. 1997 ISBN 0-7923-4383-2
260. M.L. Dalla Chiara, K. Doets, D. Mundici and J. van Benthem (eds.): *Structures and Norms in Science*. Volume Two of the Tenth International Congress of Logic, Methodology and Philosophy of Science, Florence, August 1995. 1997 ISBN 0-7923-4384-0
Set ISBN (Vols 259 + 260) 0-7923-4385-9
261. A. Chakrabarti: *Denying Existence*. The Logic, Epistemology and Pragmatics of Negative Existentials and Fictional Discourse. 1997 ISBN 0-7923-4388-3
262. A. Biletzki: *Talking Wolves*. Thomas Hobbes on the Language of Politics and the Politics of Language. 1997 ISBN 0-7923-4425-1
263. D. Nute (ed.): *Defeasible Deontic Logic*. 1997 ISBN 0-7923-4630-0
264. U. Meixner: *Axiomatic Formal Ontology*. 1997 ISBN 0-7923-4747-X
265. I. Brinck: *The Indexical 'I'*. The First Person in Thought and Language. 1997 ISBN 0-7923-4741-2
266. G. Hölmström-Hintikka and R. Tuomela (eds.): *Contemporary Action Theory*. Volume 1: Individual Action. 1997 ISBN 0-7923-4753-6; Set: 0-7923-4754-4

267. G. Hölmström-Hintikka and R. Tuomela (eds.): *Contemporary Action Theory.* Volume 2: Social Action. 1997 ISBN 0-7923-4752-8; Set: 0-7923-4754-4
268. B.-C. Park: *Phenomenological Aspects of Wittgenstein's Philosophy.* 1998 ISBN 0-7923-4813-3
269. J. Paśniczek: *The Logic of Intentional Objects.* A Meinongian Version of Classical Logic. 1998 Hb ISBN 0-7923-4880-X; Pb ISBN 0-7923-5578-4
270. P.W. Humphreys and J.H. Fetzer (eds.): *The New Theory of Reference.* Kripke, Marcus, and Its Origins. 1998 ISBN 0-7923-4898-2
271. K. Szaniawski, A. Chmielewski and J. Woleński (eds.): *On Science, Inference, Information and Decision Making.* Selected Essays in the Philosophy of Science. 1998 ISBN 0-7923-4922-9
272. G.H. von Wright: *In the Shadow of Descartes.* Essays in the Philosophy of Mind. 1998 ISBN 0-7923-4992-X
273. K. Kijania-Placek and J. Woleński (eds.): *The Lvov–Warsaw School and Contemporary Philosophy.* 1998 ISBN 0-7923-5105-3
274. D. Dedrick: *Naming the Rainbow.* Colour Language, Colour Science, and Culture. 1998 ISBN 0-7923-5239-4
275. L. Albertazzi (ed.): *Shapes of Forms.* From Gestalt Psychology and Phenomenology to Ontology and Mathematics. 1999 ISBN 0-7923-5246-7
276. P. Fletcher: *Truth, Proof and Infinity.* A Theory of Constructions and Constructive Reasoning. 1998 ISBN 0-7923-5262-9
277. M. Fitting and R.L. Mendelsohn (eds.): *First-Order Modal Logic.* 1998 Hb ISBN 0-7923-5334-X; Pb ISBN 0-7923-5335-8
278. J.N. Mohanty: *Logic, Truth and the Modalities from a Phenomenological Perspective.* 1999 ISBN 0-7923-5550-4
279. T. Placek: *Mathematical Intiutionism and Intersubjectivity.* A Critical Exposition of Arguments for Intuitionism. 1999 ISBN 0-7923-5630-6
280. A. Cantini, E. Casari and P. Minari (eds.): *Logic and Foundations of Mathematics.* 1999 ISBN 0-7923-5659-4 set ISBN 0-7923-5867-8
281. M.L. Dalla Chiara, R. Giuntini and F. Laudisa (eds.): *Language, Quantum, Music.* 1999 ISBN 0-7923-5727-2; set ISBN 0-7923-5867-8
282. R. Egidi (ed.): *In Search of a New Humanism.* The Philosophy of Georg Hendrik von Wright. 1999 ISBN 0-7923-5810-4
283. F. Vollmer: *Agent Causality.* 1999 ISBN 0-7923-5848-1
284. J. Peregrin (ed.): *Truth and Its Nature (if Any).* 1999 ISBN 0-7923-5865-1
285. M. De Caro (ed.): *Interpretations and Causes.* New Perspectives on Donald Davidson's Philosophy. 1999 ISBN 0-7923-5869-4
286. R. Murawski: *Recursive Functions and Metamathematics.* Problems of Completeness and Decidability, Gödel's Theorems. 1999 ISBN 0-7923-5904-6
287. T.A.F. Kuipers: *From Instrumentalism to Constructive Realism.* On Some Relations between Confirmation, Empirical Progress, and Truth Approximation. 2000 ISBN 0-7923-6086-9
288. G. Holmström-Hintikka (ed.): *Medieval Philosophy and Modern Times.* 2000 ISBN 0-7923-6102-4
289. E. Grosholz and H. Breger (eds.): *The Growth of Mathematical Knowledge.* 2000 ISBN 0-7923-6151-2

290. G. Sommaruga: *History and Philosophy of Constructive Type Theory*. 2000 ISBN 0-7923-6180-6
291. J. Gasser (ed.): *A Boole Anthology*. Recent and Classical Studies in the Logic of George Boole. 2000 ISBN 0-7923-6380-9
292. V.F. Hendricks, S.A. Pedersen and K.F. Jørgensen (eds.): *Proof Theory*. History and Philosophical Significance. 2000 ISBN 0-7923-6544-5
293. W.L. Craig: *The Tensed Theory of Time*. A Critical Examination. 2000 ISBN 0-7923-6634-4
294. W.L. Craig: *The Tenseless Theory of Time*. A Critical Examination. 2000 ISBN 0-7923-6635-2
295. L. Albertazzi (ed.): *The Dawn of Cognitive Science*. Early European Contributors. 2001 ISBN 0-7923-6799-5
296. G. Forrai: *Reference, Truth and Conceptual Schemes*. A Defense of Internal Realism. 2001 ISBN 0-7923-6885-1
297. V.F. Hendricks, S.A. Pedersen and K.F. Jørgensen (eds.): *Probability Theory*. Philosophy, Recent History and Relations to Science. 2001 ISBN 0-7923-6952-1
298. M. Esfeld: *Holism in Philosophy of Mind and Philosophy of Physics*. 2001 ISBN 0-7923-7003-1
299. E.C. Steinhart: *The Logic of Metaphor*. Analogous Parts of Possible Worlds. 2001 ISBN 0-7923-7004-X
300. To be published.
301. T.A.F. Kuipers: *Structures in Science Heuristic Patterns Based on Cognitive Structures*. An Advanced Textbook in Neo-Classical Philosophy of Science. 2001 ISBN 0-7923-7117-8
302. G. Hon and S.S. Rakover (eds.): *Explanation*. Theoretical Approaches and Applications. 2001 ISBN 1-4020-0017-0
303. G. Holmström-Hintikka, S. Lindström and R. Sliwinski (eds.): *Collected Papers of Stig Kanger with Essays on his Life and Work*. Vol. I. 2001 ISBN 1-4020-0021-9; Pb ISBN 1-4020-0022-7
304. G. Holmström-Hintikka, S. Lindström and R. Sliwinski (eds.): *Collected Papers of Stig Kanger with Essays on his Life and Work*. Vol. II. 2001 ISBN 1-4020-0111-8; Pb ISBN 1-4020-0112-6
305. C.A. Anderson and M. Zelëny (eds.): *Logic, Meaning and Computation*. Essays in Memory of Alonzo Church. 2001 ISBN 1-4020-0141-X
306. P. Schuster, U. Berger and H. Osswald (eds.): *Reuniting the Antipodes – Constructive and Nonstandard Views of the Continuum*. 2001 ISBN 1-4020-0152-5
307. S.D. Zwart: *Refined Verisimilitude*. 2001 ISBN 1-4020-0268-8
308. A.-S. Maurin: *IF Tropes*. 2002 ISBN 1-4020-0656-X
309. H. Eilstein (ed.): *A Collection of Polish Works on Philosophical Problems of Time and Spacetime*. 2002 ISBN 1-4020-0670-5
310. Y. Gauthier: *Internal Logic*. Foundations of Mathematics from Kronecker to Hilbert. 2002 ISBN 1-4020-0689-6
311. E. Ruttkamp: *A Model-Theoretic Realist Interpretation of Science*. 2002 ISBN 1-4020-0729-9
312. V. Rantala: *Explanatory Translation*. Beyond the Kuhnian Model of Conceptual Change. 2002 ISBN 1-4020-0827-9
313. L. Decock: *Trading Ontology for Ideology*. 2002 ISBN 1-4020-0865-1

314. O. Ezra: *The Withdrawal of Rights*. Rights from a Different Perspective. 2002
ISBN 1-4020-0886-4

315. P. Gärdenfors, J. Woleński and K. Kijania-Placek: *In the Scope of Logic, Methodology and Philosophy of Science*. Volume One of the 11th International Congress of Logic, Methodology and Philosophy of Science, Cracow, August 1999. 2002
ISBN 1-4020-0929-1; Pb 1-4020-0931-3

316. P. Gärdenfors, J. Woleński and K. Kijania-Placek: *In the Scope of Logic, Methodology and Philosophy of Science*. Volume Two of the 11th International Congress of Logic, Methodology and Philosophy of Science, Cracow, August 1999. 2002
ISBN 1-4020-0930-5; Pb 1-4020-0931-3

Previous volumes are still available.

KLUWER ACADEMIC PUBLISHERS – DORDRECHT / BOSTON / LONDON